Plant Cell Culture Protocols

METHODS IN MOLECULAR BIOLOGY™

John M. Walker, SERIES EDITOR

328. **New and Emerging Proteomic Techniques**, edited by *Dobrin Nedelkov and Randall W. Nelson*, 2006
327. **Epidermal Growth Factor:** *Methods and Protocols*, edited by *Tarun B. Patel and Paul J. Bertics*, 2006
326. **In Situ Hybridization Protocols**, *Third Edition*, edited by *Ian A. Darby and Tim D. Hewitson*, 2006
325. **Nuclear Reprogramming:** *Methods and Protocols*, edited by *Steve Pells*, 2006
324. **Hormone Assays in Biological Fluids**, edited by *Michael J. Wheeler and J. S. Morley Hutchinson*, 2006
323. **Arabidopsis Protocols**, *Second Edition*, edited by *Julio Salinas and Jose J. Sanchez-Serrano*, 2006
322. **Xenopus Protocols:** *Cell Biology and Signal Transduction*, edited by *X. Johné Liu*, 2006
321. **Microfluidic Techniques:** *Reviews and Protocols*, edited by *Shelley D. Minteer*, 2006
320. **Cytochrome P450 Protocols**, *Second Edition*, edited by *Ian R. Phillips and Elizabeth A. Shephard*, 2006
319. **Cell Imaging Techniques**, *Methods and Protocols*, edited by *Douglas J. Taatjes and Brooke T. Mossman*, 2006
318. **Plant Cell Culture Protocols**, *Second Edition*, edited by *Victor M. Loyola-Vargas and Felipe Vázquez-Flota*, 2005
317. **Differential Display Methods and Protocols**, *Second Edition*, edited by *Peng Liang, Jonathan Meade, and Arthur B. Pardee*, 2005
316. **Bioinformatics and Drug Discovery**, edited by *Richard S. Larson*, 2005
315. **Mast Cells:** *Methods and Protocols*, edited by *Guha Krishnaswamy and David S. Chi*, 2005
314. **DNA Repair Protocols:** *Mammalian Systems*, *Second Edition*, edited by *Daryl S. Henderson*, 2005
313. **Yeast Protocols:** *Second Edition*, edited by *Wei Xiao*, 2005
312. **Calcium Signaling Protocols:** *Second Edition*, edited by *David G. Lambert*, 2005
311. **Pharmacogenomics:** *Methods and Protocols*, edited by *Federico Innocenti*, 2005
310. **Chemical Genomics:** *Reviews and Protocols*, edited by *Edward D. Zanders*, 2005
309. **RNA Silencing:** *Methods and Protocols*, edited by *Gordon Carmichael*, 2005
308. **Therapeutic Proteins:** *Methods and Protocols*, edited by *C. Mark Smales and David C. James*, 2005
307. **Phosphodiesterase Methods and Protocols**, edited by *Claire Lugnier*, 2005
306. **Receptor Binding Techniques:** *Second Edition*, edited by *Anthony P. Davenport*, 2005
305. **Protein–Ligand Interactions:** *Methods and Applications*, edited by *G. Ulrich Nienhaus*, 2005
304. **Human Retrovirus Protocols:** *Virology and Molecular Biology*, edited by *Tuofu Zhu*, 2005
303. **NanoBiotechnology Protocols**, edited by *Sandra J. Rosenthal and David W. Wright*, 2005
302. **Handbook of ELISPOT:** *Methods and Protocols*, edited by *Alexander E. Kalyuzhny*, 2005
301. **Ubiquitin–Proteasome Protocols**, edited by *Cam Patterson and Douglas M. Cyr*, 2005
300. **Protein Nanotechnology:** *Protocols, Instrumentation, and Applications*, edited by *Tuan Vo-Dinh*, 2005
299. **Amyloid Proteins:** *Methods and Protocols*, edited by *Einar M. Sigurdsson*, 2005
298. **Peptide Synthesis and Application**, edited by *John Howl*, 2005
297. **Forensic DNA Typing Protocols**, edited by *Angel Carracedo*, 2005
296. **Cell Cycle Control:** *Mechanisms and Protocols*, edited by *Tim Humphrey and Gavin Brooks*, 2005
295. **Immunochemical Protocols,** *Third Edition*, edited by *Robert Burns*, 2005
294. **Cell Migration:** *Developmental Methods and Protocols*, edited by *Jun-Lin Guan*, 2005
293. **Laser Capture Microdissection:** *Methods and Protocols*, edited by *Graeme I. Murray and Stephanie Curran*, 2005
292. **DNA Viruses:** *Methods and Protocols*, edited by *Paul M. Lieberman*, 2005
291. **Molecular Toxicology Protocols**, edited by *Phouthone Keohavong and Stephen G. Grant*, 2005
290. **Basic Cell Culture Protocols**, *Third Edition*, edited by *Cheryl D. Helgason and Cindy L. Miller*, 2005
289. **Epidermal Cells**, *Methods and Applications*, edited by *Kursad Turksen*, 2005
288. **Oligonucleotide Synthesis**, *Methods and Applications*, edited by *Piet Herdewijn*, 2005
287. **Epigenetics Protocols**, edited by *Trygve O. Tollefsbol*, 2004
286. **Transgenic Plants:** *Methods and Protocols*, edited by *Leandro Peña*, 2005
285. **Cell Cycle Control and Dysregulation Protocols:** *Cyclins, Cyclin-Dependent Kinases, and Other Factors*, edited by *Antonio Giordano and Gaetano Romano*, 2004
284. **Signal Transduction Protocols**, *Second Edition*, edited by *Robert C. Dickson and Michael D. Mendenhall*, 2004
283. **Bioconjugation Protocols**, edited by *Christof M. Niemeyer*, 2004
282. **Apoptosis Methods and Protocols**, edited by *Hugh J. M. Brady*, 2004

METHODS IN MOLECULAR BIOLOGY™

Plant Cell Culture Protocols

Second Edition

Edited by

Victor M. Loyola-Vargas

and

Felipe Vázquez-Flota

*Unidad de Bioquímica y Biología Molecular de Plantas,
Centro de Investigacion Cientifica de Yucatán, Yucatán, México*

HUMANA PRESS ✹ TOTOWA, NEW JERSEY

© 2006 Humana Press Inc.
999 Riverview Drive, Suite 208
Totowa, New Jersey 07512

www.humanapress.com

All rights reserved. No part of this book may be reproduced, stored in a retrieval system, or transmitted in any form or by any means, electronic, mechanical, photocopying, microfilming, recording, or otherwise without written permission from the Publisher. Methods in Molecular Biology™ is a trademark of The Humana Press Inc.

All papers, comments, opinions, conclusions, or recommendations are those of the author(s), and do not necessarily reflect the views of the publisher.

This publication is printed on acid-free paper. ∞
ANSI Z39.48-1984 (American Standards Institute)
Permanence of Paper for Printed Library Materials.
Cover illustration: Cover illustration: Figure 3 from Chapter 5, "Callus and Suspension Culture Induction, Maintenance, and Characterization," by Alejandrina Robledo-Paz, María Nélida Vázquez-Sánchez, Rosa María Adame-Alvarez, and Alba Estela Jofre-Garfias.

Production Editor: Jennifer Hackworth

Cover design by Patricia F. Cleary

For additional copies, pricing for bulk purchases, and/or information about other Humana titles, contact Humana at the above address or at any of the following numbers: Tel.: 973-256-1699; Fax: 973-256-8341; E-mail: orders@humanapr.com; or visit our Website: www.humanapress.com

Photocopy Authorization Policy:
Authorization to photocopy items for internal or personal use, or the internal or personal use of specific clients, is granted by Humana Press Inc., provided that the base fee of US $30.00 per copy is paid directly to the Copyright Clearance Center at 222 Rosewood Drive, Danvers, MA 01923. For those organizations that have been granted a photocopy license from the CCC, a separate system of payment has been arranged and is acceptable to Humana Press Inc. The fee code for users of the Transactional Reporting Service is: [1-58829-547-8/06 $30.00].

Printed in the United States of America. 10 9 8 7 6 5 4 3 2 1

eISBN: 1-59259-959-1
ISSN: 1064-3745

Library of Congress Cataloging-in-Publication Data
Plant cell culture / edited by Victor M. Loyola-Vargas and Felipe
  Vázquez-Flota. – 2nd. ed.
    p. cm. – (Methods in molecular biology)
  Includes bibliographical references (p.).
  ISBN 1-58829-547-8 (alk. paper)
   1. Plant cell culture–Laboratory manuals. 2. Plant tissue culture–Laboratory manuals.
  I. Loyola-Vargas, Victor M. II. Vázquez-Flota, Felipe. III. Methods in molecular biology (Clifton, N.J.)

QK725.P5535 2005
571.6'382–dc22

2005046025

Preface

Cell culture methodologies have become standard procedures in most plant laboratories. Full courses on the subject are taught as part of the curricula of several programs in biological and agronomical sciences. However, the apparent simplicity of cell culture technology should not lead to mistakenly considering it as trivial. It would be very difficult to understand modern plant biotechnology without in vitro cell cultures, and even today, the application of recombinant DNA technology to important crops, such as peppers, is hampered by the poor embryogenic or morphogenetic response of in vitro cultured tissues.

This second edition of *Plant Cell Culture Protocols* follows a similar plot as its predecessor. It also pursues similar goals; that is, to provide an updated step-by-step guide to the most common and applicable techniques and methods for plant tissue and cell culture. A total of 30 chapters, divided into six sections, have been included. Topics selected cover from general methodologies, such as culture induction, growth and viability evaluation, and contamination control, to such highly specialized techniques as chloroplast transformation, passing through the laborious process of protoplast isolation and culture. Most of the protocols are currently used in the research programs of the authors, or represent important parts of business projects aimed at the generation of improved plant materials. Two appendixes have also been included; the first discusses common principles for the formulation of culture media and also lists the composition of the eight most commonly used media formulations. The second appendix compiles a list of useful internet sites for cell culture scientists. A total of more than 100 sites have been selected, based on the quality of the information offered in them, as well as on their user friendliness. We anticipate that some these sites will be included among the reader's favorites (if they aren't already).

We sincerely hope that readers will find this new edition of *Plant Cell Culture Protocols*, which resides in the highly useful Methods in Molecular Biology series, to be a major resource of information for their research projects; this has been its real purpose.

We would like to thank the authors of each chapter for responding to our constant requests for the materials, despite their reckless work schedules. They made it possible for us to carry this job to its completion.

Finally, we should express our profound gratitude to Professor John Walker for his trust in our experience to complete this project. We have certainly enjoyed the opportunity to interact with colleagues from all over the world.

Victor M. Loyola-Varga
Felipe Vázquez-Flota

Contents

Preface v
Contributors xi

PART I. INTRODUCTION
1 An Introduction to Plant Cell Culture: *Back to the Future*
   Victor M. Loyola-Vargas and Felipe Vázquez-Flota 3
2 History of Plant Tissue Culture
   Trevor A. Thorpe 9

PART II. CELL CULTURE AND PLANT REGENERATION: THE FUNDAMENTALS
3 Pathogen and Biological Contamination Management:
   The Road Ahead
   Alan C. Cassells and Barbara M. Doyle 35
4 Growth Measurements: *Estimation of Cell Division
   and Cell Expansion*
   Gregorio Godoy-Hernández and Felipe A. Vázquez-Flota 51
5 Callus and Suspension Culture Induction, Maintenance,
   and Characterization
   **Alejandrina Robledo-Paz, María Nélida
   Vázquez-Sánchez, Rosa María Adame-Alvarez,
   and Alba Estela Jofre-Garfias** 59
6 Measurement of Cell Viability in In Vitro Cultures
   **Lizbeth A. Castro-Concha, Rosa María Escobedo,
   and María de Lourdes Miranda-Ham** 71
7 Cryopreservation of Embryogenic Cell Suspensions
   by Encapsulation-Vitrification
   Qiaochun Wang and Avihai Perl 77
8 Somatic Embryogenesis in *Picea* Suspension Cultures
   Claudio Stasolla 87
9 Indirect Somatic Embryogenesis in Cassava for Genetic
   Modification Purposes
   **Krit Raemakers, Isolde Pereira, Herma Koehorst van Putten,
   and Richard Visser** 101

10 Direct Somatic Embryogenesis in *Coffea canephora*
   **Francisco Quiroz-Figueroa, Miriam Monforte-González,
   Rosa M. Galaz-Ávalos, and Victor M. Loyola-Vargas** 111

PART III. PLANT PROPAGATION IN VITRO
11 A New Temporary Immersion Bioreactor System
   for Micropropagation
   **Manuel L. Robert, José Luis Herrera-Herrera,
   Gastón Herrera-Herrera, Miguel Ángel Herrera-Alamillo,
   and Pedro Fuentes-Carrillo** 121
12 Protocol to Achieve Photoautotrophic Coconut Plants
   Cultured In Vitro With Improved Performance Ex Vitro
   **Gabriela Fuentes, Carlos Talavera, Yves Desjardins,
   and Jorge M. Santamaría** 131
13 Use of Statistics in Plant Biotechnology
   Michael E. Compton 145
14 An Efficient Method for the Micropropagation of *Agave* Species
   **Manuel L. Robert, José Luis Herrera-Herrera, Eduardo Castillo,
   Gabriel Ojeda, and Miguel Angel Herrera-Alamillo** 165
15 Micropropagation of Endangered Plant Species
   Zhihua Liao, Min Chen, Xiaofen Sun, and Kexuan Tang 179
16 Clonal Propagation of Softwoods
   Trevor A. Thorpe, Indra S. Harry, and Edward C. Yeung 187

PART IV. APPLICATIONS FOR PLANT PROTOPLASTS
17 Isolation, Culture, and Plant Regeneration From Leaf
   Protoplasts of *Passiflora*
   **Michael R. Davey, Paul Anthony, J. Brian Power,
   and Kenneth C. Lowe** 201
18 Isolation, Culture, and Plant Regeneration From *Echinacea
   purpurea* Protoplasts
   **Zeng-guang Pan, Chun-zhao Liu, Susan J. Murch,
   and Praveen K. Saxena** 211
19 Production of Cybrids in Brassicaceae Species
   **Maksym Vasylenko, Olga Ovcharenko, Yuri Gleba,
   and Nikolay Kuchuk** 219
20 Guard Cell Protoplasts: *Isolation, Culture,
   and Regeneration of Plants*
   Gary Tallman 233

Contents

21 Production of Interspecific Hybrid Plants in *Primula*
    Juntaro Kato and Masahiro Mii 253

PART V. PROTOCOLS FOR GENOMIC MANIPULATION
22 *Agrobacterium*-Mediated Transformation of Petunia Leaf Discs
    Ingrid M. van der Meer 265
23 Transformation of Wheat Via Particle Bombardment
    Indra K. Vasil and Vilma Vasil 273
24 Chloroplast Transformation
    Xiao-Mei Lu, Wei-Bo Yin, and Zan-Min Hu 285
25 The Biochemical Basis for the Resistance to Aluminum
    and Their Potential as Selection Markers
    César De los Santos-Briones and Teresa Hernández-Sotomayor 305
26 Transformation of Maize Via *Agrobacterium tumefaciens*
    Using a Binary Co-Integrate Vector System
    Zuo-yu Zhao and Jerry Ranch 315

PART VI. ACCUMULATION OF METABOLITES IN PLANT CELLS
27 Capsaicin Accumulation in *Capsicum* spp. Suspension Cultures
    Neftalí Ochoa-Alejo 327
28 Isolation and Purification of Ribosome-Inactivating Proteins
    **Sang-Wook Park, Balakrishnan Prithiviraj, Ramarao
    Vepachedu, and Jorge M. Vivanco** 335
29 *Catharanthus roseus* Shoot Cultures for the Production
    of Monoterpenoid Indole Alkaloids
    **Elizabeta Hernández-Domínguez, Freddy Campos-Tamayo,
    Mildred Carrillo-Pech, and Felipe Vázquez-Flota** 349
30 Methods for Regeneration and Transformation in *Eschscholzia
    californica*: A Model Plant to Investigate Alkaloid Biosynthesis
    Benjamin P. MacLeod and Peter J. Facchini 357

Appendix A: The Components of the Culture Media 369
Appendix B: Plant Biotechnology and Tissue Culture
    Resources on the Internet 379

Index 385

Contributors

Rosa María Adame-Alvarez • *Departamento de Ingeniería Genética, Unidad Irapuato, Centro de Investigación y de Estudios Avanzados del I.P.N., Irapuato, Guanajuato, México*
Paul Anthony • *Plant Sciences Division, School of Biosciences, University of Nottingham, Sutton Bonington Campus, Loughborough, UK*
Freddy Campos-Tamayo • *Unidad de Bioquímica y Biología Molecular de Plantas, Centro de Investigación Científica de Yucatán, Yucatán, México*
Mildred Carrillo-Pech • *Unidad de Bioquímica y Biología Molecular de Plantas, Centro de Investigación Científica de Yucatán, Yucatán, México*
Alan C. Cassells • *Department of Zoology, Ecology and Plant Science, The National University of Ireland, Cork, Ireland*
Eduardo Castillo • *Unidad de Biotecnología, Centro de Investigación Científica de Yucatán, Yucatán, México*
Lizbeth A. Castro-Concha • *Unidad de Bioquímica y Biología Molecular de Plantas, Centro de Investigación Científica de Yucatán, Yucatán, México*
Min Chen • *School of Life Sciences, Institute of Modern Biopharmaceuticals, Southwest China Normal University, Chongqing, P.R. China*
Michael E. Compton • *School of Agriculture, University of Wisconsin–Platteville, Platteville, WI*
Michael R. Davey • *Plant Sciences Division, School of Biosciences, University of Nottingham, Sutton Bonington Campus, Loughborough, UK*
César De los Santos-Briones • *Unidad de Bioquímica y Biología Molecular de Plantas, Centro de Investigación Científica de Yucatán, Yucatán, México*
Yves Desjardins • *Centre de Recherche en Horticulture, Département de Phytologie, Faculté des Sciences de l'Agriculture et de l'Alimentation, Université Laval, Ste-Foy, Québec, G1K7P4, Canada*
Barbara M. Doyle • *Department of Zoology, Ecology, and Plant Science, The National University of Ireland, Cork, Ireland*

ROSA MARÍA ESCOBEDO • *Unidad de Bioquímica y Biología Molecular de Plantas, Centro de Investigación Científica de Yucatán, Yucatán, México*

PETER J. FACCHINI • *Department of Biological Sciences, University of Calgary, Calgary, Alberta, Canada*

GABRIELA FUENTES • *Centro de Investigación Científica de Yucatán, Unidad de Biotecnología, Calle 43, No. 130, Col. Chuburná de Hidalgo C.P. 97200, Mérida, Yucatán, México*

PEDRO FUENTES-CARRILLO • *Unidad de Biotecnología, Centro de Investigación Científica de Yucatán, Yucatán, México*

ROSA M. GALAZ-ÁVALOS • *Unidad de Bioquímica y Biología Molecular de Plantas, Centro de Investigación Científica de Yucatán, Yucatán, México*

YURI GLEBA • *Institute of Cell Biology and Genetic Engineering, Kiev, Ukraine.*

GREGORIO GODOY-HERNÁNDEZ • *Unidad de Bioquímica y Biología Molecular de Plantas, Centro de Investigación Científica de Yucatán, Yucatán, México*

INDRA S. HARRY • *SemBioSys, Calgary, Alberta, Canada*

ELIZABETA HERNÁNDEZ-DOMÍNGUEZ • *Unidad de Bioquímica y Biología Molecular de Plantas, Centro de Investigación Científica de Yucatán, Yucatán, México*

TERESA HERNÁNDEZ-SOTOMAYOR • *Unidad de Bioquímica y Biología Molecular de Plantas, Centro de Investigación Científica de Yucatán, Yucatán, México*

JOSÉ LUIS HERRERA-HERRERA • *Unidad de Biotecnología, Centro de Investigación Científica de Yucatán, Col. Chuburná de Hidalgo, Mérida, Yucatán, México*

GASTÓN HERRERA-HERRERA • *Unidad de Biotecnología, Centro de Investigación Científica de Yucatán, Col. Chuburná de Hidalgo, Mérida, Yucatán, México*

MIGUEL ANGEL HERRERA-ALAMILLO • *Unidad de Biotecnología, Centro de Investigación Científica de Yucatán, Yucatán, México*

ZAN-MIN HU • *Institute of Genetics and Developmental Biology, Chinese Academy of Sciences, Beijing, P.R. China*

ALBA ESTELA JOFRE-GARFIAS • *Departamento de Ingeniería Genética, Unidad Irapuato, Centro de Investigación y de Estudios Avanzados del I.P.N, Irapuato, Guanajuato, México*

Contributors

JUNTARO KATO • *Department of Biology, Aichi-University of Education, Kariya, Aichi, Japan*

HERMA KOEHORST VAN PUTTEN • *Laboratory of Plant Breeding, Graduate School of Experimental Plant Sciences, Wageningen University, Wageningen, The Netherlands*

NIKOLAY KUCHUK • *Institute of Cell Biology and Genetic Engineering, Kiev, Ukraine*

ZHIHUA LIAO • *State Key Laboratory of Genetic Engineering, School of Life Sciences, Fudan-SJTU-Nottingham Plant Biotechnology Research and Development Center, Morgan-Tan International Center for Life Sciences, Fudan University, Shanghai, P.R. China; and School of Life Sciences, Institute of Modern Biopharmaceuticals, Southwest China Normal University, Chongqing, P.R. China*

CHUN-ZHAO LIU • *State Key Laboratory of Biochemical Engineering, Institute of Process Engineering, Chinese Academy of Sciences, P.R. China*

KENNETH C. LOWE • *School of Biology, University of Nottingham, University Park, Nottingham, UK*

VÍCTOR M. LOYOLA-VARGAS • *Unidad de Bioquímica y Biología Molecular de Plantas, Centro de Investigación Científica de Yucatán, Yucatán, México*

BENJAMIN P. MACLEOD • *Department of Biological Sciences, University of Calgary, Calgary, Alberta, Canada*

MASAHIRO MII • *Faculty of Horticulture, Chiba University, Matsudo, Chiba, Japan*

MARÍA DE LOURDES MIRANDA-HAM • *Unidad de Bioquímica y Biología Molecular de Plantas, Centro de Investigación Científica de Yucatán, Mérida, Yucatán, México*

XIAO-MEI LU • *Institute of Genetics and Developmental Biology, Chinese Academy of Sciences, Beijing, P.R. China; and Tomato Genetics Lab, Northeast Agriculture University, Harbin, P.R. China*

MIRIAM MONFORTE-GONZÁLEZ • *Unidad de Bioquímica y Biología Molecular de Plantas, Centro de Investigación Científica de Yucatán, Yucatán, México*

SUSAN J. MURCH • *Department of Plant Agriculture, University of Guelph, Guelph, Canada*

NEFTALÍ OCHOA-ALEJO • *Departamento de Ingeniería Genética, Unidad Irapuato, Centro de Investigación y de Estudios Avanzados del I.P.N., Guanajuato, México.*

GABRIEL OJEDA • *Unidad de Biotecnología, Centro de Investigación Científica de Yucatán, Yucatán, México*
OLGA OVCHARENKO • *Institute of Cell Biology and Genetic Engineering, Kiev, Ukraine*
ZENG-GUANG PAN • *Department of Plant Agriculture, University of Guelph, Guelph, Canada*
SANG-WOOK PARK • *Department of Horticulture and Landscape Architecture, Colorado State University, Fort Collins, Colorado, CO*
ISOLDE PEREIRA • *Laboratory of Plant Breeding, Graduate School of Experimental Plant Sciences, Wageningen University, Wageningen, The Netherlands*
AVIHAI PERL • *Department of Fruit Tree Breeding and Molecular Genetics, Agricultural Research Organization, The Volcani Center, Bet Dagan, Israel*
J. BRIAN POWER • *Plant Sciences Division, School of Biosciences, University of Nottingham, Sutton Bonington Campus, Loughborough, UK*
BALAKRISHNAN PRITHIVIRAJ • *Department of Horticulture and Landscape Architecture, Colorado State University, Fort Collins, CO*
FRANCISCO R. QUIROZ-FIGUEROA • *Unidad de Bioquímica y Biología Molecular de Plantas, Centro de Investigación Científica de Yucatán, Yucatán, México*
KRIT RAEMAKERS • *Laboratory of Plant Breeding, Graduate School of Experimental Plant Sciences, Wageningen University, Wageningen, The Netherlands*
JERRY RANCH • *Pioneer Hi-Bred International Inc., Johnston, IA*
MANUEL L. ROBERT • *Unidad de Biotecnología, Centro de Investigación Científica de Yucatán, Yucatán, México*
ALEJANDRINA ROBLEDO-PAZ • *Programa de Semillas. Colegio de Postgraduados. Montecillo, Edo. de México, México*
JORGE M. SANTAMARÍA • *Centro de Investigación Científica de Yucatán, Unidad de Biotecnología, Col. Chuburná de Hidalgo, Mérida, Yucatán, México*
PRAVEEN K. SAXENA • *Department of Plant Agriculture, University of Guelph, Guelph, Canada*
CLAUDIO STASOLLA • *Department of Plant Science, University of Manitoba, Winnipeg, Canada*
XIAOFEN SUN • *State Key Laboratory of Genetic Engineering, School of Life Sciences, Fudan-SJTU-Nottingham Plant Biotechnology Research and Development Center, Morgan-Tan International Center for Life Sciences, Fudan University, Shanghai, P.R. China*

Contributors

CARLOS TALAVERA • *Centro de Investigación Científica de Yucatán, Unidad de Biotecnología, 130, Col. Chuburná de Hidalgo, Mérida, Yucatán, México*
GARY TALLMAN • *Department of Biology, Willamette University, Salem, OR*
KEXUAN TANG • *State Key Laboratory of Genetic Engineering, School of Life Sciences, Fudan-SJTU-Nottingham Plant Biotechnology Research and Development Center, Morgan-Tan International Center for Life Sciences, Fudan University; and Plant Biotechnology Research Center, School of Agriculture and Biology, Shanghai Jiao Tong University, Shanghai, P.R. China*
TREVOR A. THORPE • *Department of Biological Science, University of Calgary, Calgary, Alberta, Canada*
INGRID M. VAN DER MEER • *Business Unit Bioscience, Plant Research International B.V., Wageningen, The Netherlands.*
INDRA K. VASIL • *Laboratory of Plant Cell and Molecular Biology, University of Florida, Gainesville, FL*
VILMA VASIL • *Laboratory of Plant Cell and Molecular Biology, University of Florida, Gainesville, FL*
MAXIM VASYLENKO • *Institute of Cell Biology and Genetic Engineering, Kiev, Ukraine*
MARÍA NÉLIDA VÁZQUEZ-SÁNCHEZ • *Departamento de Ingeniería Genética, Unidad Irapuato, Centro de Investigación y de Estudios Avanzados del I.P.N., Irapuato, Guanajuato, México*
FELIPE VÁZQUEZ-FLOTA • *Unidad de Bioquímica y Biología Molecular de Plantas, Centro de Investigación Científica de Yucatán, Yucatán, México*
RAMARAO VEPACHEDU • *Department of Horticulture and Landscape Architecture, Colorado State University, Fort Collins, CO*
RICHARD VISSER • *Laboratory of Plant Breeding, Graduate School of Experimental Plant Sciences, Wageningen University, Wageningen, The Netherlands*
JORGE M. VIVANCO • *Department of Horticulture and Landscape Architecture, Cell and Molecular Biology Program, Colorado State University, Fort Collins, CO*
QIAOCHUN WANG • *Institute of Horticulture, Sichuan Academy of Agricultural Science, Chengdu, Sichuan, P.R, China*
EDWARD C. YEUNG • *Department of Biological Sciences, University of Calgary, Calgary, Alberta, Canada*
WEI-BO YIN • *Institute of Genetics and Developmental Biology, Chinese Academy of Sciences, Beijing, P.R. China*
ZUO-YU ZHAO • *Pioneer Hi-Bred International, Inc., Johnston, IA*

I

INTRODUCTION

1

An Introduction to Plant Cell Culture

Back to the Future

Victor M. Loyola-Vargas and Felipe Vázquez-Flota

Summary

Plant cell, tissue, and organ culture is a set of techniques designed for the growth and multiplication of cells and tissues using nutrient solutions in an aseptic and controlled environment. This technology explores conditions that promote cell division and genetic reprogramming in in vitro conditions. Mainly developed in the early 1960s, plant tissue culture has turned into a standard procedure for modern biotechnology and today one can recognize five major areas where in vitro cell cultures are currently applied: large-scale propagation of elite materials, generation of genetic modified fertile individuals, as a model system for fundamental plant cell physiology aspects, preservation of endangered species, and metabolic engineering of fine chemicals. This chapter reviews the recent advances in such areas.

Key Words: Aseptic culture; genetic modified organisms; large-scale propagation; metabolic engineering; plant cell culture; techniques.

1. Introduction

Multicellular organisms vary greatly in their complexity, developmental patterns, and reproductive strategies. Nevertheless, and despite such differences, all of them share the common feature of having a single-celled phase at the very beginning of their life cycles. This single cell will become a new individual cell after several rounds of division, resulting in the development of structures made of specialized cells. For this reason, the unicellular stage is recognized as totipotent; this attribute will be present only in the cells produced during the initial rounds of division. Once the early stages of cell division have been completed, and different sets of cells have been committed to form specialized structures, totipotency is almost lost. The knowledge of the events that govern the transformation of that single cell into a complete and functional individual lies at the core of the understanding of life itself.

Biologists have known for many years that somatic plant cells are more amenable to be reprogrammed than cells isolated from animal tissues. If exposed to the proper conditions, even the most specialized plant cells can be induced to express genes that otherwise would only be expressed at the onset of the process of morphogenesis. As a result of such a change in the pattern of gene expression, new cell types, different from the original one, may be formed. Plant cell and tissue culture technology explores conditions that promote cell division and genetic re-programming in in vitro conditions.

Plant tissue culture refers to the set of techniques designed for the growth and multiplication of cells, tissues, and organs using nutrient solutions in an aseptic and controlled environment (*see* Chapters 2, 4–7 and **Fig. 1**). Throughout the years, since the early 1960s to the mid-1980s (*see* Chapter 2), tissue culture has turned into a basic asset to modern biotechnology, and facilities for in vitro cell cultures are found today in practically each plant biology laboratory, serving different purposes, from the fundamental biochemical aspects to the massive propagation of selected individuals. Browsing through the most influential journals on plant sciences, one can recognize five major areas where in vitro cell cultures are being currently applied: large-scale propagation of elite materials, generation of genetic modified fertile individuals, as a model system for fundamental plant cell physiology aspects, preservation of endangered species, and metabolic engineering of fine chemicals.

2. Large-Scale Propagation of Elite Materials

The commercial technology for plant massive multiplication is primarily based on micropropagation (*see* Chapters 11, 14, 16). Micropropagation protocols are aimed at the rapid multiplication of plantlets true-to-type to the original material. Meristematic tissues, located either on terminal or axillary buds, are induced to proliferate in response to hormonal treatments. Hypocotyls are also frequently used as the original explant, because bud formation can be easily induced in them as compared with more mature tissues. Culture conditions, mainly nitrogen source, light regime, and temperature, can play critical roles in favoring bud development into vitroplants. Although the propagation protocols for a number of plant species were developed in the 1960s, transition of in vitro to ex vitro conditions even now frequently represents a bottleneck step. Therefore, successful commercial propagation protocols also require procedures for the establishment of plants to field conditions in addition to the efficient multiplication of plantlets (*see* Chapter 12). The massive production of

Introduction

Fig. 1. (**A**) Callus from *Catharanthus roseus*. (**B**) Suspension culture from *Coryphanta* spp. (**C**) Tumors from *C. roseus*. (**D**) Hairy roots from *C. roseus*. (**E**) Regeneration of plantlets from *C. roseus* callus. (**F**) Protoplasts from *Coffea arabica*. (**G**) Micropropagation of *Agave tequilana*. (**H**) Somatic embryogenesis of *Coffea canephora*. (**I**) Root culture from *Psacalium decompositum*. **A**, **C**, **D**, **E**, **F**, **H**, and **I** from the authors' laboratories. **B** from the laboratory of Dr. Lourdes Miranda. **G** from the laboratory of Dr. Manuel Robert. All of them are from Centro de Investigación Cientifica de Yucatán.

embryos, and their later development into entire plants, can also provide a methodology for the propagation of selected materials (*see* Chapter 8).

3. Generation of Genetic Modified Fertile Individuals

The events required to genetically transform an organism occur at the cellular level. Plant transformation protocols require of both a reliable DNA delivery methodology, as well as the efficient regeneration of fertile individuals from the modified cells. Techniques to introduce foreign genes into plant genomes include the *Agrobacterium* system and the bombardment of DNA-covered microprojectiles (*see* Chapters 22–24, 26). Selective markers, such as antibiotic resistance, chromophores, or fluorochromes, are incorporated to distinguish the transformed tissues from those untransformed. Genetically modified plants will rise from individual cells and, because DNA insertion is a random process, an efficient regeneration procedure will increase the probability of recovering a transgenic plant. For this reason, the use of tissues with a high-morphogenetic or embryogenic potential is recommended (*see* Chapters 9 and 10). Protoplasts can also be used (*see* Chapters 17–21); however, they may require a considerable amount of labor before regenerating a new plant, although with better odds of obtaining actual transformants.

4. In Vitro Cultures as a Model for the Study of Plant Cell Physiology

In vitro cultures represent an advantageous system for the study of different processes in plant cell physiology. Conditions can be strictly controlled, allowing monitoring of the effects of a single factor on a given process. One of the best examples of the use of cell cultures with such purposes may be the triggering of the defense mechanisms in response to elicitation. A number of genes involved in different aspects of such response, including those in perception of the stimulus, as well as in the signaling pathway, have been isolated and characterized in cell cultures from different species. Elicitation of secondary metabolism in cell cultures has also led to the identification of the biosynthetic enzymes involved in the process, in addition to the isolation of both structural and regulatory genes (*see* Chapter 30). Cell mechanisms for resistance to metals (*see* Chapter 25), salinity or drought among others, can be analyzed without having the interference of tissue organization. Furthermore, in vitro cultures submitted to morphogenetic conditions provide an optimum system for the study of the biochemical and molecular aspects associated with plant differentiation (*see* Chapter 10).

5. Preservation of Endangered Species and Phytogenetic Resources

Every year, an important number of plant species disappear, partly as a result of the loss of natural areas. Plants with a complex reproductive biology are particularly endangered given the reduction of their natural habitats, along with the small sizes of their populations and their prolonged life cycle. Furthermore, endangered, asexually propagated plants have to deal with the reduction of their genetic variability, which increases their susceptibility to an abrupt environmental change or to the introduction of new elements into their ecosystem. In vitro culture represents an alternative to preserve and regenerate endangered species' populations through micropropagation techniques (*see* Chapter 15). New sources of genetic variation can be generated that arise from somaclonal variation, which is inherent to callus or cell suspension cultures (*see* Chapter 6).

Introduction

It is not only tropical and exotic species that are endangered. The use of improved plant varieties have resulted in the diminished utilization of traditional varieties of several crops, such as maize, potato, and tomato. Quite often, these traditional varieties, which may have been bred for hundreds of years, and are still cultivated by farmers of small communities, are adapted to very specific environments or conditions isolated by distance or geographical conditions. Besides their cultural value, they may represent an unexplored source for resistance genes to pathogens, insects, and drought. In vitro culture provides the technology for the preservation of such phytogenetic resources, which may not be adapted to flourish in nurseries or under greenhouse conditions. In vitro cultures may also be used to preserve extended collections of germoplasm in reduced areas under strictly controlled environments. This approach is particularly valuable in the case of plants, which are vegetatively propagated. Terminal or axillary buds cultured in vitro may also be preserved by cryogenic techniques, thereby minimizing the excessive tissue manipulation required (*see* Chapter 7). The preservation of valuable tropical genetic resources, deposited in germoplasm banks and maintained by means of in vitro techniques, represents a growing trend in tissue culture applications.

6. Metabolic Engineering of Fine Chemicals

Once the technology for culturing plant cells in a manner similar to fungal and bacterial cells was available, the production of fine chemicals (natural products or secondary metabolites) was among the first applications to be pursued. However, differing from fungi and bacteria, the pattern of natural products yielded by plant cells in culture frequently showed variations from those of organized tissues. Despite many years and numerous attempts by several laboratories around the world, in vitro cell cultures have not turned out to be efficient factories of natural products. Nevertheless, cell cultures proved to be an invaluable source for enzymes and genes involved in the synthesis of these metabolites (*see* Chapter 30), as well as for establishing the relationship between cell differentiation and secondary metabolism. Early approaches to promote the production of secondary metabolites included the induction of cell lines from highly productive tissues or individuals, the systematic screening of numerous cell lines for strains with a high-biosynthetic potential, and the formulation of culture media composition. Because very little was known about the regulation of secondary metabolism, such approaches were based on following a series of trial and error assays, rather than on a solid biochemical rationalization. The better understanding of the tight regulation governing secondary metabolism pathways as well as its close relationship with branches of the primary metabolism, can now be applied through metabolic engineering strategies in order to promote the accumulation of valuable natural products through use of in vitro cultures. Metabolic engineering is aimed to improve cell processes, by means of recombinant DNA technology, for commercial purposes. Genes coding for enzymes involved in limiting steps in a pathway may be over-expressed in cell cultures favoring carbon flux through it (*see* Chapters 27–30). Alternatively, new enzymatic activities can be introduced, resulting in the formation of new compounds. Recently, the identification of regulatory genes, controlling the coordinated activation of a set of enzymes involved in secondary metabolism, has created new possibilities for the genetic manipulation of the whole pathway, by means of a single gene.

7. Future Perspectives

Plant cell cultures have turned into an invaluable tool to plant scientists and today in vitro culture techniques are standard procedures in most laboratories. However, the usefulness of this technology goes beyond academic laboratories. Massive propagation of plants represents an economically rewarding enterprise as a number of companies report significant profits every year. Plant biotechnology has experienced new and exciting advances thanks to the development of genomics, proteomics, and more recently, metabolomics. These approaches will, without doubt, accelerate the discovery, isolation, and characterization of genes, thus conferring new agronomic traits to crops. However, as mentioned in Chapters 22–26, successful genetic engineering programs for the development of new plant varieties, require the stable insertion of those genes into the plant genome and their stable transmission to the offspring. Nevertheless, in some cases, regeneration of whole plants from transgenic cells or tissues may not be necessary. The production of fine chemicals in genetically modified cells cultured in large volumes may prove advantageous over production in field grown plants that would occupy considerable extensions of land, which could, in turn, be used for agriculture. Therefore, technologies for cell culture in such volume conditions should be established. The requirements for the development of such methodologies seem enough to keep cell culture scientists busy for a number of years.

2

History of Plant Tissue Culture

Trevor A. Thorpe

Summary

Plant tissue culture, or the aseptic culture of cells, tissues, organs, and their components under defined physical and chemical conditions in vitro, is an important tool in both basic and applied studies as well as in commercial application. It owes its origin to the ideas of the German scientist, Haberlandt, at the beginning of the 20th century. The early studies led to root cultures, embryo cultures, and the first true callus/tissue cultures. The period between the 1940s and the 1960s was marked by the development of new techniques and the improvement of those already in use. It was the availability of these techniques that led to the application of tissue culture to five broad areas, namely, cell behavior (including cytology, nutrition, metabolism, morphogenesis, embryogenesis, and pathology), plant modification and improvement, pathogen-free plants and germplasm storage, clonal propagation, and product (mainly secondary metabolite) formation, starting in the mid-1960s. The 1990s saw continued expansion in the application of the in vitro technologies to an increasing number of plant species. Cell cultures have remained an important tool in the study of basic areas of plant biology and biochemistry and have assumed major significance in studies in molecular biology and agricultural biotechnology. The historical development of these in vitro technologies and their applications are the focus of this chapter.

Key Words: Cell behavior; cell suspensions; clonal propagation; organogenesis; plantlet regeneration; plant transformation; protoplasts; somatic embryogenesis; vector-dependent/independent gene transfer.

1. Introduction

Plant tissue culture, also referred to as cell, in vitro, axenic, or sterile culture, is an important tool in both basic and applied studies, as well as in commercial application *(1)*. Plant tissue culture is the aseptic culture of cells, tissues, organs and their components under defined physical and chemical conditions in vitro. The theoretical basis for plant tissue culture was proposed by

Gottlieb Haberlandt in his address to the German Academy of Science in 1902 on his experiments on the culture of single cells *(2)*. He opined that, to my knowledge, no systematically organized attempts to culture isolated vegetative cells from higher plants have been made. Yet the results of such culture experiments should give some interesting insight to the properties and potentialities that the cell, as an elementary organism, possesses. Moreover, it would provide information about the interrelationships and complementary influences to which cells within a multicellular whole organism are exposed (from the English translation, *[3]*). He experimented with isolated photosynthetic leaf cells and other functionally differed cells and was unsuccessful, but nevertheless he predicted that one could successfully cultivate artificial embryos from vegetative cells. He, thus, clearly established the concept of totipotency, and further indicated that the technique of cultivating isolated plant cells in nutrient solution permits the investigation of important problems from a new experimental approach. On the basis of that 1902 address and his pioneering experimentation before and later, Haberlandt is justifiably recognized as the father of plant tissue culture. Other studies led to the culture of isolated root tips *(4,5)*. This approach of using explants with meristematic cells produced the successful and indefinite culture of tomato root tips *(6)*. Further work allowed for root culture on a completely defined medium. Such root cultures were used initially for viral studies and later as a major tool for physiological studies *(7)*. Success was also achieved with bud cultures *(8,9)*.

Embryo culture also had its beginning early in the first decade of the last century with barley embryos *(10)*. This was followed by the successful rescue of embryos from nonviable seeds of a cross between *Linum perenne* ↔ *Linum austriacum* *(11)*, and for full embryo development in some early-ripening species of fruit trees *(12)*; thus providing one of the earliest applications of in vitro culture. The phenomenon of precocious germination was also encountered *(13)*.

The first true plant tissue cultures were obtained by Gautheret *(14,15)* from cambial tissue of *Acer pseudoplatanus*. He also obtained success with similar explants of *Ulmus campestre, Robinia pseudoacacia*, and *Salix capraea* using agar-solidified medium of Knop's solution, glucose and cysteine hydrochloride. Later, the availability of indole acetic acid and the addition of B vitamins allowed for the more or less simultaneous demonstrations with carrot root tissues *(16,17)*, and with tumor tissue of a *Nicotiana glauca* ↔ *Nicotiana langsdorffii* hybrid *(18)*, which did not require auxin, that tissues could be continuously grown in culture; and even made to differentiate roots and shoots *(19,20)*. However, all of the initial explants used by these pioneers included meristematic tissue. Nevertheless, these findings set the stage for the dramatic increase in the use of in vitro cultures in the subsequent decades. Greater detail on the early pioneering events in plant tissue culture can be found in White

(21), Bhojwani and Razdan *(22)*, and Gautheret *(23)*. This current article is based on an earlier review by the author *(24)* (used with permission from Elsevier).

2. The Development and Improvement of Techniques

The 1940s, 1950s, and 1960s proved an exciting time for the development of new techniques and the improvement of those already available. The application of coconut water (often incorrectly referred to as coconut milk) allowed for the culture of young embryos *(25)* and other recalcitrant tissues, including monocots. Callus cultures of numerous species, including a variety of woody and herbaceous dicots and gymnosperms, as well as crown gall tissues, were established as well *(23)*. It was recognized at this time that cells in culture underwent a variety of changes, including loss of sensitivity to applied auxin or habituation *(26,27)*, as well as variability of meristems formed from callus *(27,28)*. Nevertheless, it was during this period that most of the in vitro techniques used today were largely developed.

Studies by Skoog et al. *(29)* showed that the addition of adenine and high levels of phosphate allowed nonmeristematic pith tissues to be cultured and to produce shoots and roots, but only in the presence of vascular tissue. Further studies using nucleic acids led to the discovery of the first cytokinin (kinetin), as the breakdown product of herring sperm DNA *(30)*. The availability of kinetin further increased the number of species that could be cultured indefinitely, but perhaps most importantly, led to the recognition that the exogenous balance of auxin and kinetin in the medium influenced the morphogenic fate of tobacco callus *(31)*. A relative high level of auxin to kinetin-favored rooting, the reverse led to shoot formation and intermediate levels to the proliferation of callus or wound parenchyma tissue. This morphogenic model has been shown to operate in numerous species *(32)*. Native cytokinins were subsequently discovered in several tissues, including coconut water *(33)*. The formation of bipolar somatic embryos (carrot) was first reported independently by Reinert *(34,35)* and Steward *(36)* in addition to the formation of unipolar shoot buds and roots.

The culture of single cells (and small cell clumps) was achieved by shaking callus cultures of *Tagetes erecta* and tobacco, and subsequently placing them on filter paper resting on well-established callus, giving rise to the so-called nurse culture *(37,38)*. Later, single cells could be grown in medium in which tissues had already been grown (i.e., conditioned medium) *(39)*. As well, single cells incorporated in a 1-mm layer of solidified medium formed some cell colonies *(40)*. This technique is widely used for cloning cells and in protoplast culture *(22)*. Finally, in 1959, success was achieved in the culture of mechanically isolated mature differentiated mesophyll cells of *Macleaya cordata* *(41)*, and later in the induction of somatic embryos from the callus *(42)*. The first large-scale culture of plant cells was obtained from cell suspensions of Ginkgo, holly, Lolium and rose in simple sparged 20-L carboys *(43)*. The utilization of coconut water as an additive to fresh medium, instead of using conditioned medium, finally led to realization of Haberlandt's dream of producing a whole plant (tobacco) from a single cell by Vasil and Hildebrandt *(44)*, thus demonstrating the totipotency of plant cells.

The earliest nutrient media used for growing plant tissues in vitro were based on the nutrient formulations for whole plants, for which they were many *(21)*; but Knop's solution and that of Uspenski and Uspenskia were used the most, and provided less than 200 mg/L of total salts. Based on studies with carrot and Virginia creeper tissues, the concentration of salts was increased twofold *(45)*, and was further increased ca. 4 g/L, based on work with Jerusalem artichoke *(46)*. However, these changes did not provide optimum growth for tissues, and complex addenda, such as yeast extract, protein hydrolysates, and coconut water, were frequently required. In a different approach, based on an examination of the ash of tobacco callus, Murashige and Skoog (MS) *(47)* developed a new medium. The concentration of some salts was 25 times that of Knop's solution. In particular, the levels of NO_3^- and NH_4^+ were very high and the arrays of micronutrients were increased. MS formulation allowed for a further increase in the number of plant species that could be cultured, many of them using only a defined medium consisting of macro- and micronutrients, a carbon source, reduced N, B vitamins, and growth regulators *(48)*. The MS salt formulation is now the most widely used nutrient medium in plant tissue culture.

Plantlets were successfully produced by culturing shoot tips with a couple of primordia of *Lupinus* and *Tropaeolum* *(9)*, but the importance of this finding was not recognized until later when this approach to obtain virus-free orchids, its potential for clonal propagation was realized *(49)*. The potential was rapidly exploited, particularly with ornamentals *(50)*. Early studies had shown that cultured root tips were free of viruses *(51)*. It was later observed that the virus titer in the shoot meristem was very low *(52)*. This was confirmed when virus-free *Dahlia* plants were obtained from infected plants by culturing their shoot tips *(53)*. Virus elimination was possible because vascular tissues, within which the viruses move, do not extend into the root or shoot apex. The method was further refined *(54)*, and now routinely used.

Techniques for in vitro culture of floral and seed parts were developed during this period. The first attempts at ovary culture yielded limited growth of the ovaries accompanied by rooting of pedicels in several species *(56)*. Compared to studies with embryos, successful ovule culture is very limited. Studies with both ovaries and ovules have been geared mainly to an understanding of factors regulating embryo and fruit development *(56)*. The first continuously growing tissue cultures from an endosperm were from immature maize *(57)*. Plantlet regeneration via organogenesis was later achieved in *Exocarpus cupressiformis* *(58)*.

In vitro pollination and fertilization was pioneered using *Papaver somniferum* *(59)*. The approach involves culturing excised ovules and pollen grains together in the same medium and has been used to produce interspecific and intergeneric hybrids *(60)*. Earlier, cell colonies were obtained from *Ginkgo* pollen grains in culture *(61)*, and haploid callus was obtained from whole anthers of *Tradescantia reflexa* *(62)*. However, it was the finding of Guha and Maheshwari *(63,64)* that haploid plants could be obtained from cultured anthers of *Datura innoxia* that opened the new area of androgenesis. Haploid plants of tobacco were also obtained *(65)*, thus confirming the totipotency of pollen grains.

Plant protoplasts or cells without cell walls were first mechanically isolated from plasmolysed tissues well over 100 yr ago, and the first fusion was achieved in 1909 *(23)*. Nevertheless, this remained an unexplored technology until the use of a fungal cellulase by Cocking *(66)* ushered in a new era. The commercial availability of cell wall degrading enzymes led to their wide use and the development of protoplast technology in the 1970s. The first demonstration of the totipotency of protoplasts was by Takebe et al. *(67)*, who obtained tobacco plants from mesophyll protoplasts. This was followed by the regeneration of the first interspecific hybrid plants (*N. glauca* ↔ *N. langsdorffii*) *(68)*.

Braun *(69)* showed that in sunflower *Agrobacterium tumefaciens* could induce tumors, not only at the inoculated sites, but at distant points. These secondary tumors were free of bacteria and their cells could be cultured without auxin *(70)*. Further experiments showed that crown gall tissues, free of bacteria, contained a tumor-inducing principle (TIP), which was probably a macromolecule *(71)*. The nature of the TIP was worked out in the 1970s *(72)*, but Braun's work served as the foundation for *Agrobacterium*-based transformation. It should also be noted that the finding by Ledoux *(73)* that plant cells could take up and integrate DNA remained controversial for more than a decade.

3. The Recent Past

Based on the availability of the various in vitro techniques discussed in **Subheading 2.**, it is not surprising that, starting in the mid-1960s, there was a dramatic increase in their application to various problems in basic biology, agriculture, horticulture, and forestry through the 1970s and 1980s. These applications can be divided conveniently into five broad areas, namely: (1) cell behavior, (2) plant modification and improvement, (3) pathogen-free plants and germplasm storage, (4) clonal propagation, and (5) product formation *(1)*.

Detailed information on the approaches used can be gleaned from Bhojwani and Razdan *(22)*, Vasil *(74)*, and Vasil and Thorpe *(75)*, among several sources.

3.1. Cell Behavior

Included under this heading are studies dealing with cytology, nutrition, primary, and secondary metabolism, as well as morphogenesis and pathology of cultured tissues *(1)*. Studies on the structure and physiology of quiescent cells in explants, changes in cell structure associated with the induction of division in these explants and the characteristics of developing callus, and cultured cells and protoplasts have been carried out using light and electron microscopy *(76–79)*. Nuclear cytology studies have shown that endoreduplication, endomitosis, and nuclear fragmentation are common features of culture cells *(80,81)*.

Nutrition was the earliest aspect of plant tissue culture investigated, as indicated earlier. Progress has been made in the culture of photoautotrophic cells *(82,83)*. In vitro cultures, particularly cell suspensions, have become very useful in the study of both primary and secondary metabolism *(84)*. In addition to providing protoplasts from which intact and viable organelles were obtained for study (e.g., vacuoles) *(85)*, cell suspensions have been used to study the regulation of inorganic nitrogen and sulfur assimilation *(86)*, carbohydrate metabolism *(87)*, and photosynthetic carbon metabolism *(88,89)*; thus clearly showing the usefulness of cell cultures for elucidating pathway activity. Most of the work on secondary metabolism was related to the potential of cultured cells to form commercial products, but has also yielded basic biochemical information *(90,91)*.

Morphogenesis or the origin of form is an area of research with which tissue culture has long been associated; and one to which tissue culture has made significant contributions both in terms of fundamental knowledge and application *(1)*. Xylogenesis or tracheary element formation has been used to study cytodifferentiation *(92–94)*. In particular the optimization of the *Zinnia* mesophyll single cell system has dramatically improved our knowledge of this process. The classical findings of Skoog and Miller *(31)* on the hormonal balance for organogenesis has continued to influence research on this topic; a concept supported more recently by transformation of cells with appropriately modified *Agrobacterium* T-DNA *(95,96)*. However, it is clear from the literature that several additional factors, including other growth active substances, interact with auxin and cytokinin to bring about *de novo* organogenesis *(97)*. In addition to bulky explants, such as cotyledons, hypocotyls, and callus *(97)*, thin (superficial) cell layers *(98,99)* have been used in traditional morphogenic studies, as well as to produce *de novo* organs and plantlets in hundreds of plant species *(50,100)*. As well, physiological and biochemical studies on organogenesis have been carried out *(97,101,102)*. The third area of morphogenesis, somatic embryogenesis, also developed in this period with over 130 species reported to form the bipolar structures by the early 1980s *(103,104)*. Successful culture was achieved with cereals, grasses, legumes, and conifers, previously considered to be recalcitrant groups. The development of a single cell to embryo system in carrot *(105)* has allowed for an in depth study of the process.

Cell cultures have continued to play an important role in the study of plant-microbe interaction, not only in tumorigenesis *(106)*, but also on the biochemistry of virus multiplication *(107)*, phytotoxin action *(108)*, and disease resistance, particularly as affected by phytoalexins *(109)*. Without doubt the most important studies in this area dealt with *Agrobacteria*, and although aimed mainly at plant improvement (*see* **Subheading 3.2.**) provided good fundamental information *(96)*.

3.2. Plant Modification and Improvement

During this period, in vitro methods were used increasingly as an adjunct to traditional breeding methods for the modification and improvement of plants. The technique of controlled in vitro pollination on the stigma, placenta, or ovule has been used for the production of interspecific and intergeneric hybrids, overcoming sexual self-incompatibility, and the induction of haploid plants *(109)*. Embryo, ovary, and ovule cultures have been used in overcoming embryo inviability, monoploid production in barley and in overcoming seed dormancy and related problems *(110,112)*. In particular, embryo rescue has played a most important role in producing interspecific and intergeneric hybrids *(113)*.

By the early 1980s, androgenesis had been reported in some 171 species, many of which were important crop plants *(114)*. Gynogenesis was reported in some 15 species, in some of which androgenesis was not successful *(115)*. The value of these haploids was that they could be used to detect mutations and for recovery of unique recombinants, because there is no masking of recessive alleles. As well, the production of double haploids allowed for hybrid production and their integration into breeding programs.

Cell cultures have also played an important role in plant modification and improvement, as they offer advantages for isolation of variants *(116)*. Although tissue culture-produced variants have been known since the 1940s (e.g., habituation), it was only in the 1970s that attempts were made to utilize them for plant improvement. This somaclonal variation is dependent on the natural variation in a population of cells, either pre-existing or culture-induced, and is usually observed in regenerated plantlets *(117)*. The variation may be genetic or epigenetic and is not simple in origin *(118,119)*. The changes in the regenerated plantlets have potential agricultural and horticultural significance, but this potential has not yet been realized. It has also been possible to produce a wide spectrum of mutant cells in culture *(120)*. These include cells showing biochemical differences, antibiotic, herbicide, and stress resistance. In addition, auxotrophs, autotrophs, and those with altered developmental systems have been selected in culture; usually the application of the selective agent in the presence of a mutagen is required. However, in only a few cases has it been possible to regenerate plants with the desired traits (e.g., herbicide-resistant tobacco) *(121)*, and methyl tryptophan-resistant *Datura innoxia (122)*.

By 1985, nearly 100 species of angiosperms could be regenerated from protoplasts *(123)*. The ability to fuse plant protoplasts by chemical (e.g., with polyethylene glycol [PEG]) and physical means (e.g., electrofusion) allowed for production of somatic hybrid plants; the major problem being the ability to regenerate plants from the hybrid cells *(124,125)*. Protoplast fusion has been used to produce unique nuclear-cytoplasmic combinations. In one such example, *Brassica campestris* chloroplasts coding for atrazine resistance (obtained from protoplasts) were transferred into *B. napus* protoplasts with *Raphanus sativus* cytoplasm (which confers cytoplasmic male sterility from its mitochondria). The selected plants contained *B. napus* nuclei, chloroplasts from *B. campestris* and mitochondria from *R. sativus*, had the desired traits in a *B. napus* phenotype, and could be used for hybrid seed production *(126)*. Unfortunately, only a few such examples exist to date.

Genetic modification of plants has been achieved by direct DNA transfer via vector-independent and vector-dependent means since the early 1980s. Vector-independent methods with protoplasts include electroporation *(127)*, liposome fusion *(128)*, and microinjection *(129)*, as well as high-velocity microprojectile bombardment (biolistics) *(130)*. This latter method can be executed with cells, tissues, and organs. The use of *Agrobacterium* in vector-mediated transfer has progressed very rapidly since the first reports of stable transformation *(131,132)*. Although the early transformations utilized protoplasts, regenerable organs such as leaves, stems, and roots have been subsequently used *(133,134)*. Much of the research activity utilizing these tools has focused on engineering important agricultural traits for the control of insects, weeds, and plant diseases.

3.3. Pathogen-Free Plants and Germplasm Storage

Although these two uses of in vitro technology may appear unrelated, a major use of pathogen-free plants is for germplasm storage and the movement of living material across international borders *(1)*. The ability to rid plants of viruses, bacteria, and fungi by culturing meristem-tips has been widely used since the 1960s. The approach is particularly needed for virus-infected material, because bactericidal and fungicidal agents can be used successfully in ridding plants of bacteria and fungi *(22)*. Meristem-tip culture is often coupled with thermotherapy or chemotherapy for virus eradication *(135)*.

Traditionally, germplasm has been maintained as seed, but the ability to regenerate whole plants from somatic and gametic cells and shoot apices has led to their use for storage *(22,135)*. Three in vitro approaches have been developed, namely use of growth retarding compounds (e.g., maleic hydrazide, B995, and abscisic acid [ABA]) *(136)*, low nonfreezing temperatures (1–9°C) *(22)*, and cryopreservation *(135)*. In this last approach, cell suspensions, shoot apices, asexual embryos, and young plantlets, after treatment with a cryoprotectant, is frozen and stored at the temperature of liquid nitrogen (ca. −196°C) *(135,137)*.

3.4. Clonal Propagation

The use of tissue culture technology for the vegetative propagation of plants is the most widely used application of the technology. It has been used with all classes of plants *(138,139)*, although some problems still need to be resolved (e.g., hyperhydricity, abberant plants). There are three ways by which micropropagation can be achieved. These are enhancing axillary bud breaking, production of adventitious buds directly or indirectly via callus, and somatic embryogenesis directly or indirectly on explants *(50,138)*. Axillary bud breaking produces the smallest number of plantlets, but they are generally genetically true-to-type; whereas somatic embryogenesis has the potential to produce the greatest number of plantlets, but is induced in the lowest number of plant species. Commercially, numerous ornamentals are produced, mainly via axillary bud breaking *(140)*. As well, there are many lab-scale protocols for other classes of plants, including field and vegetable crops, fruit, plantation, and forest trees, but cost of production is often a limiting factor in their use commercially *(141)*.

3.5. Product Formation

Higher plants produce a large number of diverse organic chemicals, which are of pharmaceutical and industrial interest. The first attempt at the large-scale culture of plant cells for the production of pharmaceuticals took place in the 1950s at the Charles Pfizer Co. The failure of this effort limited research in this area in the United States, but work elsewhere in Germany and Japan in particular, led to development, so that by 1978 the industrial application of cell cultures was considered feasible *(142)*. Furthermore, by 1987, there were 30 cell culture systems that were better producers of secondary metabolites than the respective plants *(143)*. Unfortunately, many of the economically important plant products are either not formed in sufficiently large quantities or not at all by plant cell cultures. Different approaches have been taken to enhance yields of secondary metabolites. These include cell cloning and the repeated selection of high-yielding strains from heterogenous cell populations *(142,144)* and by using enzyme linked immunosorbent assay (ELISA) and radioimmunoassay techniques *(145)*. Another approach involves selection of mutant cell lines that overproduce the desired product *(146)*. As well, both abiotic factors—such as ultraviolet (UV) irradiation, exposure to heat or cold and salts of heavy metals—and biotic elicitors of plant and microbial origin, have been shown to enhance secondary product formation *(147,148)*. Lastly, the use of immobilized cell technology has also been examined *(149,150)*.

Central to the success of producing biologically active substances commercially is the capacity to grow cells on a large scale. This is being achieved using stirred tank reactor systems and a range of air-driven reactors *(141)*. For many systems, a two-stage (or two-phase) culture process has been tried *(151,152)*. In the first stage, rapid cell growth and biomass accumulation are emphasized, whereas the second stage concentrates on product synthesis with minimal cell division or growth. However, by 1987 the naphthoquinone, shikonin, was the only commercially produced secondary metabolite by cell cultures *(153)*.

4. The Present

During the 1990s, continued expansion in the application of in vitro technologies to an increasing number of plant species was observed. Tissue culture techniques are being used with all types of plants, including cereals and grasses *(154)*, legumes *(155)*, vegetable crops *(156)*, potato *(157)* and other root and tuber crops *(158)*, oilseeds *(159)*, temperate *(160)* and tropical *(161)* fruits, plantation crops *(162)*, forest trees *(163)*, and, of course, ornamentals *(164)*. As can be seen from these articles, the application of in vitro cell technology went well beyond micropropagation, and embraced all the in vitro approaches that were relevant or possible for the particular species, and the problem(s) being addressed. However, only limited success has been achieved in exploiting somaclonal variation *(165)*, or in the regeneration of useful plantlets from mutant cells *(166)*; also, the early promise of protoplast technology has remained largely unfulfilled *(167)*. Substantial progress has been made in extending cryopreservation technology for germplasm storage *(168)* and in artificial seed technology *(169)*. Some novel approaches for culturing cells such as on rafts, membranes, and glass rods, as well as manipulation of the culture environment by use of non-ionic surfactants have been successfully developed *(170)*.

Cell cultures have remained an important tool in the study of plant biology. Thus progress is being made in cell biology, for example, in studies of the cytoskeleton *(171)*, on chromosomal changes in cultured cells *(172)*, and in cell-cycle studies *(173,174)*. Better physiological and biochemical tools have allowed for a re-examination of neoplastic growth in cell cultures during habituation and hyperhydricity, and relate it to possible cancerous growth in plants *(175)*. Cell cultures have remained an extremely important tool in the study of primary metabolism; for example, the use of cell suspensions to develop in vitro transcription systems *(176)*, or the regulation of carbohydrate metabolism in transgenics *(177)*. The development of medicinal plant cell culture techniques has led to the identification of more than 80 enzymes of alkaloid biosynthesis (reviewed in **ref.** *178*). Similar information arising from the use of cell cultures for molecular and biochemical studies on other areas of secondary metabolism, is generating research activity on metabolic engineering of plant secondary metabolite production *(179)*.

Cell cultures remain an important tool in the study of morphogenesis, even though the present use of developmental mutants, particularly of *Arabidopsis*, is adding valuable information on plant development (*see* **ref.** *180*). Molecular, physiological, and biochemical studies have allowed for an indepth understanding of cytodifferentiation, mainly tracheary element formation *(181)*, organogenesis *(182,183)*, and somatic embryogenesis *(184–186)*.

Advances in molecular biology are allowing for the genetic engineering of plants, through the precise insertion of foreign genes from diverse biological systems. Three major breakthroughs have played major roles in the development of this transformation technology *(187)*. These are the development of shuttle vectors for harnessing the natural gene transfer capability of *Agrobacterium (188)*, the methods to use these vectors for the direct transformation of regenerable explants obtained from plant organs *(189)*, and the development of selectable markers *(190)*. For species not amenable to *Agrobacterium*-mediated transformation, physical, chemical, and mechanical means are used to get the DNA into the cells. With these latter approaches, particularly biolistics *(191)*, it has become possible to transform virtually any plant species and genotype.

The initial wave of research in plant biotechnology has been driven mainly by the seed and agri-chemical industries, and has concentrated on agronomic traits of direct relevance to these industries, namely the control of insects, weeds, and plant diseases *(192)*. At present, over 100 species of plants have been genetically engineered, including nearly all the major dicotyledonous crops and an increasing number of monocotyledonous ones, as well as some woody plants. Current research is leading to routine gene transfer systems for all important crops; for example, the production of golden rice *(193)*. In addition, technical improvements are further increasing transformation efficiency, extending transformation to elite commercial germplasm and lowering transgenic plant production costs. The next wave in agricultural biotechnology is already in progress with biotechnological applications of interest to the food processing, speciality chemical, and pharmaceutical industries.

The current emphasis and importance of plant biotechnology can be gleamed from the last two International *Congresses on Plant Tissue and Cell Culture and Biotechnology* held in Israel in June 1998, and in the United States in June 2002. The theme of the Israeli Congress was *Plant Biotechnology* and *In Vitro Biology in the 21st Century* and the theme of the last Congress was *Plant Biotechnology 2002 and Beyond*. The proceedings for these two congresses *(194, 195)* were developed through a scientific program that focused on the most important developments, both basic and applied, in the areas of plant tissue culture and molecular biology and their impact on plant improvement and biotechnology. They clearly show where tissue culture is today and where it is heading (i.e., as an equal partner with molecular biology), as a tool in basic plant biology and in various areas of application. In fact, progress in applied plant biotechnology is fully matching and is without doubt stimulating fundamental scientific progress, which remains the best hope for achieving sustainable and environmentally stable agriculture *(196)*. Indeed, the advancements made in the last 100 yr with in vitro technology have gone well beyond what Haberlandt and the other pioneers could have imagined.

References

1. Thorpe, T. A. (1990) The current status of plant tissue culture, in *Plant Tissue Culture: Applications and Limitations* (Bhojwani, S. S., ed.), Elsevier, Amsterdam, pp. 1–33.
2. Haberlandt, G. (1902) Kulturversuche mit isolierten Pflanzenzellen. Sitzungsber. Akad. Wiss. Wien. *Math.-Naturwiss. Kl., Abt. J.* **111,** 69–92.
3. Krikorian, A. D. and Berquam, D. L. (1969) Plant cell and tissue cultures: the role of Haberlandt. *Botan. Rev.* **35,** 59–67.
4. Kotte, W. (1922) Kulturversuche mit isolierten Wurzelspitzen. *Beitr. Allg. Bot.* **2,** 413–434.
5. Robbins, W. J. (1922) Cultivation of excised root tips and stem tips under sterile conditions. *Bot. Gaz.* **73,** 376–390.
6. White, P. R. (1934) Potentially unlimited growth of excised tomato root tips in a liquid medium. *Plant Physiol.* **9,** 585–600.
7. Street, H. E. (1969) Growth in organized and unorganized systems, in *Plant Physiology* (Steward, F. C., ed.), Vol. 5B, Academic Press, New York, pp. 3–224.
8. Loo, S. W. (1945) Cultivation of excised stem tips of asparagus in vitro. *Am. J. Bot.* **32,** 13-17.
9. Ball, E. (1946) Development in sterile culture of stems tips and subjacent regions of *Tropaeolum malus* L. and of *Lupinus albus* L. *Am. J. Bot.* **33,** 301–318.
10. Monnier, M. (1995) Culture of zygotic embryos, in *In Vitro Embryogenesis in Plants* (Thorpe, T. A., ed.), Kluwer Academic, Dordrecht, The Netherlands, pp. 117–153.
11. Laibach, F. (1929) Ectogenesis in plants. Methods and genetic possibilities of propagating embryos otherwise dying in the seed. *J. Hered.* **20,** 201–208.

12. Tukey, H. B. (1934) Artificial culture methods for isolated embryos of deciduous fruits. *Proc. Am. Soc. Hortic. Sci.* **32,** 313–322.
13. LaRue, C. D. (1936) The growth of plant embryos in culture. *Bull. Torrey Bot. Club* **63,** 365–382.
14. Gautheret, R. J. (1934) Culture du tissus cambial. *C.R. Hebd. Seances Acad. Sc.* **198,** 2195–2196.
15. Gautheret, R. J. (1935) *Recherches sur la culture des tissus végétaux.* Ph.D. Thesis, Paris.
16. Gautheret, R. J. (1939) Sur la possibilité de réaliser la culture indéfinie des tissus de tubercules de carotte. *C.R. Hebd. Seances Acad. Sc.* **208,** 118–120.
17. Nobécourt, P. (1939) Sur la pérennité et l'augmentation de volume des cultures de tissues végétaux. *C.R. Seances Soc. Biol. Ses Fil.* **130,** 1270–1271.
18. White, P. R. (1939) Potentially unlimited growth of excised plant callus in an artificial nutrient. *Am. J. Bot.* **26,** 59–64.
19. Nobécourt, P. (1939) Sur les radicelles naissant des cultures de tissus végétaux. *C.R. Seances Soc. Biol. Ses Fil.* **130,** 1271–1272.
20. White, P .R. (1939) Controlled differentiation in a plant tissue culture. *Bull. Torrey Bot. Club* **66,** 507–513.
21. White, P. R. (1963) *The Cultivation of Animal and Plant Cells*, 2nd ed., Ronald Press, New York.
22. Bhojwani, S. S. and Razdan, M. K. (1983) *Plant Tissue Culture: Theory and Practice. Developments in Crop Science*, Vol. 5. Elsevier, Amsterdam.
23. Gautheret, R. J. (1985) History of plant tissue and cell culture: A personal account, in *Cell Culture and Somatic Cell Genetics of Plants* (Vasil, I. K., ed.), Vol. 2, Academic Press, New York, pp. 1–59.
24. Thorpe, T. A. (2000) History of plant cell culture. Chap. 1, in *Plant Tissue Culture: Techniques and Experiments*, (Smith, R. H., ed.) 2nd ed., Academic Press, California, pp. 1–32. (With permission from Elsevier).
25. Van Overbeek, J., Conklin, M. E., and Blakeslee, A. F. (1941) Factors in coconut milk essential for growth and development of very young Datura embryos. *Science* **94,** 350–351.
26. Gautheret, R. J. (1942) Hétéro-auxines et cultures de tissus végétaux. *Bull. Soc. Chim. Biol.* **24,** 13–41.
27. Gautheret, R. J. (1955) Sur la variabilité des propriétés physiologiques des cultures de tissues végétaux. *Rev. Gén. Bot.* **62,** 5–112.
28. Nobécourt, P. (1955) Variations de la morphologie et de la structure de cultures de tissues végétaux. *Ber. Schweiz. Bot. Ges.* **65,** 475–480.
29. Skoog, F. and Tsui, C. (1948) Chemical control of growth and bud formation in tobacco stem segments and callus cultured *in vitro. Am. J. Bot.* **35,** 782–787.
30. Miller, C., Skoog, F., Von Saltza, M. H., and Strong, F. M. (1955) Kinetin, a cell division factorfrom desoxyribonucleic acid. *J. Am. Chem. Soc.* **77,** 1392.
31. Skoog, F. and Miller, C. O. (1957) Chemical regulation of growth and organ formation in plant tissue cultures in vitro. *Symp. Soc. Exp. Biol.* **11,** 118–131.
32. Evans, D. A., Sharp, W. R., and Flick, C. E. (1981) Growth and behavior of cell cultures: Embryogenesis and organogenesis, in *Plant Tissue Culture: Methods*

and Applications in Agriculture (Thorpe, T. A., ed.), Academic Press, New York, pp. 45–113.
33. Letham, D. S. (1974) Regulators of cell division in plant tissues. The cytokinins of coconut milk. *Physiol. Plant.* **32**, 66–70.
34. Reinert, J. (1958) Utersuchungen die Morphogenese an Gewebekulturen. *Ber. Dtsch. Bot. Ges.* **71**, 15.
35. Reinert, J. (1959) Uber die Kontrolle der Morphogenese und die Induktion von Adventivembryonen an Gewebekulturen aus Karotten. *Planta* **53**, 318–333.
36. Steward, F. C., Mapes, M. O., and Mears, K. (1958) Growth and organized development of cultured cells. II. Organization in cultures grown from freely suspended cells. *Am. J. Bot.* **45**, 705–708.
37. Muir, W.H., Hildebrandt, A.C., and Riker, A.J. (1954) Plant tissue cultures produced from single isolated plant cells. *Science* **119**, 877–878.
38. Muir, W. H., Hildebrandt, A. C., and Riker, A. J. (1958) The preparation, isolation and growth in culture of single cells from higher plants. *Am. J. Bot.* **45**, 585–597.
39. Jones, L. E., Hildebrandt, A. C., Riker, A. J., and Wu, J. H. (1960) Growth of somatic tobacco cells in microculture. *Am. J. Bot.* **47**, 468–475.
40. Bergmann, L. (1959) A new technique for isolating and cloning cells of higher piarits. *Nature* **184**, 648–649.
41. Kohlenbach, H. W. (1959) Streckungs-und Teilungswachstum isolierter Mesophyllzellen von *Macleaya cordata* (Wild.) R. Br. *Naturwissenschaften* **46**, 116–117.
42. Kohlenbach, H. W. (1966) Die Entwicklungspotenzen explantierter und isolierter Dauerzellen. I. Das Strechungs- und Teilungswachstum isolierter Mesophyllzellen von *Macleaya cordata Z. Pflanzenphysiol.* **55**, 142–157.
43. Tulecke, W. and Nickell, L. G. (1959) Production of large amounts of plant tissue by submerged culture. *Science* **130**, 863–864.
44. Vasil, V. and Hildebrandt, A. C. (1965) Differentiation of tobacco plants from single, isolated cells in micro cultures. *Science* **150**, 889–892.
45. Heller, R. (1953) Recherches sur la nutrition minerale des tissus végétaux cultivé *in vitro. Ann. Sci. Nat. Bot. Biol. Veg.* **14**, 1–223.
46. Nitsch, J. P. and Nitsch, C. (1956) Auxin-dependent growth of excised Helianthus tuberosus tissues. *Am. J. Bot.* **43**, 839–851.
47. Murashige, T. and Skoog, F. (1962) A revised medium for rapid growth and bioassays with tobacco tissue cultures. *Physiol. Plant.* **15**, 473–497.
48. Gamborg, O. L., Murashige, T., Thorpe, T. A., and Vasil, I. K. (1976) Plant tissue culture media. *In Vitro* **12**, 473–478.
49. Morel, G. (1960). Producing virus-free cymbidium. *Am. Orchid Soc. Bull.* **29**, 495–497.
50. Murashige, T. (1974) Plant propagation through tissue culture. *Annu. Rev. Plant Physiol.* **25**, 135–166.
51. White, P. R. (1934) Multiplication of the viruses of tobacco and Aucuba mosaics in growing excised tomato root tips. *Phytopathology* **24**, 1003–1011.

52. Limasset, P. and Cornuet, P. (1949) Recherche du virus de la mosaïque du tabac dans les méristèmes des plantes infectées. *C.R. Hebd. Seances Acad. Sci.* **228**, 1971–1972.
53. Morel, G. and Martin, C. (1952) Guérison de dahlias atteints d'une maladie á virus. *C.R. Hebd. Seances Acad. Sc.* **235**, 1324–1325.
54. Quack, F. (1961) Heat treatment and substances inhibiting virus multiplication in meristem culture to obtain virus-free plants. *Adv. Hortic. Sci. Their Appl., Proc. Int. Hortic. Congr. 15th, 1958* **1**, 144–148.
55. LaRue, C. D. (1942) The rooting of flowers in culture. *Bull. Torrey Bot. Club* **69**, 332–341.
56. Rangan, T. S. (1982) Ovary, ovule and nucellus culture, in *Experimental Embryology of Vascular Plants* (Johri, B. M., ed.), Springer-Verlag, Berlin, pp. 105–129.
57. LaRue, C. D. (1949) Culture of the endosperm of maize. *Am. J. Bot.* **36**, 798.
58. Johri, B. M. and Bhojwani, S. S. (1965) Growth responses of mature endosperm in cultures. *Nature* **208**, 1345–1347.
59. Kanta, K., Rangaswamy, N. S., and Maheshwari, P. (1962) Test-tube fertilization in flowering plants. *Nature* **194**, 1214–1217.
60. Zenkteler, M., Misiura, E., and Guzowska, I. (1975) Studies on obtaining hybrid embryos in test tubes, in *Form, Structure and Function in Plants* (Mohan Ram, H. Y., Shaw, J. J., and Shaw, C. K., eds.), Sarita Prakashan, Meerut, India, pp. 180–187.
61. Tulecke, W. (1953) A tissue derived from the pollen of Ginkgo biloba. *Science* **117**, 599–600.
62. Yamada, T., Shoji, T., and Sinoto, Y. (1963) Formation of calli and free cells in a tissue culture of Tradescantia reflexa. *Bot. Mag.* **76**, 332–339.
63. Guha, S. and Maheshwari, S. C. (1964) *In vitro* production of embryos from anthers of Datura. *Nature* **204**, 497.
64. Guha, S. and Maheshwari, S. C. (1966) Cell division and differentiation of embryos in the pollen grains of *Datura in vitro*. *Nature* **212**, 97–98.
65. Bourgin, J.P. and Nitch, J.P. (1967) Obtention de Nicotiana haploides à partir de'étamines cultivées in vitro. *Ann. Physiol. Vég.* **9**, 377–382.
66. Cocking, E. C. (1960) A method for the isolation of plant protoplasts and vacuoles. *Nature* **187**, 927–929.
67. Takebe, I., Labib, C., and Melchers, G. (1971) Regeneration of whole plants from isolated mesophyll protoplasts of tobacco. *Naturwissenschaften* **58**, 318–320.
68. Carlson, P. S., Smith, H. H., and Dearing, R. D. (1972) Parasexual interspecific plant hybridization. *Proc. Natl. Acad. Sci. USA* **69**, 2292–2294.
69. Braun, A. C. (1941) Development of secondary tumor and tumor strands in the crown-gall of sunflowers. *Phytopathology* **31**, 135–149.
70. Braun, A. C. and White, P. R. (1943) Bacteriological sterility of tissues derived from secondary crown-gall tumors. *Phytopathology* **33**, 85–100.
71. Braun, A. C. (1950) Thermal inactivation studies on the tumor inducing principle in crown-gall. *Phytopathology* **40**, 3.

72. Zaenen, I., van Larebeke, N., Touchy, H., Van Montagu, M., and Schell, J. (1974) Super-coiled circular DNA in crown-gall inducing Agrobacterium strains. *J. Mol. Biol.* **86,** 109–127.
73. Ledoux, L. (1965) Uptake of DNA by living cells. *Prog. Nucleic Acid Res. Mol. Biol.* **4,** 231–267.
74. Vasil, I. K. (ed.) (1994) *Cell Culture and Somatic Cell Genetics of Plants*, Vol. 1, Laboratory Procedures and Their Applications. Academic Press, New York.
75. Vasil, I. K. and Thorpe, T. A. (eds.) (1994) *Plant Cell and Tissue Culture*, Kluwer Acad. Publ., Dordrecht, The Netherlands.
76. Yeoman, M. M. and Street, H. E. (1977) General cytology of cultured cells, in *Plant Tissue and Cell Culture* (Street, H. E., ed.), Blackwell Scientific, Oxford, pp. 137–176.
77. Lindsey, K. and Yeoman, M. M. (1985) Dynamics of plant cell cultures, in *Cell Culture and Somatic Cell Genetics of Plants* (Vasil, I. K., ed.), Vol. 2, Academic Press, New York, pp. 61–101.
78. Fowke, L.C. (1986) Ultrastructural cytology of cultured plant tissues, cells, and protoplasts, in *Cell Culture and Somatic Cell Genetics of Plants* (Vasil, I. K., ed.), Vol. 3, Academic Press, New York, pp. 323–342.
79. Fowke, L. C. (1987) Investigations of cell structure using cultured cells and protoplasts, in *Plant Tissue and Cell Culture* (Green, C. E., Somers, D. A., Hackett, W. P., and Biesboer, D. D., eds.), A. R. Liss, New York, pp. 17–31.
80. D'Amato, F. (1978) Chromosome number variation in cultured cells and regenerated plants, in *Frontiers of Plant Tissue Culture 1978* (Thorpe, T. A., ed.), Intl. Assoc. Plant Tissue Culture, Univ. of Calgary Printing Services, pp. 287–295.
81. Nagl, W., Pohl, J., and Radler, A. (1985) The DNA endoreduplication cycles, in *The Cell Division Cycle in Plants* (Bryant, J. A. and Francis, D., eds.), Cambridge University Press, Cambridge, pp. 217–232.
82. Yamada, Y., Fumihiko, S., and Hagimori, M. (1978) Photoautotropism in green cultured cells, in *Frontiers of Plant Tissue Culture 1978* (Thorpe, T. A., ed.), Intl. Assoc. Plant Tissue Culture, Univ. of Calgary Printing Services, pp. 453–462.
83. Hüsemann, W. (1985) Photoautotrophic growth of cells in culture, in *Cell Culture and Somatic Cell Genetics of Plants* (Vasil, I. K., ed.), Vol. 2, Academic Press, New York, pp. 213–252.
84. Neumann, K.-H., Barz, W., and Reinhard, E. (eds.) (1985) *Primary and Secondary Metabolism of Plant Cell Cultures,* Springer-Verlag, Berlin.
85. Leonard, R. T. and Rayder, L. (1985) The use of protoplasts for studies on membrane transport in plants, in *Plant Protoplasts* (Fowke, L. C. and Constabel, F., eds.), CRC Press, Boca Raton, Florida, pp. 105–118.
86. Filner, P. (1978) Regulation of inorganic nitrogen and sulfur assimilation in cell suspension cultures, in *Frontiers of Plant Tissue Culture 1978* (Thorpe, T. A., ed.), Intl. Assoc. Plant Tissue Culture, Univ. of Calgary Printing Services, pp. 437–442.
87. Fowler, M. W. (1978) Regulation of carbohydrate metabolism in cell suspension cultures, in *Frontiers of Plant Tissue Culture 1978* (Thorpe, T. A., ed.), Intl. Assoc. Plant Tissue Culture, Univ. of Calgary Printing Services, pp. 443–452.

88. Bender, L., Kumar, A., and Neumann, K.-H. (1985) On the photosynthetic system and assimilate metabolism of Daucus and Arachis cell cultures, in *Primary and Secondary Metabolism of Plant Cell Cultures* (Neumann, K.-H., Barz, W., and Reinhard, E., eds.), Springer-Verlag, Berlin, pp. 24–42.
89. Herzbeck, H. and Husemann, W. (1985) Photosynthetic carbon metabolism in photoautotrophic cell suspension cultures of *Chenopodium rubrum* L. in *Primary and Secondary Metabolism of Plant Cell Culture* (Neumann, K. -H., Barz, W., and Reinhard, E., eds.), Springer-Verlag, Berlin, pp. 15–23.
90. Constabel, F. and Vasil, I. K. (eds.) (1987) *Cell Culture and Somatic Cell Genetics of Plants*, Vol. 4. Academic Press, New York.
91. Constabel, F. and Vasil, I. K. (eds.) (1988) *Cell Culture and Somatic Cell Genetics of Plants*, Vol. 5. Academic Press, New York.
92. Roberts, L. W. (1976) *Cytodifferentiation in Plants: Xylogenesis as a Model System*, Cambridge University Press, Cambridge.
93. Phillips, R. (1980) Cytodifferentiation. *Int. Rev. Cytol., Suppl.* **11A**, 55–70.
94. Fukuda, H. and Komamine, A. (1985) Cytodifferentiation, in *Cell Culture and Somatic Cell Genetics of Plants* (Vasil, I. K., ed.), Vol. 2, Academic Press, New York, pp. 149–212.
95. Schell, J., van Montague, M., Holsters, M., et al. (1982) Plant cells transformed by modified Ti plasmids: A model system to study plant development, in *Biochemistry of Differentiation and Morphogenesis*, Springer-Verlag, Berlin, pp. 65–73.
96. Schell, J.S. (1987) Transgenic plants as tools to study the molecular organization of plant genes. *Science* **237**, 1176–1183.
97. Thorpe, T.A. (1980) Organogenesis in vitro: Structural, physiological, and biochemical aspects. *Int. Rev. Cytol., Suppl.* **11A,** 71–111.
98. Tran Thanh Van, K., and Trinh, H. (1978) Morphogenesis in thin cell layers: Concept, methodology and results, in *Frontiers of Plant Tissue Culture 1978* (Thorpe, T. A., ed.), Intl. Assoc. Plant Tissue Culture, Univ. of Calgary Printing Services, pp. 37–48.
99. Tran Thanh Van, K. (1980) Control of morphogenesis by inherent and exogenously applied factors in thin cell layers. *Int. Rev. Cytol. Suppl.* **11A**, 175–194.
100. Murashige, T. (1979) Principles of rapid propagation, in *Propagation of Higher Plants Through Tissue Culture: A Bridge Between Research and Application* (Hughes, K. W., Henke, R., and Constantin, M., eds.), Tech. Information Center, U.S. Dept. of Energy, pp. 14–24.
101. Brown, D.C.W. and Thorpe, T.A. (1986) Plant regeneration by organogenesis, in *Cell Culture and Somatic Cell Genetics of Plants* (Vasil, I. K., ed.), Vol. 3, Academic Press, New York, pp. 49–65.
102. Thompson, M. R. and Thorpe, T. A. (1990) Biochemical perspectives in tissue culture for crop improvement, in *Biochemical Aspects of Crop Improvement* (Khanna, K. R., ed.), CRC Press, Boca Raton, Florida, pp. 327–358.
103. Ammirato, P .V. (1983) Embryogenesis, in *Handbook of Plant Cell Culture* (Evans, D. A., Sharp, W. R., Ammirato P. V., and Yamada, Y., eds.), Vol. 1, MacMillan, New York, pp. 82–123.
104. Thorpe, T. A. (1988) *In vitro somatic embryogenesis*. ISI Atlas of Science: Animal and Plant Sciences, pp. 81–88.

105. Nomura, K. and Komamine, A. (1985) Identification and isolation of single cells that produce somatic embryos at a high frequency in a carrot suspension culture. *Plant Physiol.* **79,** 988–991.
106. Butcher, D. N. (1977) Plant tumor cells, in *Plant Tissue and Cell Culture* (Street, H. E., ed.), Blackwell Scientific, Oxford, pp. 429–461.
107. Rottier, P. J. M. (1978) The biochemistry of virus multiplication in leaf cell protoplasts, in *Frontiers of Plant Tissue Culture 1978* (Thorpe, T. A., ed.), Intl. Assoc. Plant Tissue Culture, Univ. of Calgary Printing Services, pp. 255–264.
108. Earle, E. D. (1978) Phytotoxin studies with plant cells and protoplasts, in *Frontiers of Plant Tissue Culture 1978* (Thorpe, T. A., ed.), Intl. Assoc. Plant Tissue Culture, Univ. of Calgary Printing Services, pp. 363–372.
109. Miller, S. A. and Maxwell, D. P. (1983) Evaluation of disease resistance, in *Handbook of Plant Cell Culture* (Evans, D. A., Sharp, W. R., Ammirata, P. V., and Yamada, Y. eds.), Vol. 1, Macmillan, New York, pp. 853–879.
110. Yeung, E. C., Thorpe, T. A., and Jensen, C .J. (1981) In vitro fertilization and embryo culture, in *Plant Tissue Culture*: Methods and Applications in Agriculture (Thorpe, T. A., ed.), Academic Press, New York, pp. 253–271.
111. Zenkteler, M. (1984) In vitro pollination and fertilization, in *Cell Culture and Somatic Cell Genetics of Plants* (Vasil, I. K., ed.), Vol. 1, Academic Press, New York, pp. 269–275.
112. Raghavan, V. (1980) Embryo culture. *Int. Rev. Cytol. Suppl.* **11B,** 209–240.
113. Collins, G. B. and Grosser, J. W. (1984) Culture of embryos, in *Cell Culture and Somatic Cell Genetics of Plants* (Vasil, I. K., ed.), Vol. 1, Academic Press, New York, pp. 241–257.
114. Hu, H., and Zeng, J. Z. (1984) Development of new varieties of anther culture, in *Handbook of Plant Cell Culture* (Ammirato, P. V., Evans, D. A., Sharp, W. R., and Yamada, Y., eds.), Vol. 3, Macmillan, New York, pp. 65–90.
115. San, L. H. and Gelebart, P. (1986) Production of gynogenetic haploids, in *Cell Culture and Somatic Cell Genetics of Plants* (Vasil, I. K., ed.), Vol. 3, Academic Press, New York, pp. 305–322.
116. Flick, C. E. (1983) Isolation of mutants from cell culture, in *Handbook of Plant Cell Culture* (Ammirato, P. V., Evans, D. A., Sharp, W. R., and Yamada, Y., eds.), Vol. I, Macmillan, New York, pp. 393–441.
117. Larkin, P. l., and Scowcroft, W. R. (1981) Somaclonal variation -a novel source of variability from cell culture for plant improvement. *Theor. Appl. Genet.* **60,** 197–214.
118. Larkin, P. J., Brettell, R. I. S., Ryan, S. A., Davies, P. A., Pallotta, M. A., and Scowcroft, W. R. (1985) Somaclonal variation: impact on plant biology and breeding strategies, in *Biotechnology in Plant Science* (Day, P., Zaitlin, M., and Hollaender, A., eds.), Academic Press, New York, pp. 83–100.
119. Scowcroft, W. R., Brettell, R. I. S., Ryan, S. A., Davies, P. A., and Pallotta, M. A. (1987) Somaclonal variation and genomic flux, in *Plant Tissue and Cell Culture* (Green, C. E., Somers, D. A., Hackett, W. P., and Biesboer, D. D., eds.), A. R. Liss, New York, pp. 275–286.

120. Jacobs, M., Negrutiu, I., Dirks, R., and Cammaerts, D. (1987) Selection programmes for isolation and analysis of mutants in plant cell cultures, in *Plant Tissue and Cell Culture* (Green, C. E., Somers, D. A., Hackett, W. P., and Biesboer, D. D., eds.), A. R. Liss, New York, pp. 243–264.
121. Hughes, K. (1983) Selection for herbicide resistance, in *Handbook of Plant Cell Culture* (Ammirato, P. V., Evans, D. A., Sharp, W. R., and Yamada, Y., eds.), Vol. 1, Macmillan, New York, pp. 442–460.
122. Ranch, J.P., Rick, S., Brotherton, J.E., and Widholm, J. (1983) Expression of 5-methyltryptophan resistance in plants regenerated from resistant cell lines of Datura innoxia. *Plant Physiol.* **71,** 136–140.
123. Binding, H. (1986) Regeneration from protoplasts, in *Cell Culture and Somatic Cell Genetics of Plants* (Vasil, I. K., ed.), Vol. 3, Academic Press, New York, pp. 259–274.
124. Evans, D. A., Sharp, W. R., and Bravo, J. E. (1984) Cell culture methods for crop improvement, in *Handbook of Plant Cell Culture* (Sharp, W. R., Evans, D. A., Ammirato, P. V., and Yamada, Y., eds.), Vol. 2, Macmillan, New York, pp. 47–68.
125. Schieder, 0. and Kohn, H. (1986) Protoplast fusion and generation of somatic hybrids, in *Cell Culture and Somatic Cell Genetics of Plants* (Vasil, I. K., ed.), Vol. 3, Academic Press, New York, pp. 569–588.
126. Chetrit, P., Mathieu, C., Vedel, F., Pelletier, G., and Primard, C. (1985) Mitochondrial DNA polymorphism induced by protoplast fusion in Cruciferae. *Theor. Appl. Genet.* **69,** 361–366.
127. Potrykus, I., Shillito, R. D., Saul, M., and Paszkowski, J. (1985) Direct gene transfer: State of the art and future potential. *Plant Mol. Biol. Rep.* **3,** 117–128.
128. Deshayes, A., Herrera-Estrella, L., and Caboche, M. (1985) Liposome-mediated transformation of tobacco mesophyll protoplasts by an *Escherichia coli* plasmid. *EMBO J.* **4,** 2731–2739.
129. Crossway, A., Oakes, J. V., Irvine, J. M., Ward, B., Knauf, V. C., and Shewmaker, C. K. (1986) Integration of foreign DNA following microinjection of tobacco mesophyll protoplasts. *Mol. Gen. Genet.* **202,** 179–185.
130. Klein, T. M., Wolf, B. D., Wu, R., and Sanford, J. C. (1987) High-velocity microprojectiles for delivering nucleic acids into living cells. *Nature* **327,** 70–73.
131. DeBlock, M., Herrera-Estrella, L., van Montague, M., Schell, J., and Zambryski, P. (1984) Expression of foreign genes in regenerated plants and in their progeny. *EMBO J.* **3,** 1681–1689.
132. Borsch, R. B., Fraley, R. T., Rogers, S. G., Sanders, F. R., Lloyd, A., and Boffmann, N. (1984) Inheritance of functional foreign genes in plants. *Science* **223,** 496–498.
133. Gasser, C. S. and Fraley, R. T. (1989) Genetically engineering plants for crop improvement. *Science* **244,** 1293–1299.
134. Uchimiya, H., Handa, T., and Brar, D. S. (1989) Transgenic plants. *J. Biotech.* **12,** 1–20.

135. Kartha, K.K. (1981) Meristem culture and cryopreservation methods and applications, in *Plant Tissue Culture: Methods and Applications in Agriculture* (Thorpe, T. A., ed.), Academic Press, New York, pp. 181–211.
136. Dodds, J. (1989) Tissue culture for germplasm management and distribution, in Strengthening Collaboration in *Biotechnology: International Agricultural Research and the Private Sector* (Cohen, J. I., ed.), Bureau of Science and Technology, AID, Washington, D.C. pp. 109–128.
137. Withers, L. A. (1985) Cryopreservation of cultured cells and meristems, in *Cell Culture and Somatic Cell Genetics of Plants* (Vasil, I. K., ed.), Vol. 2, Academic Press, New York, pp. 253–316.
138. Murashige, T. (1978) The impact of plant tissue culture on agriculture, in *Frontiers of Plant Tissue Culture 1978* (Thorpe, T. A., ed.), Intl. Assoc. Plant Tissue Culture, Univ. of Calgary Printing Services, pp. 15–26.
139. Conger, B. V., ed. (1981) *Cloning Agricultural Plants Via In Vitro Techniques.* CRC Press, Boca Raton, Florida.
140. Murashige, T. (1990) Plant propagation by tissue culture: practice with unrealized potential, in *Handbook of Plant Cell Culture* (Ammirato, P. V., Evans, D. A., Sharp, W. R., and Bajaj, Y. P. S., eds.), Vol. 5, McGraw-Hill, New York, pp. 3–9.
141. Zimmerman, R. H. (1986) Regeneration in woody ornamentals and fruit trees, in *Cell Culture and Somatic Cell Genetics of Plants* (Vasil, I. K., ed.), Vol. 3, Academic Press, New York, pp. 243–258.
142. Zenk, M. H. (1978) The impact of plant cell culture on industry, in *Frontiers of Plant Tissue Culture 1978* (Thorpe, T. A., ed.), Intl. Assoc. Plant Tissue Culture, Univ. of Calgary Printing Services, pp. 1–13.
143. Wink, M. (1987) Physiology of the accumulation of secondary metabolites with special reference to alkaloids, in *Cell Culture and Somatic Cell Genetics of Plants* (Constabel, F. and Vasil, I. K., eds.), Vol. 4, Academic Press, New York, pp. 17–42.
144. Dougall, D. K. (1987) Primary metabolism and its regulation, in *Plant Tissue and Cell Culture* (Green, C. E., Somers, D. A., Hackett, W. P., and Biesboer, D. D., eds.), A. R. Liss, New York, pp. 97–117.
145. Kemp, H. A. and Morgan, M. R. A. (1987) Use of immunoassays in the detection of plant cell products, in *Cell Culture and Somatic Cell Genetics of Plants* (Constabel, F. and Vasil, I. K., eds.), Vol. 4, Academic Press, New York, pp. 287–302.
146. Widholm, J. M. (1987) Selection of mutants which accumulate desirable secondary products, in *Cell Culture and Somatic Cell Genetics of Plants* (Constabel, F. and Vasil, I. K., eds.), Vol. 4, Academic Press, New York, pp. 125–137.
147. Eilert, U. (1987) Elicitation: Methodology and aspects of application, in *Cell Culture and Somatic Cell Genetics of Plants* (Constabel, F. and Vasil, I. K., eds.), Vol. 4, Academic Press, New York, pp. 153–196.
148. Kurz, W.G.W. (1988) Semicontinuous metabolite production through repeated elicitation of plant cell cultures: A novel process, in *Plant Biotechnology* (Mabry, T. J., ed.), IC2 Institute, Austin, pp. 93–103.
149. Brodelius, P. (1985) The potential role of immobilisation in plant cell biotechnology. *Trends Biotechnol.* **3,** 280–285.

150. Yeoman, M. M. (1987) Techniques, characteristics, properties, and commercial potential of immobilized plant cells, in *Cell Culture and Somatic Cell Genetics of Plants* (Constabel, F. and Vasil, I. K., eds.), Vol. 4, Academic Press, New York, pp. 197–215.
151. Fowler, M.W. (1987) Process systems and approaches for large-scale plant cell culture, in *Plant Tissue and Cell Culture* (Green, C. E., Somers, D. A., Hackett, W. P., and Biesboer, D. D., eds.), A. R. Liss, New York, pp. 459–471.
152. Beiderbeck, R. and Knoop, B. (1987) Two-phase culture, in *Cell Culture and Somatic Cell Genetics of Plants* (Constabel, F. and Vasil, I. K., eds.), Vol. 4, Academic Press, New York, pp. 255–266.
153. Fujita, Y. and Tabata, M. (1987) Secondary metabolites from plant cells—pharmaceutical applications and progress in commercial production, in *Plant Tissue and Cell Culture* (Green, C. E., Somers, D. A., Hackett, W. P., and Biesboer, D. D., eds.), A. R. Liss, New York, pp. 169–185.
154. Vasil, I. K. and Vasil, V. (1994) *In vitro* culture of cereals and grasses, in *Plant Cell and Tissue Culture* (Vasil, I. K. and Thorpe, T. A., eds.), Kluwer Acad. Publ., Dordrecht, The Netherlands, pp. 293–312.
155. Davey, M. R., Kumar, V., and Hammatt, N. (1994) *In vitro* culture of legumes, in *Plant Cell and Tissue Culture* (Vasil, I. K. and Thorpe, T. A., eds.), Kluwer Acad. Publ., Dordrecht, The Netherlands, pp. 313–329.
156. Reynolds, J. F. (1994) *In vitro* culture of vegetable crops, in *Plant Cell and Tissue Culture* (Vasil, I. K. and Thorpe, T. A., eds.), Kluwer Acad. Publ., Dordrecht, The Netherlands, pp. 331–362.
157. Jones, M. G. K. (1994) *In vitro* culture of potato, in *Plant Cell and Tissue Culture* (Vasil, I. K. and Thorpe, T. A., eds.), Kluwer Acad. Publ., Dordrecht, The Netherlands, pp. 363–378.
158. Krikorian, A. D. (1994) *In vitro* culture of root and tuber crops, in *Plant Cell and Tissue Culture* (Vasil, I. K. and Thorpe, T. A., eds.), Kluwer Acad. Publ., Dordrecht, The Netherlands, pp. 379–411.
159. Palmer, C. E. and Keller, W. A. (1994) *In vitro* culture of oilseeds, in *Plant Cell and Tissue Culture* (Vasil, I. K. and Thorpe, T. A., eds.), Kluwer Acad. Publ., Dordrecht, The Netherlands, pp. 413–455.
160. Zimmerman, R. H. and Swartz, H. J. (1994) *In vitro* culture of temperate fruits, in *Plant Cell and Tissue Culture* (Vasil, I. K. and Thorpe, T. A., eds.), Kluwer Acad. Publ., Dordrecht, The Netherlands, pp. 457–474.
161. Grosser, J. W. (1994) *In vitro* culture of tropical fruits, in *Plant Cell and Tissue Culture* (Vasil, I. K. and Thorpe, T. A., eds.), Kluwer Acad. Publ., Dordrecht, The Netherlands, pp. 475–496.
162. Krikorian, A. D. (l994) *In vitro* culture of plantation crops, in *Plant Cell and Tissue Culture* (Vasil, I. K. and Thorpe, T. A., eds.), Kluwer Acad. Publ., Dordrecht, The Netherlands, pp. 497–537.
163. Harry, I. S. and Thorpe, T. A. (1994) *In vitro* culture of forest trees, in *Plant Cell and Tissue Culture* (Vasil, I. K. and Thorpe, T. A., eds.), Kluwer Acad. Publ., Dordrecht, The Netherlands, pp. 539–560.

164. Debergh, P. (1994) *In vitro* culture of ornamentals, in *Plant Cell and Tissue Culture* (Vasil, I. K. and Thorpe, T. A., eds.), Kluwer Acad. Publ., Dordrecht, The Netherlands, pp. 561–573.
165. Karp, A. (1994) Origins, causes and uses of variation in plant tissue cultures, in *Plant Cell and Tissue Culture* (Vasil, I. K. and Thorpe, T. A., eds.), Kluwer Acad. Publ., Dordrecht, The Netherlands, pp. 139–151.
166. Dix, P. J. (1994) Isolation and characterisation of mutant cell lines, in *Plant Cell and Tissue Culture* (Vasil, I. K. and Thorpe, T. A., eds.), Kluwer Acad. Publ., Dordrecht, The Netherlands, pp. 119–138.
167. Feher, A. and Dudits, D. (1994) Plant protoplasts for cell fusion and direct DNA uptake: culture and regeneration systems, in *Plant Cell and Tissue Culture* (Vasil, I. K. and Thorpe, T. A., eds.), Kluwer Acad. Publ., Dordrecht, The Netherlands, pp. 71–118.
168. Kartha, K. K. and Engelmann, F. (1994) Cryopreservation and geffiplasm storage: *In Plant Cell and Tissue Culture* (Vasil, I. K. and Thorpe, T. A., eds.), Kluwer Acad. Publ., Dordrecht. The Netherlands, pp. 195–230.
169. Redenbaugh, K., ed. (1993) *Synseeds: Applications of Synthetic Seeds to Crop Improvement.* CRC Press, Boca Raton, FL.
170. Lowe, K. C., Davey, M. R., and Power, J. B. (1996) Plant tissue culture: past, present and future. *Plant Tiss. Cult. Biotechnol.* **2,** 175–186.
171. Kong, L., Attree, S. M., Evans, D. E., Binarova, P., Yeung, E. C., and Fowke, L. C. (1998) Somatic embryogenesis in white spruce: studies of embryo development and cell biology, in *Somatic Embryogenesis in Woody Plants* (Jain, S. M. and Gupta, P. K., eds.), Vol. 4, Kluwer Acad. Publ., Dordrecht, The Netherlands, pp. 1–28.
172. Kaeppler, S. M. and Phillips, R. L. (1993) DNA methylation and tissue culture-induced variation in plants. *In Vitro Cell. Dev. Biol.* **29P,** 125–130.
173. Komamine, A., Ito, M., and Kawahara, R. (1993) Cell culture systems as useful tools for investigation of developmental biology in higher plants: analysis of mechanisms of the cell cycle and differentiation using plant cell cultures, in *Advances in Developmental Biology and Biotechnology of Higher Plants* (Soh, W. Y., Liu, J. R., and Komamine, A., eds.), Proceedings First Asia Pacific Conference on Plant Cell and Tissue Culture, held in Taedok Science Town, Taejon, Korea, 5-9 Sept. 1993,. The Korean Society of Plant Tissue Culture, pp. 289–310.
174. Trehin, C., Planchais, S., Glab, N., Perennes, C., Tregear, J., and Bergounioux, C. (1998) Cell cycle regulation by plant growth regulators: involvement of auxin and cytokinin in the re-entry of Petunia protoplasts into the cell cycle. *Planta* **206,** 215–224.
175. Gaspar, T. (1995) The concept of cancer in in vitro plant cultures and the implication of habituation to hormones and hyperhydricity. *Plant Tiss. Cult. Biotechnol.* **1,** 126–136.
176. Suguira, M. (1997) In vitro transcription systems from suspension-cultured cells. *Annu. Rev. Plant Physiol. Plant Mol. Biol.* **48,** 383–398.

177. Stitt, M. and Sonnewald, U. (1995) Regulation of carbohydrate metabolism in transgenics. *Annu. Rev. Plant Physiol. Plant Mol. Biol.* **46,** 341–368.
178. Kutchin, T. M. (1998) Molecular genetics of plant alkaloid biosynthesis, in *The Alkaloids* (Cordell, G., ed.), Vol. 50, Academic Press, San Diego, pp. 257–316.
179. Verpoorte, R., van der Heijden, R., ten Hoopen, H. J. G., and Memclink, J. (1998) Metabolic engineering for the improvement of plant secondary metabolite production. *Plant Tiss. Cult. Biotechnol.* **4,** 3–20.
180. *The Plant Cell*, Special Issue, July 1997.
181. Fukuda, H. (1997) Xylogenesis: initiation, progression, and cell death. *Annu. Rev. Plant Physiol. Plant Mol. Biol.* **47,** 299–325.
182. Thorpe, T.A. (1993) Physiology and biochemistry of shoot bud formation *in vitro*, in *Advances in Developmental Biology and Biotechnology of Higher Plants* (Soh, W. Y., Liu, J. R., and Komamine, A., eds.) Proceedings First Asia Pacific Conference on Plant Cells and Tissue Culture, held in Taedok Science Town, Taejon, Korea, 5–9 Sept. 1993, The Korean Society of Plant Tissue Culture, pp. 210–224.
183. Joy IV, R. W. and Thorpe, T. A. (1999) Shoot morphogenesis: Structure, physiology, biochemistry and molecular biology, in *Morphogenesis in Plant Tissue Cultures* (Soh W. Y., and Bhojwani, S. S., eds.), Kluwer Acad. Publ., Dordrecht, The Netherlands, pp. 171–214.
184. Nomura, K. and Komamine, A. (1995). Physiological and biochemical aspects of somatic embryogenesis, in *In Vitro Embryogenesis in Plants* (Thorpe, T. A., ed.), Kluwer Acad. Publ., Dordrecht, The Netherlands, pp. 249–265.
185. Dudits, D., Györgyey, J., Bögre, L., and Bakó, L. (1995) Molecular biology of somatic embryogenesis, in In *Vitro Embryogenesis in Plants* (Thorpe, T. A., ed.), Kluwer Acad. Publ., Dordrecht, The Netherlands, pp. 267–308.
186. Thorpe, T. A. and Stasolla, C. (2001) Somatic embryogenesis, in *Current Trends in the Embryology of Angiosperms* (Bhojwani S. S. and Soh, W. Y., eds.), Kluwer Acad. Publ., Dordrecht, The Netherlands, pp. 279–236.
187. Hinchee, M. A. W., Corbin, D. R., Armstrong, C. L., et al. (1994) Plant transformation, in *Plant Cell and Tissue Culture* (Vasil, I. K., and Thorpe, T. A., eds.), Kluwer Acad. Publ., Dordrecht, The Netherlands, pp. 231–270.
188. Fraley, R. T., Rogers, S. G., Borsch, R. B., et al. (1985) The SEV system: a new disarmed Ti plasmid vector system for plant transformation. *Bio/Technol.* **3,** 629–635.
189. Horsch, R. B., Fry, J., Hoffman, N., et al. (1985) A simple and general method for transferring genes into plants. *Science* **227,** 1229–1231.
190. Cloutier, S. and Landry, B. S. (1994) Molecular markers applied to plant tissue culture. *In Vitro Cell. Dev. Biol.* **31P,** 32–39.
191. Sanford, J. C. (2000) The development of the biolistic process. *In Vitro Cell. Dev. Biol. Plant* **36,** 303–308.
192. Fraley, R. (1992) Sustaining the food supply. *Bio/Technol.* **10,** 40–43.
193. Potrykus, I. (2001) The Golden Rice tale. *In Vitro Cell. Dev. Biol.-Plant* **37,** 93–100.

194. Altman, A., Ziv, M., and Izhar, S. (eds.) (1999) *Plant Biotechnology and In Vitro Biology in the 21st Century*. Proceedings of the IXth International Congress of the International Association for Plant Tissue Culture and Biotechnology, Jerusalem, Israel, 14–19 June, 1998. Kluwer Acad. Publ., Dordrecht, The Netherlands.
195. Vasil, I. K. (ed.) (2003) *Plant Biotechnology 2002 and Beyond*. Proceedings of the 10th IAPTC&B Congress, June 23–28, 2002, Orlando, FL, USA. Kluwer Acad. Publ., Dordrecht, The Netherlands.
196. Schell, J. (1995) Progress in plant sciences is our best hope to achieve an economically rewarding, sustainable and environmentally stable agriculture. *Plant Tiss. Cult. Biotechnol.* **1,** 10–12.

II

CELL CULTURE AND PLANT REGENERATION:
THE FUNDAMENTALS

3

Pathogen and Biological Contamination Management

The Road Ahead

Alan C. Cassells and Barbara M. Doyle

Summary

Multiplication of certified pathogen-free stock plants in vitro makes an important contribution to the production of disease-free planting material for vegetatively propagated crops. Meristem culture is extensively used to eliminate pathogens and contaminants from microbially contaminated plants prior to micropropagation. The approach to pathogen and contamination management differs. It is essential to avoid the release of pathogen-contaminated microplants and to ensure this the plant pathogen-testing guidelines and protocols issued by the Food and Agriculture Organisation (FAO) and its regional representative organizations should be followed. Where in vitro methods are used to eliminate pathogens, the progeny plants should be established in vivo under quarantine conditions and tested under FAO guidelines before being used as stock plants for in vitro multiplication. At establishment of microplants in vitro (stage 1), cultures should be culture-indexed for the presence of microbial contaminants. If pathogen- and contaminant-free cultures are established, then the risk is that of managing laboratory contamination by common environmental microorganisms based on Hazard Analysis Critical Control Points (HACCP) principles.

International plant health certification organizations are conservative and rely on established pathogen indexing protocols. They are reluctant to accept DNA-based tests and do not accept testing of in vitro cultures. Given that in time both of these restrictions may be relaxed, micropropagators may look forward to availing more of diagnostic service providers using polymerase chain reaction-based multiplex assays for pathogen-indexing and advances in diagnostic kits for environmental microorganisms in support of laboratory contamination management; with the caveat that molecular tests for pathogens may continue to require confirmation by inoculation of indicator plant species.

Key Words: Environmental microorganisms; HACCP; meristem culture; molecular beacons; multiplex assays; PCR.

1. Introduction

Biotic contamination in tissue culture is caused by the entry into culture of pathogens and other microbial contaminants on or in the explants used to initiate the cultures, and by entry of environmental microorganisms and micro-arthropods during the in vitro culture stages (*see* **Fig. 1**). Biotic contaminants associated with the explants used to establish cultures are usually pathogens of the crop or opportunistic colonizers of the plant *(1)*. The latter may be organisms abundant in the environment of the plants that were not excluded during apical meristem explant preparation or microorganisms that escaped surface sterilization by persistence in biofilms, in surface niches or as endophytes. Such contaminants may remain latent in the explants and in regenerated plants, to transmit disease or contamination through the progeny plants, or they may emerge during culture when suppressive media factors are removed or the medium is diluted, to overrun the cultures. When clean (aseptic) cultures are established, biotic contamination can occur during the in vitro stages. In this case the contaminants are environmental microorganisms, human-associated microorganisms, and micro-arthropods (*see* **Fig. 1**; *[2–4]*). Typically, these microorganisms are non-fastidious and their presence can be seen by visual inspection of the cultures.

The distinction between plant pathogens (phytopathogens) and environmental contaminants in tissue culture is in many instances a case of role reversal. Phytopathogens may be latent in tissues in vitro (i.e., be symptomless) whereas environmental organisms may rapidly overgrow the explant, killing the tissues by becoming vitropathogens. The risk with phytopathogen contamination is that of clonal propagation and possible international distribution and environmental release of pathogens; the risk from contamination with vitropathogens is loss of the cultures. A further distinction is that vitropathogens are mainly non-fastidious environmental fungi and bacteria whereas phytopathogens are mainly intracellular non-cultivable viruses and viroids or fastidious inter- or intra-cellular bacteria. The identification of phytopathogens may require diagnostics and diagnostic protocols, which in the case of plant health certification schemes are prescribed by international regulatory authorities *(5)*.

The consequences of the distinction both in the character of the contaminants and in the sequence of their occurrence in tissue culture, is that pathogen and biological contamination management, arguably, are best managed by two strategies. Phytopathogens and opportunistic endophytes and surface contaminants should be eliminated at the stage of establishment of the cultures and contamination management in the laboratory, caused by vitropathogens, is best managed by good working practice *(6)*. This approach will be discussed further here with emphasis on emerging "omics" (genomic, proteomic, and metabolomic) identification methods.

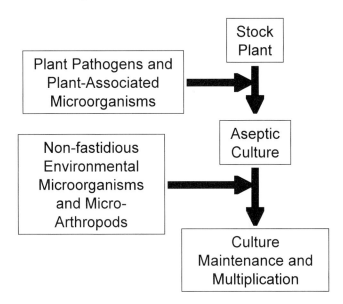

Fig. 1. A scheme illustrating a two-phase strategy for managing contamination in tissue culture where the focus is on pathogen-indexing and pathogen and contaminant elimination before or at the establishment of the cultures and good laboratory working practice is used to exclude environmental organisms during the in vitro stages.

2. Diagnostics and Diagnosis

In plant health certification, the requirement is to be able to distinguish between non-pathogenic and pathogenic isolates and to distinguish between races of pathovars, where pathovars are defined as "a strain or set of strains with the same or similar characteristics differentiated at infrasubspecific level from other strains of the same species or subspecies on the basis of distinctive pathogenicity to one or more host plants." Classification of a taxon as a pathovar does not exclude recognition of difference in biochemical, serological, or other non-pathogenic characters between that and other pathovars of the same species or subspecies, but implies that at the infrasubspecific level other differences are considered to have less significance in comparison to differences in pathogenicity. Usually pathovars are distinguished in terms of proved differences in host range. However, clear differences in symptomatology on the same plant species (e.g., *Xanthomonas campestris pv. oryzae* and *X. campestris pv. oryzicola*) can warrant different pathovar designation (www.ag.uidaho.cdu/bacteriology/pathovar.html). Consequently, to confirm the identity of a pathovar or pathovar race putatively identified by any test method where the pathogenicity gene(s) or gene product(s) are not detected by the method may require inoculation of an indicator host plant.

Table 1
Characteristics of CCP Biomarkers

Indicator organism	Type	Characteristics	Quick test or reference
Pseudomonas fluorescens	Eubacteriaceae	Gram-negative, motile rods that oxidise glucose *(1)*	*(7)*
Erwinia spp.	Eubacteriaceae	Gram-negative, motile more than 4 peritrichous flagellae, grows anaerobically *(1)*	*(7)*
Klebsiella spp.	Eubacteriaceae		*(7)*
Serratia spp.	Eubacteriaceae	Gram-negative, motile rods, highly mucoid colonies, produces red pigment, facultative anaerobes, oxidase negative, nitrate positive *(1)*	api 20E[a]
Agrobacterium spp.	Eubacteriaceae	Gram-negative, motile, non-spore forming rods, grows on D-1 medium *(1)*	*(7)*
Bacillus spp.	Eubacteriaceae	Gram-positive, motile rods, produces large; gray-white colonies with irregular margins on blood agar; produces endospores when stressed; motile/non-motile; catalase positive *(1)*	*(7)*
Rhodotorula spp.	Basidiomycetes	Unicellular (blasto-) conida globose to elongate which may be encapsulated; pseudohyphae are absent or rudimentary; colonies cream to pink, coral red, orange or yellow in color	api 20C AUX[a]
Penicillium spp.	Deuteromycetes	Conidiophores arising from the mycelium singly or less commonly in groups or fused, branched near the apex, ending in spore producing cells, the spores are produced sequentially and form chains	*(43)*
Staplylococcus spp.	Euacteriaceae	Gram-positive, non-motile, spherical in clusters, anaerobic fermentation of glucose with acid production, catalase, nitrate, and coagulase positive.	api STAPH[a]
Streptococcus spp.	Euacteriaceae	Gram-positive, non-motile, spherical in chains, aerobic action on glucose catalase negative	api STREP[a]
Candida albicans	Saccharomycetales	Produces blastospores, pseudohyphae, and chlamydospores on cornmeal Tween-80 agar at 25°C after 72 h	api ID32C[a]

[a]www.biomerieux.fr
For a complete list of commercial diagnostic suppliers for bacterial and yeast biochemical detrection kits *see* http://www.aoac.org/testkits/microbiologykits.htm.

In tissue culture in general, diagnostics are only required for pathovar identification; however, if good laboratory management practice is followed there will be an interest in identifying those laboratory contaminants, at least to ge-

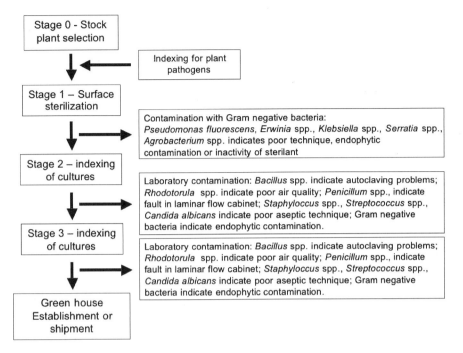

Fig. 2. A flow diagram showing the stages in micropropagation that lists microorganisms that indicate failure of equipment and aseptic techniques at the respective stages.

nus level, which may be indicators of sources of contamination (*see* **Fig. 2 *[6]***). The latter are mainly cultivable environmental microorganisms (yeasts and bacteria) and can be identified by classical microbiological methods or by biochemical test kits (e.g., *see* www.biomereiux.fr and **Table 1**). Historically, plant associated-bacteria have been identified based on the Gram stain, morphology, biochemical tests, and growth and colony characteristics on selective media *(7,8)*. Fungal including yeast contaminants are normally identified based on morphology *(9,10)*. Microarthropod contaminants are identified microscopically *(11)*. The following methods are used for pathogen and contaminant identification.

2.1. Conventional Methods

2.1.1. Microscopy Methods

Light microscopy is used extensively in the identification of fungal and bacterial plant pathogens and culture contaminants. It is also used to characterize and distinguish yeasts from bacteria and in bacterial identification, in combination with simple biochemical tests and growth on selective media *(7,8)*. Gram-, capsule-, and flagella-staining are important in the identification of bacteria. The critical control point (CCP) indicator yeasts, *Rhodotorula* spp. and *Candida albicans* are identified microscopically (**Table 1**).

Thin-section electron microscopy (EM) can be used to study inter- and intracellular pathogens *(12)* but is not used in routine diagnosis. Leaf-dip electron microscopy, especially when combined with serological techniques (e.g., immuno-electron microscopy and immunogold electron microscopy) is a sensitive technique but is limited by the high capital and running costs *(13)*.

2.1.2. Biochemical Kits

Kits for bacterial and yeast identification based on miniaturised biochemical tests are available from a number of companies (*see* www.aoac.org/testkits/microbiologykits.htm). These kits are useful for the identification of laboratory contaminants (common environmental microorganisms) but as identification is based on a comparison of the biochemical profile of the unknown with a reference database, their use in plant pathogen identification is limited as the latter tend not to be on the databases. They involve some basic tests on the microorganism (e.g., Gram stain, oxidase test) to categorize the organism for the purpose of selecting the appropriate test kit *(14)*. More useful for plant bacterial pathogen identification is fatty acid profiling which involves extraction, derivatization, separation, and identification of the fatty acids as some fatty acids indicate taxonomic relationships *(15)*.

2.1.3. Identification Based on Selective Culture Media

Growth on selective media, following staining and microscopic examination, is a relatively low cost traditional strategy to identify bacteria to genus level, which can be used for identification of CCP indicator organisms (*see* **Fig. 2** and **Table 1**). These methods are also applicable to the identification of bacterial plant pathogens but in the latter case require inoculation of indicator plants to confirm the pathogenicity and identify the pathovar *(7,8)*.

2.1.4. Serological Diagnostics

Enzyme-linked immunosorbent assay (ELISA) kits are currently the most widely used diagnostics for the identification of viral plant pathogens and are commercially available (for example, *see* www.agdia.com). ELISA kits are also available commercially for a limited number of plant bacterial and fungal pathogens. It is essential that the negative/positive threshold is set correctly and that negative results are validated by inoculation of indicator plants for certification pathogens *(16)*. Antibodies are also used in ISM (*see* **Subheading 2.1.1.**) and in latex agglutination tests *(17)*. The development of protein array (protein chip) technology offers the prospect of antibody arrays for the detection of multiple pathogens in the same sample *(18)*. This technology should be compared with multiplex DNA-based assays (*see* **Subheading 2.1.5.1.**).

2.1.5. "Omics" Methods

2.1.5.1. GENOMIC FINGERPRINTING

Increasing information is available on sequence data for pathogens (e.g., GenBank; www.ncbi.nih.gov/Genbank/). This facilitates primer design for the detection of pathovars where the diagnostician can consult publicly available databases and use the "Basic Local Alignment Search Tool" (BLAST) to compare the primer sequences against previous deposits (http://www.ncbi.nlm.nih.gov/blast/ *[19,20]*). Exploiting unique sequences for pathovar identification (e.g., *argK-tox* gene of *P. syringae pv. phaseolicola*) has an advantage in terms of specificity but a disadvantage in terms of sensitivity for the detection of single copy sequences *(19)*. More commonly, ribosomal DNA is the target for polymerase chain reaction (PCR) amplification. It is composed of both transcribed and non-transcribed sequences. The internal transcribed spacer (ITS) regions (*see* **Fig. 3**) are nonconserved and can be targeted in specific amplification events. This approach can identify pathogens to the species or subspecific level *(21)* but may not identify to the pathovar level which may require inoculation of indicator plants (**Fig. 4**).

The advances in technology for pathogen detection at the molecular level allow for more manageable one-tube multiplex assays for the simultaneous detection of a number of pathogens to be carried out *(22)*, eliminating, in many cases, the need for subsequent time-consuming gel-electrophoretic analysis of PCR end products followed by confirmation of the amplicon (PCR amplification product) *(19)*. A multiplex assay for the simultaneous detection of two viruses in potato tubers (Potato Leafroll Virus and Potato Virus Y) using differently colored molecular beacons has been described by Klerks et al. *(22)*. An added benefit of these one-tube assays is the elimination of further possible points of entry for DNA contamination during the detection protocol.

According to Bailey et al. *(23)* PCR offers the most sensitive means for detecting plant pathogens. But even with the current advances in molecular technology, sometimes detection of pathogens, for example, in asymptomatic material may prove difficult. Latency is potentially a problem and as a consequence identifying the source of the contaminant(s) can prove difficult *(6)*. In many cases experienced diagnosticians are still needed for diagnosis and detection *(19)*. Many service companies claim that a minimum 5–8 yr of experience (which includes specialist training) is needed to perform these PCR type assays (*see* www.mdpl.co.in/faq.asp) for what appears to be simple one-

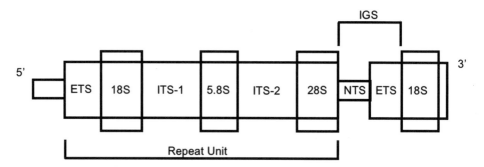

Fig. 3. Schematic representation of basic repeat unit of nrDNA. (Reproduced with permission from **ref. 42**.) ETS = external transcribed spacer; ITS = internal transcribed spacer; NTS = non-transcribed spacer; IGS = intergenic spacer.

tube assays where plant extract is added to off-the-shelf beads and the appropriate signals are read (molecular beacons).

The obvious advantages of DNA-based diagnostics for virus detection over host indexing or dot-blot assays using labeled probes, include efficiency and reliability. Gosalvez et al. *(24)* highlighted the sensitivity of reverse transcriptase (RT)-PCR (625× more sensitive) for the detection of Melon Necrotic Spot Virus in melon leaves compared to dot blot assays.

Random amplified polymorphic DNA (RAPD) analysis (an example of classical PCR) has been used for the successful detection of many plant pathogens. RAPD assays are reliant on the use of random primers (usually 10-mer) where no prior knowledge of the genome is required. They can be used to determine polymorphisms and generate molecular markers which can lead to the development of specific primers. Rigotti et al. *(25)* reported on the application of this technology for the detection of *Botrytis cinerea* in strawberry.

PCR-based diagnostics are becoming the "gold standard" for plant pathogen identification *(19)*. PCR-based assays can be examined under classical- or real-time. Both share a common reliance on primer specificity for the successful identification of a particular pathogen but there are a number of significant differences between the two systems. One major difference between classical and real-time PCR is the method employed for the detection of the PCR product. Whereas classical PCR relies on Southern blot hybridization to unambiguously determine the nature of the amplicon, real-time PCR can utilize in the main three fluorescent methods: TaqMan® probes, fluorescent resonance energy transfer (FRET) probes, and molecular probes *(19)*. In real-time PCR the intensity of the fluorescence of the sample is proportional to the amount of PCR product that has accumulated during the reaction *(26)*. This method of detection of the PCR product obviates the need for further gel electrophoretic

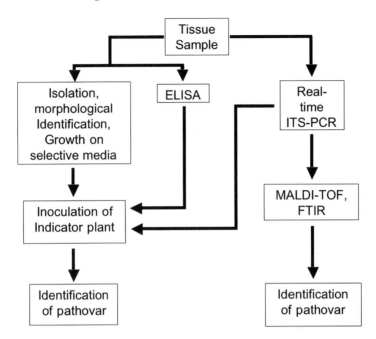

Fig. 4. Strategy for use of conventional and "omics" methods for pathogen identification comparing conventional with "omics" strategies. Real-time polymerase chain reaction can identify the pathogen group, specific pathogenicity factors can be potentially identified by spectroscopy (e.g., MALDI-TOF and FTIR, respectively).

analysis. A frequent problem encountered with both types of PCR is inhibition of the amplification step usually as a result of the presence of contaminating material (e.g., polysaccharides or secondary metabolites in the reaction mix) *(27)*. Many protocols have been published on ways to "clean-up" the target sample but this can result in a lower yield overall of extracted DNA. One way around this is to use BIO-PCR which concomitantly increases the sensitivity of the reaction by first incorporating an enrichment step into the procedure *(28)*. A further advantage here is that only viable cells are detected. Another alternative for increasing the sensitivity of the PCR reaction is to use immunomagnetic separation *(19,27)* whereby the target organism is extracted by binding to magnetic beads that have been pre-coated with anti-rabbit immunoglobulin-G washed with anti-target serum.

A method which combines PCR and ELISA was described by Bailey et al. *(23)* for the identification to the species level for *Phytophthora* and *Pythium*. In this report the ITS1 region (*see* **Fig. 3**) was sequenced for a number of species from both genera and through sequence alignment a number of differences were discovered which allowed for the preparation of particular capture probes which were complementary in sequence to the target DNA molecule(s). These probes were biotin-labeled and immobilized in a microtiter plate coated with streptavidin. The DIG (digoxygenin-11-UTP)-labeled amplicons that subsequently bound to the capture probes were immunologically detected.

Aside from using the organism's chromosomal DNA as a target in PCR another target for amplification are plasmid sequences. But this procedure can be problematic as these plasmid sequences are interchangeable between bacteria (horizontal transfer). According to Schaad and Frederick *(19)*, the use of larger plasmids (megaplasmids) as a target, where the probability of horizontal transfer is low, is preferred.

Denaturing gradient gel electrophoresis (DGGE) and temperature-gradient gel electrophoresis (TGGE) are often used in microbial ecology for analyzing microbial populations *(29)*. These techniques allow for the separation of amplicons of the same length but which differ with respect to their nucleotide sequences. Watanabe et al. *(29)* compared bacterial universal primers (frequently used in PCR/DGGE) with 16S rDNA sequences. An improvement in the universal primers with the incorporation of the base inosine followed by DGGE analyses allowed for greater levels of detection of environmental samples than when compared to previous studies using PCR/DGGE.

2.1.5.2. Proteomic Fingerprinting

Matrix-assisted laser desorption/ionization time-of-flight (MALDI-TOF) based on analysis of whole intact bacteria is a rapid method of bacterial identification based on biomarker protein analysis *(30)*. Intact bacteria may be analyzed. More information, however, can be obtained by fragmentation of the bacteria and analysis of the total proteome *(31)*. This method is also applicable to fungi and viruses. The taxonomic relationship of the organism can be obtained from stable membrane protein biomarkers, it may also be possible to detect specific proteinaceous pathogenicity markers to identify to pathovar level *(32)*.

2.1.5.3. Metabolomic Fingerprinting

There are emerging techniques for the rapid identification of microorganisms based on metabolomic analysis by micro-spectroscopy (e.g., yeast can be identified by Fourier-Transformed Infra-Red [FT-IR] spectroscopy) *(33)*. For a review of this application *see* **ref.** *34*.

2.2. Indexing for Plant Pathogens

In selecting stock plants for tissue culture, the main question is not whether they are infected on the surface with pest or microorganisms that, in principle, can be eliminated by surface sterilization of the explant, but whether there are inter- or intra-cellular endophytes that may enter the cultures to cause contamination. Where stock plants are showing systemic disease symptoms, they should preferably be rejected and disease escapes sought. If the latter are not available, then meristem culture should be used to introduce the plants into culture *(35)*. There is, however, the risk that phytopathogens (as opposed to vitropathogens) may be latent in the plant and in the established cultures and so the stock plants must be formally indexed for specified pathogens in the case of crops for certification. It is advisable that non-certified crops also be indexed for major inter- and intra-cellular phytopathogens of the crop. Indexing must comply with the current rules/guidelines of the regional certification authority (e.g., The European and Mediterranean Plant Protection Organization in the European Union [*see* guidelines for indexing at www.eppo.org] and the North American Plant Protection Organization in the United States and Canada [*see* guidelines for indexing at www.nappo.org]). In general, these organizations are conservative and require that traditional indexing methods be used. These are primarily based on transmission by mechanical inoculation or grafting to indicator plants and ELISA.

Aside from the officially approved diagnostic tests, the certification agencies specify the conditions (age of plant, tissue to be tested) under which the tests must be carried out. One of the limitations is that the certification authorities do not yet approve the testing of in vitro cultures *(16)*. Consequently, where meristem culture is used to eliminate phytopathogens, certification may require that a sample of production be established and grown in vivo for indexing.

2.3. Establishment of Aseptic Cultures

Although a case can be made for searching for pathogen-free stock plants and for rigorous indexing of the selected stock plants prior to their use in tissue culture, in practice this is only mandatory where micropropagation is used in official certification schemes *(36)*. Commonly, the selection of the explant that is used to introduce the plant into culture is made on the basis of the genetic stability of the tissue in vitro *(35)*. Leaf and petiole explants can be used from plants, typically represented by *Saintpaulia* and *Begonia*, whose tissues are considered to give rise to genetically stable adventitious shoots whereas cultures of potato, which are unstable in adventitious regeneration, are usually initiated from apical explants and multiplied as nodal cultures. In either case, or where the cultures are being set up for experimental purposes, the criterion of visible expression of contamination is generally used as an indicator of non-aseptic status. Many studies have shown that this is not a reliable strategy. Aside from the symptomless transmission of pathogens, there are also reported cases of microbial contaminants been "suppressed" by components of the plant tissue culture media. These may be clonally propagated or transmitted in subculture to emerge and over run the cultures when the medium is changed (e.g., removal of additives such as phloroglucinol and alteration in the hormone content or reduction of sugar or medium salts). It is strongly recommended that cultures be indexed for microbial contamination on a range of common fungal and bacteriological media. It may also be beneficial to add inhibitors of phenolic production and bacterial substrates, such as casein hydrolysate, to the Stage 1 medium to encourage microbial growth *(37)*.

Meristem culture has long been recognized as an efficient strategy to eliminate pathogens from plants and, perhaps unknowingly, to eliminate many biotic contaminants. In general, this is a reliable strategy but there are examples where, either because the apical explant was too large or because the contaminant/pathogen was highly invasive of the apical tissue, this approach has not been successful *(12)*. In these cases, attempts can be made to find escapes, or the plants can be heat-treated or treated with antibiotics in vivo prior to explantation. Alternatively, the plants can be heat treated or exposed to antibiotics in vitro *(12,38)*.

If explants, other than apical meristems are used, care should be taken as epiphytic bacteria and yeasts may form resistant biofilms, may be protected in leaf or stem crevices, in damaged hairs or glands or may have entered the plant opportunistically through wounds and so on, thereby avoiding the action of surface sterilants. Common problems are the lack of a wetting agent in the sterilizing solution or breakdown of the active principle *(39)*. Leaf-surface inhibiting bacterial genera are CCP indicator organisms for failure to establish contaminant-free (aseptic) cultures (*see* **Fig. 2**). They can be identified to genera by classical plant bacteriological methods; by use of commercial biochemical test kits in-house or sent for fatty acid profiling or PCR analysis to service companies (addresses can be found using web search engines such as yahoo.com).

3. Laboratory Contamination Management

HACCP has been used very effectively in the food industry to control contamination and is equally applicable in the tissue culture laboratory *(6)*. It is important when designing tissue culture laboratories, especially commercial laboratories, to compartmentalize activities so that barriers are in place to prevent the entry of contamination from the external environment and between the sterile and nonsterile areas. The staff should receive an adequate training in aseptic technique and staff hygiene standards should be maintained as in the food industry. Jewellery and watches should be removed before entering the clean areas, hair and ears should be covered, overshoes should be worn, and face masks used. It is particularly important that workers wash their hands/gloves between shifts. The equipment should be regularly checked to ensure that the specifications for sterilization and air-cleaning are maintained.

Biomarkers for autoclave functioning, the efficiency of laminar-flow cabinets and laboratory air quality are given in **Fig. 2** *(6)*. These are all common bacteria/yeasts and can be identified by traditional methods or by using biochemical test kits (**Table 1**). Growth room air quality can also be determined by monitoring these microorganisms *(40)*. A high risk in growth rooms is infestation by micro-arthropods and cultures should be regularly inspected for microcolony trails which reflect the transmission of bacteria by the micro-arthropods as they move about the cultures. All contaminated cultures should be autoclaved before disposal and care should be taken to avoid storing fungal cultures which may act as initial sources of infection. The movement of micro-arthropods in the growthroom can be slowed by placing (transparent) film on the culture shelves which prevents the pest dropping down from upper shelves. Chemical treatments are available but these can be toxic to the plant tissues *(4)*.

It has been known for some time that the cultures can be contaminated by the laboratory workers as indicated by the presence of human-associated microorganisms in the cultures. It has also been reported that human pathogens may be transmitted via plant cultures to infect workers *(41)*. Contamination with human-associated bacteria can be used to monitor the aseptic technique of micropropagation workers.

4. Future Trends

It is not surprising that the guidelines for indexing plants in certified stock production and for international plant movement, rely on traditional methods as these guidelines are drawn up by committees of representatives of individual countries and consensus must be obtained before changes can be introduced that are acceptable as a basis for international plant trade. Nevertheless, European Union draft legislation approves the use of PCR for indexing for *Ralstonia solanacearum* in potato and it is anticipated that the use of "omics" methods will become more widely approved. There still remains the requirement that specified in vivo plant tissues be used for indexing. Further studies of the virus-titer of in vitro material may validate the indexing of in vitro cultures for viral pathogens in the future *(16)*.

There are no restrictions on the use of "omics" methods for pathogen-indexing of in vitro cultures for non-certification purposes. Here, however, the issues of sensitivity, specificity and cost arise. Although primers to ITS regions may allow identification to species or subspecific level there remains the need to confirm pathogenicity. Where pathovars are distinguished by the production of specific metabolites (toxins) or proteins (e.g., glucanases and so on) "omics" methods have the potential to identify them. Where pathogenicity is caused by quantitative rather than qualitative gene expression this may not be possible and inoculation indicator plants will still be required. Identification to species/

subspecies level will also be required where the same pathogenicity factors occur in different species (*see* **Fig. 4**).

As mentioned above, setting up an "omics" protocol may involve considerable capital costs for equipment, for the training of personnel and for protocol development. An alternative to this is to use one of the many service providers (e.g., www.lgcpromochem-atcc.com; www.ncppb.com; www.ukncc.co.uk; www.cabri.org.) These organizations charge from approx $100 for a fatty acid analysis to several hundred dollars for a genomic fingerprint.

Monitoring of laboratory contamination for CCP indicator organisms by biochemical test kits or microscopic/growth on selective media costs little for the materials and, as argued previously, should be regarded as part of a HACCP system.

References

1. Cassells, A. C. and Tahmatsidou, V. (1996) The influence of local plant growth conditions on non-fastidious bacterial contamination of meristem-tips of Hydrangea cultured *in vitro*. *Plant Cell Tiss. Org. Cult.* **47,** 15–26.
2. Leifert, C., Morris, C., and Waites W. M. (1994) Ecology of microbial saprophytes and pathogens in tissue cultured and field grown plants. *CRC Crit. Rev. Plant Sci.* **13,** 139–183.
3. Leifert, C., Nicholas, J. R., and Waites, W. M. (1990) Yeast contaminants of micropropagated plant cultures. *J. Appl. Bacteriol.* **69,** 471–476.
4. Pype, J., Everaert, K., and Debergh, P. C. (1997) Contamination by microarthropods, in *Pathogen and Microbial Contamination Management in Micropropagation* (Cassells, A. C., ed.), Kluwer Academic Publishers, Dordrecht, pp. 259–266.
5. Krczal, G. (1998) Virus certification of ornamental plants—the European strategy, in *Plant Virus Disease Control APS Press*, (Hadidi, A., Khetarpal, R. K., and Koganezawa, H., eds.), St. Paul, pp. 277–287.
6. Leifert, C. and Cassells, A. C. (2001) Microbial hazards in plant tissue and cell cultures. *In Vitro Cell Dev. Biol.* **37,** 133–138.
7. Lelliott, R. A. and Stead, D. E. (1987) *Methods for the Diagnosis of Bacterial Diseases of Plants*, Blackwell Scientific Publications, Oxford.
8. Schaad, N. W., Jones, J. B., and Chun, W. (2001) *Laboratory Guide for Identification of Plant Pathogenic Bacteria*, APS Press, St. Louis.
9. Agrios, G.N. (1997) *Plant Pathology*, Academic Press, London.
10. Larone, D. H. (1987) *Medically Important Fungi. A Guide to Identification*, Elsevier Publishers, New York.
11. Krantz, G. W. (1978) *A Manual of Acarolgy*, Oregon State University, Corvallis.
12. Hull, R. (2001) *Matthew's Plant Virology*, Academic Press, New York.
13. Hari, V. and Das, P. (1998). Ultra microscopic detection of plant viruses and their gene products, in *Plant Virus Disease Control* (Hadidi, A., Khetarpal, R. K., and Koganezawa, H., eds.), APS Press, St. Paul.

14. Cassells, A. C. (1991) Problems in tissue culture: culture contamination, in *Micropropagation: Technology and Applications* (Debergh, P. C. and Zimmerman, R. H. eds.). Kluwer Dordrecht, pp. 31–44.
15. Stead, D. E., Elphinstone, J. G., Weller, S., Smith, N., and Hennessy, J. (2000) Modern methods for characterizing, identifying and detecting bacteria associated with plants. *Acta Hort.* **530,** 45–60.
16. O'Herlihy, E. A. and Cassells, A. C. (2003) Influence of *in vitro* factors on titre and elimination of model fruit tree viruses. *Plant Cell Tiss. Org. Cult.* **72,** 33–42.
17. Dijkstra, J. and de Jager, C. P. (1998) *Practical Plant Virology: Protocols and Exercises,* Springer Verlag, Berlin.
18. Lee, B. H. and Nagamune, T. (2004) Protein microarrays and their application. *Biotechnol. Bioprocess. Eng.* **9,** 69–75.
19. Schaad, N. W. and Frederick, R.D (2002) Real-time PCR and its application for rapid plant disease diagnostics. *Can. J. Plant Pathol.* **24,** 250–258.
20. Mishra, P. K., Fox, R. T. V., and Culham, A. (2003) Development of a PCR-based assay for rapid and reliable identification of pathogenic Fusaria. *FEMS Microbiol. Letts.* **218,** 329–332.
21. Singh, K. K., Mathew, R., Masih, I. E., and Paul, B. (2003) ITS region of the rDNA of Pythium rhizosaccharum sp. Nov. isolated from sugarcane roots: taxonomy and comparison with related species. *FEMS Microbiol. Letts.* **221,** 233–236.
22. Klerks, M. M., Leone, G. O. M., Verbeek, M., van den Heuvel, J. F., and Schoen, C. D. (2001) Development of a multiplex AmpliDet RNA for the simultaneous detection of Potato Leafroll Virus and Potato Virus Y in potato tubers. *J. Virol. Meth.* **93,** 115–125.
23. Bailey, A. M., Mitchell, D. J., Manjunath, K. L., Nolasco, G., and Niblett, C. L. (2002) Identification to the species level of the plant pathogens *Phytophthora* and *Pythium* by using unique sequences of the ITS1 region of ribosomal DNA as capture probes for PCR ELISA. *FEMS Microbiol. Letts.* **207,** 153–158.
24. Gosalvez, B., Navarro, J. A., Lorca, A., Botella, F., Sanchez-Pina, M. A., and Pallas, V. (2003). Detection of melon necrotic spot virus in water samples and melon plants by molecular methods. *J. Virol. Meth.* **113,** 87–93.
25. Rigotti, S., Gindro, K., Richter, H., and Viret, O. (2002). Characterization of molecular markers for specific and sensitive detection of Botrytis cinerea Pers.: Fr. in Strawberry (Fragaria x ananassa Duch.) using PCR. *FEMS Microbiol. Letts.* **209,** 169–174.
26. Filion, M., St. Arnaud, M., and Jabaji-Hare, S. H. (2003) Direct quantification of fungal DNA from soil substrate using real-time PCR. *J. Microbiol. Meth.* **53,** 67–76.
27. Langrell, S.R.H. and Barbara, D.J. (2001) Magnetic capture hybridisation for improved PCR detection of Nectria galligena from lignified apple extracts. *Plant Molec. Biol. Rep.* **19,** 5–11.
28. Ozakman, M. and Schaad, N. W. (2003) A real-time BIO-PCR assay for the detection of Ralstonia solanacearum race 3, biovar 2, in asymptomatic potato tubers. *Can. J. Plant Pathol.* **25,** 232–239.

29. Watanabe, K., Kodama, Y., and Harayama, S. (2001) Design and evaluation of PCR primers to amplify bacterial 16S ribosomal DNA fragments used for community fingerprinting. *J. Microbiol. Meth.* **44**, 253–262.
30. Lay, J. O. (2001) MALDI-TOF mass spectrometry of bacteria. *Mass Spectrom. Revs.* **20**, 172–194.
31. Wang, Z. P., Dunlop, K., Long, S. R., and Li, L. (2002) Mass spectrometric methods for generation of protein mass database used for bacterial identification. *Anal. Chem.* **74**, 3174–3182.
32. Nedelkov, D., Rasooly, A., and Nelson, R. W. (2000) Multitoxin buiosensor-mass spectrometry analysis: a new approach for rapid, real-time, sensitive analysis of staphylococcal toxins in food. *Int. J. Food Microbiol.* **60**, 1–13.
33. Wenning, M., Seiler, H., and Scherer, S. (2002) Fourier-transformed infrared microscopy, a novel and rapid toot for identification of yeasts. *Appl. Environ. Microbiol.* **68**, 4717–4721.
34. Bull, A. T., Goodfellow, M., and Slater, J. H. (1992) Biodiversity as a source of innovation in biotechnology. *Ann. Rev. Microbiol.* **46**, 219–252.
35. George, E. F. (1996) *Plant Propagation by Tissue Culture*, Exegetics, Basingstoke.
36. Cassells, A. C. (1997) *Pathogen and Microbial Contamination Management in Micropropagation*, Kluwer Academic Publishers, Dordrecht.
37. Menard, D., Coumans, M., and Gaspar, T. H. (1985) Micropropagation du Pelargonium a partir de meristems. Meded. Fac. Landbouwett. Rijksuniv. *Gent.* **50**, 327–331.
38. Barrett, C. and Cassells, A. C. (1994) An evaluation of antibiotics for the elimination of *Xanthomonas campestris* pv. Pelargonii (Brown) from *Pelargonium x domesticum* cv. Grand Slam explants *in vitro*. *Plant Cell Tiss. Org. Cult.* **36**, 169–175.
39. Hoffman, P. N., Death, J. E., and Coates, D. (1981) The stability of sodium hypochlorite solutions, in *Disinfectants: Their Use and Evaluation of Effectiveness* (Collins, C. H., Allwood, M. C., Bloomfield, S. J. and Fox, A., eds.), Academic Press, London, pp. 77–83.
40. Gregory, P. H. (1973) *The Microbiology of Atmosphere*, Leonard Hill Books, Aylesbury.
41. Weller, R. and Leifert, C. (1996) Transmission of Trichophyton interdigitale via an intermediate plant host. *Brit. J. Dermatol.* **135**, 656–657.
42. Saar, D. E., Polans, N. O., Sorensen, P. D., and Duvall, M. R. (2001) Angiosperm DNA contamination by endophytic fungi: detection and methods of avoidance. *Plant Mol. Biol. Rep.*, **19**, 249–260.
43. Barnett, H. L. and Hunter, B. B. (1998) *Illustrated Genera of Imperfect Fungi*, 4th ed., APS Press, St. Louis.

4

Growth Measurements

Estimation of Cell Division and Cell Expansion

Gregorio Godoy-Hernández and Felipe A. Vázquez-Flota

Summary
The main parameters for the estimation of growth within in vitro cultures are reviewed. Procedures to measure these parameters are described, emphasizing in each case their convenience of use, depending on the features of the culture to evaluate.

Key Words: Cell counting; duplication time; dry weight; fresh weight; growth index; packed cell volume.

1. Introduction

The accurate, fast, and reliable determination of cell growth is of critical importance in plant cell and tissue culture. The precise assessment of growth kinetics is essential for the efficient design of bioprocess engineering. However, the measurement of growth parameters in the different types of cultures, and concomitantly the use of various containers along with the heterogenity in cell morphology, introduce diverse problems that must be addressed by using a specific methodology for each case (*1*). Because callus and cell suspension cultures represent two of the most common in vitro systems, this chapter concentrates in growth measurements in such systems.

1.1. Callus Cultures

Plant cell cultures are initiated through the formation of a mass of nondifferentiated cells called "callus." A callus culture is obtained by excising a piece of tissue, the explant, from the parent plant and placing it onto a nutrient base, solidified with agar. This nutrient base contains macronutrients (nitrogen, potassium, phosphorus, and so on), micronutrients (cobalt, copper, iron, and so on), a carbon source (usually sucrose), and various plant growth regulators. Callus tissue is not an ideal system to work with, because of their slow growth rate and high biochemical variability. Nevertheless, because they are an almost obligatory step to initiate an in vitro culture, the proper assessment of the effect of different media composition or growth regulators on their growth rate is an important parameter to define the optimal media to use. The most common growth parameters used for callus cultures include fresh weight, dry weight, and growth index.

1.2. Suspension Cultures

The success in the establishment of a cell suspension culture depends, in a great extent, on the availability of "friable" callus tissue (i.e., a tissue that, when stirred in liquid medium, rapidly disaggregate into single cells and small clusters). Such a system is much more amenable for biochemical studies and process development than calli, because they generally grow in a faster rate and allow cells to be in direct contact with the medium nutrients. Suspension or liquid cultures of plant cells are usually grown as microbial cells. However, plant cell dimensions are larger than those of bacteria or fungi, ranging from 20 to 40 µm in diameter and from 100 to 200 µm in length *(2,3)*. The central vacuole occupies a large portion of the mature cell volume. Plant cell cultures tend to contain clumps, formed by a variable number of cells. These clumps arise as a result of the failure of new cells to separate after division or from the adherence of free cells among themselves. In some cases, such clumps (also known as aggregates) may contain up to 200 cells, and reach up to 2 mm in diameter. Although cell stickiness can be overcome by modifying the culture medium, cells in culture become "sticky" in late lag phase of growth as a rule.

Methods to obtain suspensions composed largely of free cells (fine-cell suspensions) include the use of cell wall degrading enzymes and sieving. Unfortunately once established, a fine cell suspension has a tendency with time to revert to a clumped condition *(4)*.

1.3 Methods for Evaluating Growth in In Vitro Cultures

There are several methods for evaluating growth kinetics in plant cell cultures. Selected examples include: fresh cell weight, dry cell weight, settled cell volume, packed cell volume, cell counting, culture optical density, residual electrical conductivity, and pH measurements, among others *(1)*.

In cultures originated from different plant species, settled volume, packed cell volume, as well as fresh weight, all show a very good linear correlation with dry weight data. Thus, any of these estimations can be used for measuring cell growth. However, during the stationary phase of the culture, there are important deviations in this correlation. At this phase, plant cell cultures show a high degree of aggregation, cell lysis, and a marked heterogenity in cell morphology.

The measurement of cell concentration by cell counting and turbidity (optical density) has also shown a reasonably good correlation with the dry-cell weight parameter. When cultures reached the stationary phase, the same problems stated above render unsatisfactory results regarding to the correlation of these parameters to dry weight estimation. On the other hand, protein, RNA and DNA contents fail to show the same good correlation with the dry weight, which may be ascribed to changes in cell physiology along the growth cycle, which frequently results in wide variations.

Although complicated and time consuming, cell counting represents the best way to assess culture growth in suspension cultures. Nevertheless, it often shows a good correlation with other parameters, such as electric conductivity. Cell density is obtained by direct counting of cells under the microscope, using a cell counting chamber, such as the Sedgewick rafter cell (Graticulates Limited, Tonbridge England) or the Newbauer chamber (Sigma-Aldrich, St. Louis, MO). Such devices hold a fixed volume of the suspension over a defined area. The base of the chamber is divided in squares, frequently containing a 1 mm^3 (1 µL) volume. By observing the suspension with a low magnification objective, cells contained in such a volume are identified and counted.

Electric conductivity of culture medium decreases inversely to biomass gain. This is a consequence of ion uptake by cells. The monitoring of this decrease to assess cell growth offers several advantages over other methods, such as: (1) continuous and *in situ* or on-line monitoring of cell growth; (2) no sampling or wet chemical analysis is required; (3) it is economical and efficient; (4) it provides an accurate, reliable, and reproducible measurement of plant cell growth rate; and (5) it is independent of cell aggregation, growth morphology, and apparent viscosity *(1)*.

2. Materials
2.1. Biological Material
1. Callus and cell suspension cultures of any species.

2.2. Glassware
1. Glass desiccators, or any other hermetic container, with silica gel.
2. 50-mL sterile, graduated centrifuge tubes.
3. Counting cell chamber (Sedgewick rafter cell).

2.3. Instrumentation

1. Analytical balance (resolution to 0.01 mg)
2. Top loading balance (resolution 0.1 g)
3. Convection oven (resolution 0.1°C)
4. Conductivity meter (range 0–200 µs and 0–200 µs; resolution 0.1 U)
5. Universal centrifuge (with a speed range including 4000g)
6. Light microscope with ×10 and ×40 objective lenses.

2.4. Chemicals

1. Chromium trioxide (CrO_3), 8% solution. Dissolve 8 g of CrO_3 in 100 mL of water. Keep in the dark at room temperature
2. Cellulase from *Aspergillus* sp., 10% suspension. Dissolve 0.5 g of cellulase (Sigma Chemical Co, St. Louis, MO) in 5 mL 100 mM potassium phosphate buffer, pH 6.0. Keep at 4°C. Use suspension within 1 wk.
3. Macerozyme R-10, 5% suspension (Pectinase from *Rhizopus*). Dissolve 0.25 g of Macerozyme R-10 (Sigma Chemical Co) in 5 mL 100 mM potassium phosphate buffer, pH 6.0. Keep at 4°C. Use suspension within 1 wk.

3. Methods

3.1. Measuring Growth in Callus Cultures

3.1.1. Fresh-Weight Determination

1. To determine the total fresh weight, harvest all the tissue on the media surface with flat-end tweezers and weigh on a balance. To avoid tissue desiccation, open the culture jar only when you are ready to processes it.

3.1.2. Dry-Weight Determination

1. Dry weight can be estimated by drying the tissue at 60°C in a convection oven, until of constant weight (usually 16 h). Take a sample of the fresh tissue, weigh it on a pre-weighed square of aluminum foil, and evaporate the water contained in the tissue in the pre-heated oven at 60°C for 12 h.
2. Allow samples to cool down in a dessicator containing silica gel for 15–20 min and then register the sample's weight.
3. Put the tissue sample back into the oven and take a new weight register after 2 h. If no variations are detected, samples have reached constant weight. If variations higher than 10% regarding the previous register are observed, return samples to the oven for another 2-h period.
4. Alternatively, dry weight can be obtained from lyophilized tissues. Once harvested, fresh tissues should be weighed, deposited in lyophilizer flasks, and frozen at –20°C for at least 12 h. Flasks with frozen tissues are then connected to the vacuum line for 24 h and weight is registered.

3.2. Measuring Growth in Cell Suspension Cultures

Growth of suspension cultures is commonly evaluated as the settled cell volume (SCV), the packed cell volume (PCV), fresh cell weight (FCW), dry cell weight (DCW). Indirect evaluations include pH measurements and medium residual conductivity (mmhos). Finally, parameters describing growth efficiency, such as specific growth rate (μ), doubling time (dt), and growth index, can be determined.

3.2.1. Settled Cell Volume and Packed Cell Volume

Both parameters allow the quick estimation of culture growth, while maintaining sterile conditions. These measurements are useful for monitoring growth in the same flasks along a culture cycle, because suspensions may be returned to prior culture conditions. Care must be taken to maintain sterile conditions. However, volume estimation may not be an accurate way of monitoring growth, given its dependence on cell morphology (cell and clump size, cell density, and other).

SCV is determined by allowing a cell suspension to sediment in graduated tubes. It is reported as the percentage of the total volume of suspension occupied by the cell mass. The PCV is determined in a similar way, after it has been compacted by centrifugation.

3.2.1.1. SETTLED CELL VOLUME

1. Pour the cell suspension in a graduated cylinder of adequate volume.
2. Allow the suspension to settle for 30 min and record the cell volume.
3. Take a second reading 30 min later. If the variation between readings is higher than 5%, record a third measurement after another 30-min wait period.
4. The volume fraction of the suspension occupied by the cells is determined as the SCV.

3.2.1.2. PACKED CELL VOLUME

1. PCV can be determined by centrifuging 10 mL of the culture in a 15-mL graduated conical centrifuge tube at 200g for 5 min.

3.2.2. Fresh Cell Weight and Dry Cell Weight

Fresh and dry cell weight represent more precise measurements of cell growth than the sole culture volume. However, both require the manipulation of samples in nonsterile conditions. Fresh weight estimation involves less time than that required for dry weight, but it may not reflect a real measurement of biomass gain, particularly at the end of the culture period, when most of the culture growth is because of water uptake *(5)*.

1. Collect the cell mass by filtration, using a Büchner funnel under vacuum.

2. Wash the cell package with about 3 mL distilled water and retain under vacuum for a fixed time period (e.g., 30 s). Weigh immediately to reduce variations caused by water evaporation.
3. Fresh and dry weights are determined as described earlier for callus tissue.

3.2.3. Culture Cell Density

In order to obtain a reliable value of the number of cells in a suspension culture, clusters should be first disaggregated. This can be accomplished by incubating the suspension with an 8% chromium trioxide solution, or with hydrolytic enzymes, such as cellulase and pectinase. The chromium trioxide method is quicker and less complicated than the use of enzymes; however, it hinders the estimation of cell viability in the same sample. Because a careful use of enzymes maintains cells viable, the assessment of the number of living cells by the exclusion of vital stains can be performed in the same sample.

3.2.3.1. Cell Cluster Disaggregation by Chromium Trioxide

1. Take 1 mL of the cell suspension and add 2 mL of 8% chromium trioxide (CrO_3).
2. Incubate the mixture for 15 min at 70°C.
3. Vortex the mixture vigorously for 15 min.

3.2.3.2. Cell Cluster Disaggregating Hydrolytic Enzymes

1. Take 1 mL of the cell suspension and mix it with 0.5 mL of 10% cellulase and 0.5 mL 5% pectinase.
2. Incubate 30 min at 25°C with rotatory agitation (100 rpm) (*see* **Note 1**).

3.2.3.3. Cell Counting

1. Fill the counting cell chamber with the mixture, position carefully the cover glass on top of the chamber, to avoid the formation of bubbles.
2. Observe under the microscope with the ×10 objective to locate the squared field.
3. Count all the cells contained in 10 squares. Add the values of the 10 squares (do not obtain the average). This number represents the number of cells in 10 µL, so multiply by 100 to determine the cell number per milliliter *(6)* (*see* **Note 2**).
4. Depending on the culture's cell density, further dilution may be required, which should be considered in the calculation.

3.3. Parameters of Growth Efficiency

3.3.1. Growth Index

Both fresh and dry weight are measurements of tissue's absolute biomass at a given sampling time. No reference to the actual growth capacity is taken in consideration. Growth index (GI) is a relative estimation of such capacity as it correlates the biomass data at the sampling time to that of the initial condition. It is calculated as the ratio of the accumulated and the initial biomass. The

Growth Measurements

accumulated biomass corresponds to the difference between the final and the initial masses.

$$GI = \frac{W_F - W_0}{W_0}$$

Where GI represents growth index, and W_F and W_0, represent the final and initial masses, respectively (either as fresh or dry weight).

3.3.2. Specific Growth Rate

It is generally accepted that growth of a cell culture with respect to time is best described by the sigmoid curve theory. At the beginning, the cell population grows relatively slow (lag phase). As the population size approaches one-half of the carrying capacity (defined by the nutrient status of the culture medium), the culture's growth per time unit increases. The rate of growth is measured by the steepness of the curve, and it is the steepest when the population density reaches one-half of the carrying capacity (in the middle of the sigmoid). After this point, the steepness of the curve decreases, keeping this tendency as the population increases, until it reaches the carrying capacity (stationary phase).

The specific growth rate (μ) refers to the steepness of such a curve, and it is defined as the rate of increase of biomass of a cell population per unit of biomass concentration. It can be calculated in batch cultures, since during a defined period of time, the rate of increase in biomass per unit of biomass concentration is constant and measurable. This period of time occurs between the lag and stationary phases. During this period, the increase in the cell population fits a straight-line equation (**6**):

$$\ln x = \mu t + \ln x_0$$

$$\mu = \frac{\ln x - \ln x_0}{t}$$

Where x_0 is the initial biomass (or cell density), x is the biomass (or cell density) at time t, and μ is the specific growth rate.

3.3.3. Doubling Time

Doubling time (dt) is the time required for the concentration of biomass of a population of cells to double. One of the greatest contrasts between the growth of cultured plant cells and microorganisms refers to their respective growth rates. Although the pattern of growth may be the same, plant cells have doubling times or division rates measured in days, while this parameter in many microorganisms is in the order of minutes to hours. One of the fastest (and quite exceptional) doubling time recorded for a plant cell culture is 15 h for tobacco cells *(2)*. The doubling time (dt) can be calculated according to the following equation *(6)*:

$$dt = \frac{\ln 2}{\mu}$$

Where µ represents the specific growth rate.

4. Notes

1. The conductivity of the culture medium can be determined using a digital conductivity meter or equivalent.
2. A very good linear correlation between the decrease in conductivity and the increase in dry weight can be found.

References

1. Ryu, D. D. Y., Lee, S. O., and Romani, R. J. (1990) Determination of growth rate for plant cell cultures: comparative studies. *Biotechnol. Bioeng.* **35**, 305–311.
2. Fowler, M. W. (1987) Products from plant cells, in *Basic Biotechnology* (Bu'lok, J. and Kristiansen, B., eds.), Academic Press, London, pp 525–544.
3. Su, W. W. (1995) Bioprocessing technology for plant cell suspension cultures. *Appl. Biochem. Biotechnol.* **50**, 189–230.
4. Fowler, M. W. (1982) The large scale cultivation of plant cells, in *Progress in Industrial Microbiology*, Vol 16. (Bull, M. J., ed.), Elsevier, Amsterdam, pp. 207–229.
5. Trejo-Tapia, G., Arias-Castro, C., and Rodríguez-Mendiola, M. (2003) Influence of the culture medium constituents and inoculum size on the accumulation of blue pigment and cell growth of *Lavandula spica*. *Plant Cell Tiss. Org. Cult.* **72**, 7–12.
6. King, P. J. and Street, H. E. (1977) Growth patterns in cell cultures, in *Botanical Monographs*, Vol 11: Plant Tissue and Cell Culture (Street, H. E., ed.), Blackwell, Oxford.
7. Ramage, C. M. and Williams, R. R. (2002) Mineral nutrition and plant morphogenesis. *In Vitro Cell Dev. Biol., Plant* **38**, 116–124.

5

Callus and Suspension Culture Induction, Maintenance, and Characterization

Alejandrina Robledo-Paz, María Nélida Vázquez-Sánchez, Rosa María Adame-Alvarez, and Alba Estela Jofre-Garfias

Summary

Callus and cell suspension can be used for long-term cell cultures maintenance. This chapter describes procedures for the induction of somatic embryos of garlic, keeping a regeneration capacity for more than 5 yr, as well as the maintenance of a tobacco suspension culture (NT-1 cells), for more than 10 yr. Methods for plant regeneration and growth kinetics of garlic cultures are described, as well as for cell viability of NT-1 cells stained with 2,3,5 triphenyltetrazolium chloride. The packed cell volume determination as a parameter of growth is detailed.

Key Words: *Allium sativum*; callus cultures; garlic; somatic embryogenesis; suspension cultures.

1. Introduction

Plant tissue cultures involve a series of aseptic techniques that allow for the production of whole regenerated plants or the induction and maintenance of proliferating undifferentiated masses of cells called callus. These kinds of cultures are initiated by placing the appropriate type of explant onto a nutrient medium in vitro.

In 1957, Skoog and Miller (*1*) demonstrated the effect of auxins and cytokinins on the induction of different responses of tobacco pith tissue. It was established that a low concentration of auxins in combination with a high concentration of a cytokinin-induced shoot development and that a high concentration of auxins combined with a low concentration of cytokinin yielded root formation. After these experiments, there were many examples of morphogenetic responses of in vitro cultures, including callus induction. Plant suspension cultures, in which undifferentiated cells grow in liquid medium, have been used in biochemical research, for cell physiology, growth, metabolism, molecular biology, genetic engineering experiments, or for medium or large-scale secondary metabolite and other substance production.

As starting plant material (explant), very diverse parts of a plant can be used. The selection will depend on the particular species that is employed and the kind of response that is desired. Among the type of explants frequently used are leaf portions, isolated meristems, hypocotyls, and root segments, among others. In short, any part of the plant that is able to respond to the culture medium and the growing conditions can be used.

For the initiation of the culture three important considerations should be taken into account: (1) explant selection, (2) election of a suitable culture medium, and (3) appropriate environmental conditions for its development.

In this chapter callus induction and maintenance, plant regeneration of garlic, and the long-term maintenance of a tobacco suspension cell line will be described.

2. Materials

2.1. Garlic (Allium sativum L.) Somatic Embryogenesis Induction

2.1.1. Plant Material

1. Disease- and pest-free garlic cloves from different garlic cultivars.

2.1.2. Laboratory Materials and Equipment

1. Laminar flow cabinets.
2. Bunsen burner.
3. Dissecting forceps and scalpel.
4. Sterile Petri dishes 15 × 100 mm.
5. Beakers.
6. Cylinders.
7. 100- and 250-mL bottles.
8. Aluminum foil.
9. Parafilm.
10. Stereo microscope.
11. Tissue culture incubator at 22°C and 16 h photoperiod.
12. Autoclave.

Callus and Suspension Culture 61

13. pH meter.
14. Analytical balance.
15. Stirrer with hot plate.

2.1.3. Reagents, Solutions, and Culture Media

1. Sterile distilled water.
2. Ridomil bravo: 9% metalaxyl, 72% cloratonyl.
3. Benlate (benomyl 50%).
4. Agrimicin 100 (streptomycin 18.7%, oxytetracycline 2%).
5. Commercial laundry bleach (e.g., Clorox®) with 6% of active chlorine.
6. Tween-20.
7. Fungicide and bactericide solution: for each cultivar prepare 500 mL of the fungicide solution. To 300 mL distilled water add the following compounds: 3 g/L ridomil bravo; 1.2 g/L benlate, and 1.0 g/L agrimicyn. Mix well until homogeneous suspension is obtained and bring volume to 500 mL.
8. Chlorine solution: prepare 500 mL of solution for each cultivar. To 300 mL of distilled water add 150 mL of bleach (final concentration is 1.8% of active chlorine) add 4 mL of Tween-20 and bring volume to 500 mL.
9. Culture media: specific media used for all steps, from sprouting of garlic cloves to plant regeneration, are described in **Table 1**. Mix all ingredients needed for each culture medium. Adjust the pH to 5.7 with $1.0N$ KOH. Add agar and sterilize in autoclave at 1.1 kg/cm^2 at 121°C for 20 min. Pour 30 mL of medium into 15 × 100-mm Petri dishes. Media B, C, and D are based on N6 salts *(2)* supplemented with Eriksson vitamins *(3)*. Media E and F are based on on the Murashige and Skoog (MS) medium *(4)*.

2.2. NT-1 Tobacco Cell Suspension

2.2.1. Reagents and Culture Medium

1. Culture medium: NT-1 medium based on MS salts supplemented with 180 mg/L KH_2PO_4, 100 mg/L inositol, 1 mg/L thiamine, 2 mg/L 2,4-dichlorophenoxyacetic acid (2,4-D), and 30 g/L sucrose. Adjust pH to 5.7 with $1.0N$ KOH and sterilize in autoclave at 1.1 kg/cm^2 at 121°C for 20 min.
2. 0.05 M Phosphate buffer: dissolve 3.403 g of KH_2PO_4 in about 400 mL of distilled water. Adjust pH to 5.8 with KOH, and bring the volume to 500 mL.
3. 2,3,5-Triphenyltetrazolium chloride solution: dissolve 1 g of 2,3,5-Triphenyltetrazolium chloride in 100 mL of 0.5 M phosphate buffer, filter, sterilize, and store at 4°C in the dark.

Table 1
Culture Media Used for Callus Induction and Maintenance and for Plant Regeneration of Garlic

Medium	Salts	Vitamins	L-Proline	2,4-D	BA	Sucrose	Agar
A	MS (50%)	MS (50%)	0	0	0	1%	0.6%
B	N6	Erickson	6 mM	2.2 μM	0	2%	0.7%
C	N6	Erickson	6 mM	4.5 μM	0	2%	0.7%
D	N6	Erickson	6 mM	0	0	2%	0.7%
E	MS	MS	0	0	4.4 μM	2%	0.7%
F	MS	MS	0	0	0	2%	0.7%

Notes: MS, Murashige and Skoog, 1962 medium (*4*); N6 = Chu et al., 1975 medium (*2*).

2.2.2. Laboratory Materials and Equipment

1. Orbital shaker.
2. Culture room at 25°C.
3. Clinical centrifuge.
4. Microcentrifuge.
5. Optical microscope.
6. Adjustable micropipets.
7. 10-mL Sterile transfer pipets with wide mouth.
8. 250-mL Erlenmeyer flasks.
9. 15-mL Centrifuge tubes.
10. 1.5-mL Microcentrifuge tubes.
11. Aluminum foil.

3. Methods
3.1. Garlic Callus Induction
3.1.1. Clove Disinfestation

Selection and prepare the vegetative material: garlic cloves to be used should not be dormant. Freshly harvested cloves are under dormancy and will not respond to the culture medium for the establishment of an in vitro culture. A method to estimate the dormancy degree of cloves at a certain time is to cut them longitudinally. The first leaf should be evident (*see* **Fig. 1**) and if this leaf occupies more than 50% of the length of the clove it is suitable for use in the callus induction procedure. If this degree is not achieved then the cloves must be brought out of dormancy stage. To do so, cloves should be maintained at 4°C for 4–5 wk. Once the first leaf is in the appropriate developmental stage the cloves are selected by their size. It is very important that the basal portion (from which roots are formed) do not present any evidence of damage because initial explants are generated from this part of the clove (5).

Selected cloves are detached from the bulb, manually peeled, and placed into a beaker where the fungicide-antibiotic solution is added to them. The beaker is capped with aluminum foil to prevent fermentation and left at room temperature for 18 h. After 18 h transfer the garlic cloves to the chlorine solution for 20 min and, finally, rinse three to four times with sterile distilled water.

3.1.2. Callus Initiation and Maintenance

Each garlic clove is placed vertically, with the basal part in the culture medium, in 100-mL bottles with 20 mL of medium A, and incubated in a culture chamber at 20°C for a photoperiod of 16 h of light from white fluorescent lamps at 50 $\mu mol/m^2 \cdot s$.

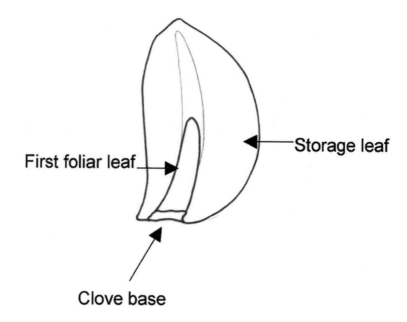

Fig. 1. Diagram of a longitudinal cut of a garlic clove showing the first foliar leaf, the storage leaf, and the basal plate from which roots emerge.

Roots should emerge after 3 d of incubation. The apical portion of each root, about 3-mm long, is dissected in a laminar flow cabinet using a sterile scalpel and forceps (*see* **Note 1**).

Twenty apical tips are placed horizontally on each Petri dish containing 30 mL of culture medium B or C. Dishes are sealed with Parafilm and maintained in a culture incubator at 22°C in darkness for 4 wk.

At the end of this period cultures are transferred to fresh medium B or C and are grown, under the same incubation conditions, for an additional 4 wk (*see* **Fig. 2A**). At this time it is possible to observe callus formation in the meristematic region of the root tip.

With the purpose of increasing the amount of callus formed, explants could be sub-cultured to fresh B or C medium at intervals of 4 wk. It is important to note that it is possible to obtain a larger callus mass if explants are maintained for 2 wk in medium D. This type of calli growing can be maintained indefinitely if they are transferred to fresh medium B or C at 4-wk intervals.

3.1.3. Plant Regeneration

If plant regeneration is desired, calli can be transferred to 100-mL bottles with 25 mL of medium D, and maintained at 22°C under a 16 h light photope-

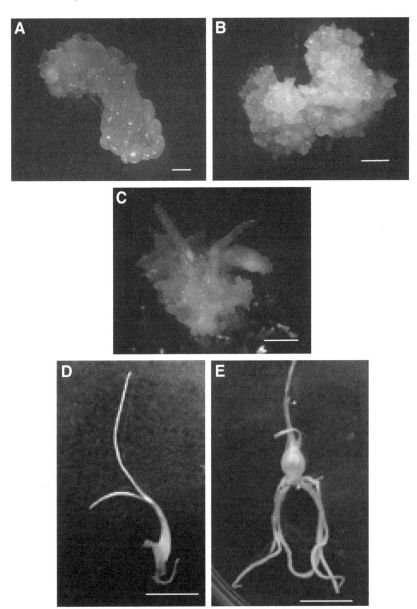

Fig. 2. Callus induction, somatic embryogenesis and plant regeneration from root-tip explants of garlic (*Allium sativum* L.). (**A**) Callus after 8 wk in culture (bar = 1 mm). (**B**) Callus showing somatic embryos after 4 wk of being transferred to medium D (bar = 2 mm). (**C**) Callus in a further stage of development, with emerging shoots (bar = 2 mm). (**D**) Regenerated garlic plant where bulb is starting to develop (bar = 1 cm). (**E**) Plant with a well developed bulb, suitable to be transferred to soil (bar = 1 cm)

riod, at 50 μmol/m^2·s. After 3 wk in this medium, small somatic embryos are developed in the surface of those calli (*see* **Fig. 2B**).

Further growth of these embryos is obtained when they are transferred to 250-mL bottles with 30 mL of medium E, and maintained under the same incubation conditions of light and temperature as before (*see* **Fig. 2C**). Small bulbs initiate and develop from these calli (*see* **Fig. 2D**) and roots start to emerge 4–6 wk later (*see* **Fig. 2E** and **Note 2**).

Because it is difficult to know *a priori* which explants will develop embryogenic calli, it is recommended to make a selection using a stereo microscope, to ensure selection of calli with the morphogenetic capacity to regenerate. Calli obtained in medium B or C have a high multiplication rate, are very easily disaggregated, and very friable.

To propagate the type of calli that allows plant regeneration it is necessary to sub-culture them onto medium B or C, at 3–4 wk intervals, and maintain them in the dark at 22°C. Under these conditions this kind of calli can be maintained indefinitely. For garlic it is recommended that when cultures have been transferred more than two times, and are under a very active division, that calli should be sub-divided to more Petri dishes to avoid competition for the nutrients in the medium.

3.1.4. Growth Kinetics of Garlic Calli

It is important to know the rate of growth of a particular kind of callus. This information allows us to better manage the callus during the multiplication, somatic embryo induction, and plant regeneration steps.

In order to do this, 500 or 700 mg of fresh weight calli are placed into Petri dishes with 30 mL of medium B or C. Dishes are sealed and maintained in the dark at 22°C. Every week these calli are transferred to new dishes of known weight. The dishes are weighed again, after transferring the calli, to calculate the callus weight. This procedure is repeated for 4 wk. A growth curve is generated with the obtained data.

In the case of garlic, culturing calli in medium B or C, at 1-wk intervals, duplicates their original weight in approx 3 wk (**Table 2**). These calli are able to preserve their capacity to increase their biomass, as well as their capacity to regenerate embryos for up to 6 yr if they are sub-cultivated to fresh medium at 4-wk intervals and are kept in the dark (*see* **Note 3**). Even if the growth differences observed in medium B or C sub-cultured calli are very low at wk 4, the accumulation of medium C fresh weight calli is still considerably higher than that obtained in medium B.

Table 2
Garlic Cultivar GT96-1 Calli Response to Weekly Subcultures to Medium B or C

Medium	Time in weeks				
	0	1	2	3	4
B	0.5[a]	0.7	0.9	1.1	1.4
	0%[b]	40.0%	80.0%	120.0%	180.0%
C	0.7[a]	0.9	1.2	1.5	2.2
	0%[b]	28.5%	71.4%	114.2%	214.3%

[a]Fresh weight in g.
[b]Cumulative fresh weight in %.

It is important to note that garlic calli are very sensitive to higher temperatures. The growth rate and the capacity to regenerate embryos decreases if incubation temperatures rise higher than 22°C. The optimum temperatures are therefore between 20 and 22°C. The culture described here has been in culture for more than 5 yr.

3.2. Long-Term Maintenance of NT-1 Cell Culture of Tobacco

This culture has been maintained in our laboratory for more than 13 yr, after it was transferred to us in 1991. It is sub-cultured every week to fresh culture medium and incubated at 25°C in the dark in a orbital shaker at 110 rpm, as described by Paszty and Lurquin in 1987 *(6)*.

Five millilitres of a 1-wk-old culture are transferred with a 10-mL pipet to an Erlenmeyer flask containing 75 mL of NT-1 medium (*see* **Note 4**). The flask is covered with aluminium foil to avoid exposure to light.

NT-1 cell culture is a very useful culture for transient expression, using the biolistic method, in experiments in which the activity of a promoter is to be determined. Results are obtained within 1–2 d after bombardments (*see* **Note 5**) *(7)*.

3.2.1. Packed Cell Volume

One form to estimate the rate of growth of a culture is to determine the packed cell volume (PCV). In order to do this, transfer 10 mL of the suspension culture to a 15-mL graduated centrifuge tube. Spin it down in a clinical centrifuge at 200g over 5 min. PCV is the volume of the pellet and is expressed as a percentage of the volume of the culture used (*see* **Note 6**). This parameter

Table 3
Packed Cell Volume

Days after subculture	PCV (v/v%)
1	4.0
2	5.0
3	6.0
4	10.0
5	15.0
6	30.0
7	40.0

is determined daily, starting 1 d after the cells are transferred to the fresh medium and ending on d 7. Results are presented in **Table 3**. It is observed that at the beginning of transfer (1–3 d), there is a lag phase during which the division of the suspension culture is very low. Following this period a log phase is observed, and at d 7 the culture is still in this growing phase.

3.2.2. Cell Viability

One procedure to estimate the viability of the cultures is the use of 2,3,5-triphenyltetrazolium chloride (TTC) *(8)*.

Transfer 500 µL of cultures from d 1 and 4 or 250 µL of the culture from d 7 into microcentrifuge tubes. Spin down for 1 min at 9500g, eliminate supernatant, and add 1 mL of TTC solution. Incubate at room temperature for 12–20 h in the dark, and observe under a compound microscope.

Living cells reduce the reagent to an insoluble red formazan precipitate because of mitochondrial activity (*see* **Note 7**).

In **Fig. 3** stained cells from cultured NT-1 cells are shown at 1, 4, and 7 d after transfer to fresh medium. Cells from 1-d-old culture show a good staining (notice that some cells are not stained). Cells from a 4-d-old culture show a staining that is more intense than the other incubation periods.

4. Notes

1. Roots start to emerge at d 3 after cloves are put on medium A to sprout. New and useful roots emerge 1 wk later. It is possible to obtain more explants from a clove.
2. Regenerated plants can be transferred to soil. They have to be conditioned to the environment in a growth chamber at 22°C and subjected to a photoperiod of 16 h light at 50 µmol/m^2·s. This step is achieved in 7–10 d and plants are ready to be transferred to a greenhouse where they reach maturity.
3. Cultures described here were initiated 6 yr ago and still have the capacity to regenerate plants which can be transferred to soil and reach maturity in the greenhouse.

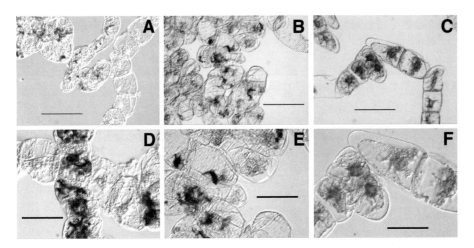

Fig. 3. Staining of NT-1 tobacco suspension cultures at different stages of incubation with TTC. (**A,D**) Stained cells from 1-d-old cultures. (**B,E**) Cell from 4-d-old cultures. (**C,F**) Cells from 7-d-old cultures. Bar in A, B, and C = 100 mm; bar in D, E, and F = 50 mm.

4. Pipets used to transfer NT-1 cell must have a wide tip so that the small clumps of cells, formed during the culture, are able to pass through.
5. Even if this suspension culture has lost the capacity to regenerate plants, it does has some advantages, such as its relative uniformity and its rapid rate of growth. Bombardment conditions should be standardized so that differences between the strength of promoters could be observed.
6. It is important to get the readings of pellet volumes soon after centrifugation, because cells start to swell if tubes are let to stand for long periods, resulting in overestimation PCV.
7. In this case the technique of Nomarski (differential interference contrast) was used to improve the images obtained in the microscope.

Acknowledgments

The authors want to thank Laura Silva-Rosales for her critical review of the manuscript and Andres Estrada and Marcelina Garcia for their technical assistance in the TTC staining and observations.

References

1. Skoog, F. and Miller, C. O. (1957) Chemical regulation of growth and organ formation in plant tissues cultured in vitro. *Symp. Soc. Exp. Biol.* **11**, 118–130.
2. Chu, C. C., Wang, C. C., Sun, S. C., et al. (1975) Establishment of an efficient medium for anther culture of rice through comparative experiments on the nitrogen sources. *Sci. Sin.* **18**, 659–688.

3. Eriksson, T. (1965) Studies on the growth requirements and growth measurements of cell cultures of *Haplopappus gracilis*. *Physiologia Plantarum.* **18,** 976–993.
4. Murashige, T. and Skoog, F. (1962) A revised medium for rapid growth and bioassays with tobacco tissue cultures. *Physiologia Plantarum.* **15,** 473–497.
5. Argüello, J. A., Bottini, R., Luna, R., de Bottini, G. A., and Racca, R. W. (1983) Dormancy in garlic (*Alliun sativum* L.) cv Rosado Paraguayo I. Levels of growth substances in "seed cloves" under storage. *Plant Cell Physiol.* **24,** 1559–1563.
6. Paszty, C. and Lurquin, P. F. (1987) Improved Plant Protoplast Plating/Selection Technique for Quantitation of Transformation. *BioTechniques.* **5,** 716–718.
7. Russel, J. A., Rory, M. K., and Sanford, J. C. (1992) Major improvements in biolistic transformation of suspension-cultured tobacco cells. *In Vitro Cell. Dev. Biol.* **28P,** 97–105.
8. Towill, L. E. and Mazur, P. (1975) Studies on the reduction of 2,3,5-triphenyltretrazolim chloride as a viability assay for plant tissue cultures. *Can. J. Bot.* **53,** 1097–1102.

6

Measurement of Cell Viability in In Vitro Cultures

Lizbeth A. Castro-Concha, Rosa María Escobedo, and María de Lourdes Miranda-Ham

Summary

An overview of the methods for assessing cell viability in in vitro cultures is presented. The protocols of four of the most commonly used assays are described in detail, so the readers may be able to determine which assay is suitable for their own projects using plant cell cultures.

Key Words: Cell viability; Evans blue; fluorescein diacetate; MTT; TTC; tetrazolium salts.

1. Introduction

Plant cell cultures have been widely used as model systems for biochemical and physiological studies, and also in some biotechnological processes, such as cell permeabilization, cell immobilization, or cell cultivation in bioreactors *(1)*. Therefore, the accurate assessment of the number of viable cells in a population is very important to prevent the inclusion of low viable or dead cells in the calculations of results per cell or on a fresh weight basis or to indicate the maximal attainable cell density in production processes.

A cell is considered viable if it has the ability to grow and develop *(2,3)*. Viability assays are based on either the physical properties of viable cells, such as membrane integrity or cytoplasmic streaming, or on their metabolic activity, such as reduction of tetrazolium salts or hydrolysis of fluorogenic susbtrates. To assess cell membrane integrity, dyes such as Evans blue (EB *[4,5]*), methylene blue *(6)*, Trypan blue *(7)*, neutral red *(8)*, and phenosaphranin *(9)* have been used. These compounds leak through the ruptured membranes and stain the contents of dead cells and then, are accounted for via microscopic observation or spectrometric estimation.

Other methods rely on the measurement of the activity of enzymes, such as reductases and esterases. In the case of reductases, both, MTT (3-[4,5-dimethylthiazol-2yl]-2,5-diphenyl tetrazolium bromide) and TTC (2,3,5-triphenyl tetrazolium chloride), accept electrons from the electron transport chain of the mitochondria; as a result, these molecules are converted to insoluble formazan within viable cells with fully active mitochondria *(10)*. On the other hand, intracellular esterases hydrolyze a fluorogenic substrate (fluorescein diacetate), that can pass through the cell membrane, whereupon they cleave off the diacetate group producing the highly fluorescent product fluorescein. Fluorescein will accumulate in cells, which possess intact membranes, so the green fluorescence can be used as a marker of cell viability *(11)*. An alternative method to measure the activity of esterases is the use of two-dimensional fluorescence spectroscopy, which eliminates the time-consuming, and often difficult, counting of viable and nonviable cells under the miscroscope *(1)*.

Cell death is an important feature to monitor during plant–pathogen interactions, so it must be measured accurately for the proper appraisal of the cellular responses. There have been contradictory reports of the measured cell viability in elicited cell suspension cultures *(12–14)*. Hence, we have analyzed the elicitor induced cell viability changes in a tomato cell suspension using different methods. The use of the reduction of tetrazolium salts is proposed as a better indicator of metabolically active cells during the early hours of fungal elicitation *(15)*.

2. Materials

2.1. Cell Suspension Cultures

Calli were induced from young leaves of *Lycopersicon esculentum* Mill. Var. Río Grande on media with Murashige and Skoog (MS) salts *(16)* and supplemented with B_5 vitamins *(17)*, 3% (w/v) sucrose, 2.24 µM 2,4-dichlorophenoxyacetic acid, 0.049 µM kinetin, and 0.22% (w/v) gelrite, pH 5.8. They were incubated in the dark at 25°C and subcultured every 3 wk. Friable calli were transferred into 50 mL liquid medium (as previous, without gelrite) with orbital shaking (100 rpm). The cell suspensions were subcultured every 7 d by 1:5 dilution with fresh media. In order to observe an adequate response to the elicitor treatments, the number of cells in the suspension cultures was adjusted to 2.5×10^6 cells per milliliter (*see* **Note 1**).

2.2. MTT/TTC Assay

1. MTT stock solution: dissolve 50 mg in 10 mL sterile 50 mM phosphate buffer, pH 7.5 and store in the dark at 4°C.
2. TTC stock solution: dissolve 0.2 g in 10 mL sterile 50 mM phosphate buffer, pH 7.5 and store in the dark at 4°C.

3. Sterile 50 m*M* phosphate buffer, pH 7.5.
4. 50% (v/v) methanol, containing 1% (w/v) sodium dodecyl sulfate (SDS).

2.3. Evans Blue Assay

1. EB stock solution: dissolve 100 mg in 10 mL sterile water and store in the dark at 4°C.
2. 50% (v/v) methanol, containing 1% (w/v) SDS.

2.4. Fluorescein Diacetate Assay

1. Fluorescein Diacetate (FDA) stock solution: dissolve 25 mg of fluorescein diacetate in acetone. Preserve the solution in an amber bottle at –20°C.
2. 50 m*M* phosphate buffer, pH 7.5.

3. Methods
3.1. Maintenance of L. esculentum Cultures

Cells should be subcultured every 7 d as follows: in laminar flow cabinet, add 10 mL of a 7-d-old suspension culture to 40 mL of fresh medium in a 250-mL Erlenmeyer flask. Close the flasks and return to the growing conditions stated before. By repeating the above process, cell cultures can be maintained indefinitely.

3.2. MTT/ TTC Assay

1. Wash aseptically cell suspension samples (1 mL) with 50 m*M* phosphate buffer, pH 7.5. Repeat twice (*see* **Note 1**).
2. Resuspend the cells in 1 mL of the same buffer.
3. Add MTT or TTC to a final concentration of 1.25 or 2.5 m*M*, respectively.
4. Incubate samples for 8 h in the dark at 25°C.
5. Solubilize formazan salts with 1.5 mL 50% methanol, containing 1% SDS, at 60°C for a period of 30 min.
6. Centrifuge at 1875*g* for 5 min and recover the supernatant.
7. Repeat **steps 5** and **6**. Pool the supernatants.
8. Quantify absorbance at 570 nm for MTT and 485 nm for TTC (*see* **Note 2**).

3.3. Evans Blue Assay

1. Add Evans Blue (EB) stock solution to cell suspension samples (1 mL) to a final concentration of 0.025% (v/v).
2. Incubate for 15 min at room temperature.
3. Wash extensively with distilled water to remove excess and unbound dye.
4. Solubilize dye bound to dead cells in 50% (v/v) methanol with 1% (w/v) SDS at 60°C for 30 min. Repeat twice and pool the supernatants.
5. Centrifuge at 1875*g* for 15 min.
6. Dilute the supernatant to a final volume of 7 mL.
7. Quantify absorbance at 600 nm (*see* **Note 2**).

3.4. FDA Assay

1. Mix cell samples (1 mL) with 10 µL of FDA stock solution.
2. Incubate for 15 min at room temperature in the dark.
3. Adjust the volume to 4 mL with distilled water.
4. Centrifuge at 1875g for 5 min.
5. Resuspend the pellet in 1 mL 50 mM phosphate buffer, pH 7.5.
6. Freeze quickly in liquid nitrogen.
7. Thaw and dilute samples to 3 mL with phosphate buffer.
8. Homogenize with a Brinkman polytron at high speed for 10 s.
9. Centrifuge at 1875g for 20 min.
10. Dilute a 100 µL sample of the supernatant to a final volume of 2 mL.
11. Determine fluorescence at 516 nm, using a 492 nm excitation beam (*see* **Note 2**).

3.5. Microscopic Assay

1. Stain cell samples with FDA by mixing the samples (1 mL) with 10 µL of the stock solution.
2. Incubate for 15 min at room temperature in the dark.
3. Wash with 50 mM phosphate buffer, pH 7.5.
4. Centrifuge at 1875g for 5 min.
5. Resuspend in phosphate buffer (1 mL).
6. Counterstain with EB, following **steps 1–5** but using 10 µL of the EB stock solution.
7. Determine the number of blue dead cells under a bright field and yellow-green fluorescent viable cells under ultraviolet light (excitation: BP 450-490 nm and emission: LP 520 nm) in an Axioplan microscope in 10 randomized fields in a Sedgewick chamber.

4. Notes

1. Cell densities were determined by the chromium trioxide method. Briefly, 1-mL samples were taken at the indicated time and mixed thoroughly with 2 mL of 8% (w/v) chromium trioxide. Then, they were incubated for 15 min at 80°C. One milliliter samples were loaded on a Sedgewick Rafter counting cell chamber (London), which holds 100 mm^3 and 10 randomized fields were counted twice per slide under bright light microscopy (×10). The mean and standard deviation of cell density values were calculated.
2. There was a robust correlation between enzyme activities and the number of viable cells employed in each assay (MTT, $r^2 = 0.995$; TTC, $r^2 = 0.989$; EB, $r^2 = 0.979$; FDA, $r^2 = 0.996$) and an internal control of heat-killed cells was used to estimate the percentages of relative viability ($r^2 = 0.998$).

References

1. Vanková, R., Kuncová, G., Opatrná, J., Süssenbeková, H., Gaudinová, A., and Vanek, T. (2001) Two-dimensional fluorescence spectroscopy—a new tool for the determination of plant cell viability. *Plant Cell Rep.* **20**, 41–47.
2. Palta, J. P., Levitt J., and Stadeimann, E. J. (1978) Plant viability assay. *Cryobiology* **15**, 249–255.
3. Pegg D. E. (1989) Viability assays for preserved cells, tissues and organs. *Cryobiology* **26**, 212–231.
4. Smith, B. A., Reider, M. L., and Fletcher, J. S. (1982) Relationship between vital staining and subculture growth during the senescence of plant tissue culture. *Plant Physiol.* **70**, 1228–1230.
5. Baker, C. J. and Mock, M. M. (1994) An improved method for monitoring cell death in cell suspension and leaf disc assays using Evans blue. *Plant Cell. Tiss. Org. Cult.* **39**, 7–12.
6. Huang, C. N., Cornejo, M. J., Bush, D. S., and Jones, R. L. (1986) Estimating viability of plant protoplasts using double and single staining. *Protoplasma* **135**, 80–87.
7. Hou, B.H. and Lin, C.G. (1996) Rapid optimization of electroporation conditions for soybean and tomato suspension cultured cells. *Plant Physiol.* **111** (**Suppl. 2**), 166.
8. Cripen, R. W. and Perrier, J. L. (1974) The use of neutral red and Evans blue for live-dead determinations of marine plankton. *Stain Technol.* **49**, 97–104.
9. Wildholm, J. M. (1972) The use of fluorescein diacetate and phenosafranine for determining viability of cultured plant cells. *Stain Technol.* **47**, 189–194.
10. Towill, L. E. and Mazur, P. (1975) Studies on the reduction of 2,3,5-triphenyltetrazolium chloride as viability assay for plant tissue cultures. *Can. J. Bot.* **53**, 1097–1102.
11. Steward, N., Martin, R., and Engasser, J. M. (1999) A new methodology for plant cell viability assessment using intracellular esterase activity. *Plant Cell Rep.* **19**, 171–176.
12. Kodama, M., Yoshida, T., Otani, H., Kohomoto, K., and Nishimura, S. (1991) Effect of AL-toxin on viability of cultures tomato cells determined by the MTT-colorimetric assay, in *Molecular Strategies of Pathogens and Host Plants* (Patil, S., ed.), Springer-Verlag, Berlin, pp. 251.
13. Vera-Estrella, R., Blumwald, E., and Higgins, V. J. (1992) Effects of specific elicitors of *Cladosporium fulvum* on tomato suspension cells. Evidence for the involvement of active oxygen species. *Plant Physiol.* **99**, 1208–1215.
14. Sánchez, L. M., Doke, N., and Kawakita, K. (1993) Elicitor induced chemiluminiscence in cell suspension cultures of tomato, sweet pepper and tobacco plants and its inhibition by suppressors from *Phytophthora* spp. *Plant Sci.* **88**, 141–148.

15. Escobedo, R. M. and Miranda-Ham, M. L. (2003) Analysis of elicitor-induced cell viability changes in *Lycopersicon esculentum* Mill. suspension culture by different methods. *In vitro Cell. Dev. Biol.-Plant* **39,** 236–239.
16. Murashige, T. and Skoog, F. A. (1962) A revised medium for rapid growth and bioassays with tobacco tissue cultures. *Physiol. Plantarum* **15,** 473–497.
17. Gamborg, O. L., Miller, R. A., and Ojima, K. (1968) Nutrient requirements of suspension cultures of soybean root cells. *Exp. Cell Res.* **50,** 151–158.

7

Cryopreservation of Embryogenic Cell Suspensions by Encapsulation-Vitrification

Qiaochun Wang and Avihai Perl

Summary

Encapsulation-vitrification, which is a combination of encapsulation-dehydration and vitrification procedures, is a newly developed technique for cryopreservation of plant germoplasm. Here, we describe the protocol of this methodology, using grapevine (*Vitis*) as a model plant. Cell suspensions at the exponential growth stage were encapsulated with 2.5% sodium alginate solution in 0.1 M calcium chloride solution for 20 min to form beads of about 4 mm in diameter containing 25% cells. The beads were stepwise precultured in increasing sucrose concentrations of 0.25, 0.5, and 0.75 M for 3 yr, with 1 d for each step. Precultured beads were treated with a loading solution for 60 min at room temperature and then dehydrated with PVS2 at 0°C for 270 min, followed by direct immersion in liquid nitrogen for 1 h. The beads were rapidly rewarmed at 40°C in a water bath for 3 min and then diluted with 1 M sucrose solution at room temperature for 30 min. Rewarmed, washed beads were post-cultured on a recovery medium for 3 d at 25°C in the dark for survival. Surviving cells were transferred to a regrowth medium to induce cell proliferation. Embryogenic cell suspensions were re-established by suspending the cells in a cell suspension maintenance medium maintained on a gyratory shaker at 25°C in the dark. For plant regeneration, surviving cells were transferred from the recovery medium to an embryo maturation medium and maintained at 25°C under light conditions. Embryos at the torpedo stage were cultured on rooting medium until whole plantlet was developed.

Key Words: Cell suspensions; cryopreservation; embryogenesis; encapsulation; liquid nitrogen; plant regeneration; vitrification.

1. Introduction

Embryogenic cell suspensions have widely been used as an efficient system for plant clonal micropropagation and as target tissues for genetic manipulation in many transformation systems (*1*). However, the competence for embryogenesis usually declines or totally loses and somatic variation may occur

From: *Methods in Molecular Biology, vol. 318: Plant Cell Culture Protocols, Second Edition*
Edited by: V. M. Loyola-Vargas and F. Vázquez-Flota © Humana Press Inc., Totowa, NJ

in such cell suspension cultures with an increase in subculture time *(2)*. Moreover, establishment of fresh embryogenic cultures is season-dependent, laborious, and time-consuming. Cryopreservation has been considered as an ideal means for avoiding loss of embryogenic potential and preventing the occurrence of somaclonal variation: all cell divisions and other metabolic processes of plant materials that are stored in this way cease. Theoretically, plant materials can thus be stored without any changes for an indefinite period of time *(3)*.

Encapsulation-vitrification, which is a combination of encapsulation-dehydration and vitrification procedures, is one of several newly developed techniques for cryopreservation of plant germplasm. In comparison with classical techniques, such as slow cooling, encapsulation-vitrification enables cells to be cryopreserved by direct immersion in liquid nitrogen, thus avoiding the use of a sophisticated and expensive programmable freezer *(3)*. Vitrification refers to the physical process by which a highly concentrated cryoprotective solution is supercooled at very low temperatures and eventually solidifies into a metastable glass without undergoing crytallization at a glass transition temperature *(4)*. Thus, vitrified cells evade the danger of intracellular freezing and survive freezing in liquid nitrogen. Encapsulation of the cells, normally in 2–3% calcium-alginate beads, can protect the specimens during the procedure and reduce the chemcial toxicity or osmotic stress of the vitrification solution. In addition, encapsulation-vitrification is much easier to handle and allows for treatment of a large number of samples at the same time in comparison to the vitrification procedure alone.

Encapsulation-vitrification was originally developed by Tannoury et al. *(5)* for cryopreservation of carnation apices (*Dianthus caryophyllus* L.). In this protocol, encapsulated apices were precultured for 16 h in progressively more concentrated sucrose medium, then incubated for 6 h in a vitrification solution containing ethylene glycol and sucrose, and finally frozen either rapidly or slowly. Survival of cryopreserved apices reached 100 and 92% after rapid and slow freezing, respectively.

Using encapsulation-vitrification procedure, much higher recovery was produced in cryopreserved shoot tips of wasabi (*Wasabia japonica*) *(6)* and strawberry (*Fragaria* × *ananassa* Duch.) *(7)* compared with encapsulation-dehydration. Wang et al. *(8)* demonstrated that encapsulation-vitrification was much more applicable to different grapevine (*Vitis*) species and cultivars than encapsulation-dehydration. Therefore, encapsulation-vitrification would have a great potential for broad applications to cell suspensions *(3,8)*.

2. Materials

2.1. General Requirements

1. Standard tissue culture facilities.
2. 2-mL Cryogenic vial.
3. 250-mL Erlenmeyer flask.
4. Gyratory shaker.
5. 1-mL Pipet.
6. Sterile disposable 25-mL pipet.
7. 20- to 35-L Taylor-Wharton Dewar flask (beaker style wide-mouth).
8. Magenta GA7 vessels.
9. 9-cm Whatmann paper.
10. 9-cm Petri dish.

2.2. Chemicals, Plant Growth Regulators, and Media

1. Activated charcoal (AC, cat. no. C 9157).
2. Calcium chloride (Sigma, cat. no. C 4901).
3. Dimethylsulfoxide (DMSO) (Sigma, cat. no. D 8418).
4. Ethylene glycol (Sigma, cat. no. E 9129).
5. Gelrite (Sigma, cat. no. G 1910).
6. Glycerol (Sigma, cat. no. G 5516).
7. Naphthalene acetic acid (NAA) (Sigma, cat. no. N 0640).
8. 2-Naphthoxyacetic acid (NOA) (Sigma, cat. no. N 3019).
9. Nitsch and Nitsch medium (NN) *(9)* (Duchefa, cat. no. N 0224).
10. Sodium alginate (Sigma, cat. no. A 2158).
11. Sucrose (Sigma, cat. no. S 7903).
12. Woody plant medium (WPM) *(10)* (Duchefa, cat. no. M 0220).

2.3. Media and Solutions

1. 0.1 M calcium chloride solution: liquid NN medium containing 0.1 M calcium chloride, 2 M glycerol, and 0.4 M sucrose, pH 5.8.
2. Cell suspension maintenance medium: liquid NN medium supplemented with 18 g/L maltose, 1 g/L casein enzymatic hydrolysate, 4.6 g/L glycerol, and 1 mg/L NOA, pH 5.8.
3. Dilution medium: liquid NN medium containing 1 M sucrose, pH 5.8.
4. Embryo maturation medium: solid NN medium supplemented with 18 g/L maltose, 1 g/L casein enzymatic hydrolysate, 4.6 g/L glycerol, and 2.6 g/L gelrite, pH 5.8.
5. Loading solution: liquid NN containing 2 M glycerol and 0.4 M sucrose, pH 5.8.
6. Plant vitrification solution (PVS2): 0.4 M sucrose, 30% (w/v) glycerol, 15% (w/v) DMSO and 15% (w/v) ethylene glycol made up in liquid NN medium, pH 5.8 *(11)*.

7. Preculture medium: cell suspension maintanence supplemented with 0.25, 0.5, or 0.75 M sucrose, respectively, pH 5.8.
8. Recovery medium: cell suspension maintanence containing 2.5 g/L AC and solidified with 2.6 g/L gelrite, pH 5.8.
9. Rooting medium: solid WPM containing 30 g/L sucrose, 1 mg/L NAA, 2.5 g/L AC, and 2.6 g/L gelrite, pH 5.8.
10. Sodium alginate solution: 0.4 M sucrose, 2 M glycerol, and 2.5% (w/v) sodium alginate made up in liquid NN medium, pH 5.8.

3. Methods

As with any method, optimal treatment parameters at each step during the whole procedure may vary greatly with different plant species and even cultivars. In this section, we make every effort to illustrate the protocol, using grapevine (*Vitis*) as an example. The method, described in **Subheadings 3.1.–3.5.** consists of (1) maintenance of embryogenic cell suspensions, (2) encapsulation-vitrification, (3) freezing and rewarming, (4) unloading, and (5) survival, regrowth, and plant regeneration. A flow-chart of the encapsulation-vitrification protocol is illustrated in **Fig. 1**.

To maintain sterility of emberyogenic cell suspensions, all appropriate manipulations should be carried out in a laminar air-flow cabinet, using aseptic techniques and sterile materials.

3.1. Maintenance of Embryogenic Cell Suspensions

1. One gram of Calli are transferred to a 250-mL Erlenmeyer flask containing 50 mL of cell suspension maintenance medium.
2. The cell suspensions are incubated at 25°C in darkness on a gyratory shaker at 90 rpm.
3. Subculture is carried out weekly by transfer of cell suspensions to fresh cell suspension maintenance medium.

3.2. Encapsulation-Vitrification

1. Three days after subculture (*see* **Note 1**), embryogenic cell suspensions are collected using a 25-mL sterile disposable pitet to remove the entire medium from the 250-mL Erlenmeyer flask. In order to avoid loss of cells, empty the medium while pressing the mouth of the pipet against the bottom of the Erlenmeyer so that the cells do not penetrate into the pipet.
2. The cells are transferred to 2.5% sodium alginate solution by a ratio of 1 g cells to 3 mL solution (*see* **Note 2**) and mixed well.
3. The mixture is dripped with a 1-mL sterile pipet into 0.1 M calcium chloride solution at room temperature and left for 20 min to form beads of about 4 mm in diameter, each containing 25% (w/v) cells (*see* **Note 2**).
4. The beads are precultured stepwise on the preculture medium enriched with increasing sucrose concentrations of 0.25, 0.5, and 0.75 M for 3 d, with 1 d for

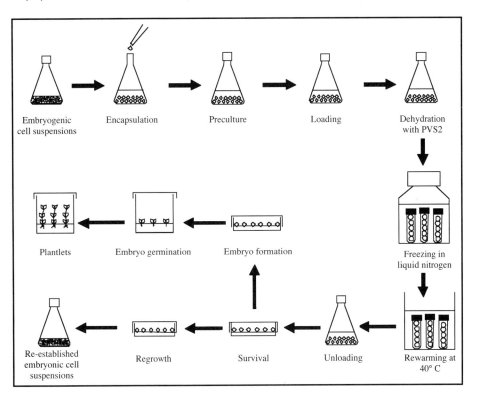

Fig. 1. A flow-chart of cryopreservation of embryogenic cell suspensions by encapsulation-vitrification procedure.

each step (*see* **Note 3**). The preculture conditions are the same as for the maintenance of embryogenic cell suspensions.

5. The precultured beads are rapidly surface-dried by blotting onto sterile 9-cm Whatmann paper for a few seconds, and then transferred to a loading solution and incubated for 60 min at room temperature in a laminar air-flow cabinet (*see* **Note 4**).
6. The beads are rapidly surface dried by blotting onto sterile 9-cm Whatmann paper for a few seconds, and then transferred to PVS2 cooled on ice and incubated for 270 min (*see* **Note 5**).

3.3. Freezing and Rewarming

1. The vitrified beads are rapidly surface dried by blotting on sterile 9-cm Whatmann paper for a few seconds. Next, 10 beads are transferred into a 2-mL cryogenic vial and directly immersed in a Taylor-Wharton Dewar Flask containing the liquid nitrogen for 1 h.

2. Cryogenic vials containing frozen cells are rapidly removed from liquid nitrogen and immetiately rewarmed at 40°C in a water bath for 3 min.

3.4. Unloading

1. Rewarmed beads are placed in the dilution medium for 30 min at room temperature (*see* **Note 6**), followed by surface drying by blotting on sterile 9-cm Whatmann paper for a few seconds.

3.5. Survival, Regrowth, and Plant Regeneration

1. The beads are post-cultured at 25°C in the dark for 3 d in 9-cm Petri dish (10 beads/dish) containing 30 mL recovery medium for survival (*see* **Note 7**).
2. For regrowth, the beads are transferred from the recovery medium to 9-cm Petri dish (10 beads/dish) containing 30 mL regrowth medium and placed at 25°C in the dark for 4 wk (*see* **Note 8**). Embryogenic cell suspensions are re-established by suspending the beads in a 50-mL Erlenmeyer flask containing 50 mL of cell suspension maintenance medium. The cultures are placed on a gyratory shaker (90 rpm) at 25°C in the dark. Subculture is carried out every week.
3. For plant regeneration, the beads are transferred from the recovery medium to 9-cm Petri dish (10 beads/dish) containing 30 mL plant regeneration medium. The cultures are maintained at 25°C under a 16 photoperiod with a light intensity of 45 µmol/s/m^2 provided by cool white fluorescent tubes. Four weeks later, single embryos at the torpedo stage are transferred to Magenta GA7 vessels containing 50 mL rooting medium. The culture conditions are the same as for embryo maturation. Subculture is done every 4 wk until a whole plantlet with a profuse root system is developed.

4. Notes

1. Growth phase at which cells are harvested at the onset of the procedure is an important factor for cryopreservation *(12)*. Cells at the exponential growth stage are more tolerant to freezing than those either at the lag or stationary stages because they have small volume and vacuoles and contain relatively little water. The growth pattern of cell suspensions differs with different plant species. For several *Vitis* species and cultivars, the exponential growth stage is reached 3 d after each subculture *(8,13)*.
2. Cell density in the beads influences significantly the regrowth and embryo formation of crypreserved cell suspensions. In general, from 25 to 50% cell density were used in encapsulated beads *(8,13–15)*. Regrowth of cryopsereved cells of *Catharanthus* was found faster in the beads containing 50% than the 25% cells *(15)*, whereas efficiency of embryo formation in *Vitis* was much higher in beads containing 25% cells *(8,13)*.
3. Preculture is required to induce resistance of cell suspensions to dehydration and subsequent freezing in liquid nitrogen. Sucrose is the most often used sugar for induction of such tolerance at this step *(3)*, although other sugars such as glucose

(16), glycerol *(17)*, maltose *(18)*, manitol *(19)*, and sorbitol *(18,20)* were used. Sugar concentration in the preculture medium ranges from 0.3 to 1 M for periods varying between 1 h *(18)* and 9 d *(15)*. Progressive increase in sucrose concentration enables to avoid the deleterious effects of direct exposure to high sucrose concentrations *(21)*. Preculture with high sugar concentrations (up to 0.75–1 M) has been shown to increase the viability of cryopreserved cell suspensions of *Catharanthus roseus* *(15)*, *Papaver somniferum* *(18)*, *Vaccinium pahalae* *(14)*, *Medicago sative* *(22)*, *Asparagus officinalis* *(23)*, and *Vitis* *(8,13)*.

4. In vitrification-based cryopreservation procedures, cells have to be dehydrated by exposure to a highly concentrated solution such as PVS2 prior to immersion in liquid nitrogen. However, direct exposure of cells to PVS2 without osmoprotection causes harmful effects on cells because of osmotic stress or chemical toxicity *(24)*, which has been described as a major limitation in determining the success of cryopreservation of cells by vitrification *(25)*. This major limitation can be overcome by the loading treatment *(24,25)*. In this treatment, cells are placed at room temperature in a solution containing cryoprotective substances such as sucrose, glycerol and ethylene glycol, for a short period varying between 10 min *(25,26)* and 60 min *(8)*, depending on plant scpeices. The most often used loading solution is composed of a basic medium containing 2 M glycerol and 0.4 M *(8,25,26)*.

5. Two plant vitrification solutions are most frequently employed: one described by Langis et al. *(27)*, which comprises 40% (w/v) ethylene glycol, 15% sorbitol, and 6% bovine serum albumin, and the other (PVS2) described by Sakai et al. *(11)*, which consists of 30% (w/v) glycerol, 15% (w/v) ethylene glycol, and 15% (w/v) DMSO 0.4 M sucrose. We applied both vitrification solutions to *Vitis* cell suspensions with much better results obtained with PVS2. Overexposure to the vitrification solution can cause chemical toxicity and excess osmotic stress *(28)*, eventually leading to reduced survival of cryopreserved cells. Dehydration can be performed either at 0°C or at room temperature. The optimal time of exposure to the vitrification solution varies with plant species, and depending on the temperature during exposure, it ranges from 20 min *(25)* to 270 min *(8)*. Dehydration at 0°C takes longer time, but usually yields a higher survival of cryopreserved cells *(8,25)*, mainly through elimination or reduction in the chemical toxicity or osmotic stress of vitrification solutions to plant cells *(29)*.

6. Unloading is designed to dilute the vitrification solutions and remove cryoprotectants from cryopreserved cell suspensions *(30)*. Because of the extremely high concentrations of the vitrification solutions, unloading is usually performed either by a one-step dilution in a high-sugar concentration solution up to 1.2 M *(25)* or by a stepwise dilution in decreasing sugar concentrations *(22)*, in order to reduce the osmotic shock. With *Vitis* cell suspensions, the best results were achieved using 1 M sucrose solution for 30 min *(8)*.

7. In many cases, recovery of cryopreserved cell suspensions was largely improved when post cultured for a few days on a solid medium before being transferred to liquid medium *(8,31)*. Addition of activated charcoal to the recovery medium

was also found beneficial to the survival of cryopreserved cell suspensions *(3,8, 13)*. The beneficial effect of activated charcoal on survival was probably achieved through reduced necrogenesis *(31)* and adsorption of toxic substances produced by frost-damaged cells *(32)*.
8. Regrowth of cryopreserved cell suspensions generally starts after a lag period that usually last from 3 d *(25)* to 10 d *(33)* and reaches the same growth pattern as that of noncryopreserved cell suspensions after two to three times of subculture *(8,13)*.

References

1. Raemakers, K., Jacobsen, E., and Visser, R. (1999) Proliferative somatic embryogenesis in woody species, in *Somatic Embryogenesis in Woody Plants*, vol. 4. (Jain, S. M., Gupta, P. K., and Newton, R. J., eds), Luwer Academic Publishers, pp. 29–59.
2. Deverno, L. L. (1995) An evaluation of somatic variation during somatic embryogenesis, in *Somatic Embryogenesis in Woody Plants*, vol. 1. (Jain, S. M., Gupta, P. K., and Newton, R. J., eds), Luwer Academic Publishers, pp. 361–377.
3. Engelmann, F. (1997) *In vitro* conservation methods, in *Biotechnology and Plant Genetic Resources* (Callow, J. A., Ford-Lloyd, B. V., and Newbury, H. J., eds), CAB International, Oxford, pp. 119–161.
4. Fahy, G. M., MacFarlande, D. R., Angell, C. A., and Meryman, H. T. (1984) Vitrification as an approch to cryopreservation. *Cryobiology* **21**, 407–426.
5. Tannoury, M., Ralambosoa, J., Kaminski, M., and Dereuddre, J. (1991) Cryopreservation by vitrification of coated shoot tips of carnation (*Dianthus caryophyllus* L.) cultured *in vitro*. *Comptes Rendus de l'Acad. des Sci*. Paris Serie III. **313**, 633–638.
6. Matsumoto, T., Sakai, A., and Yamada, K. (1994) Cryopreservation of in vitro-grown apical meristems of wasabi (*Wasabia japonica*) by vitrification and subsequent high plant regeneration. *Plant Cell Rep.* **13**, 442–446.
7. Hirai, D., Shirai, K., Shirai, S., and Sakai, A. (1998) Cryopreservation of in vitro-grown meristems of strawberry (*Fragaria* x *Ananassa* Duch.) by encapsulation-vitrification. *Euphytica* **101**, 109–115.
8. Wang, Q. C., Mawassi, M., Sahar, N., et al. (2004) Cryopreservation of grapevine (*Vitis* spp.) embryogenic cell suspensions by encapsulation-vitrification. *Plant Cell Tiss. Org Culture* **77**, 267–275.
9. Nitsch, J. P. and Nitsch, C. (1969) Haploid plants from pollen grains. *Science* **163**, 85–87.
10. Lloyd, G. and McCown, B. (1980) Commercially feasible micropropagation of mountain laurel, *Kalmia latifolia*, by use of shoot-tip culture. *Proc. Intl. Plant Prop. Soc.* **30**, 421–427.
11. Sakai, A., Kobayashi, S., and Oiyama, I. (1990) Cryopreservation of nucellar cells of navel orange (*Citrus sinensis* Osb. var. *brasiliensis* Tanaka) by vitrification. *Plant Cell Rep.* **9**, 30–33.
12. Reinhoud, P. J., Iren, F. V., and Kijne, J. K. (2000) Cryopreservation of undifferentiated plant cells, in *Cryopreservation of Tropical Plant Germplasm*, (Engelmann, E. and Takagi, H., ed.), JIRCAS, Japan, pp. 212–216.

13. Wang, Q. C., Gafny, R., Sahar, N., et al. (2002) Cryopreservation of grapevine (*Vitis vinifera* L.) embryogenic cell suspensions by encapsulation- dehydration and subsequent plant regeneration. *Plant Sci.* **162,** 551–558.
14. Shibli, R. A., Smith, M. A. L., and Shatnawi, M. A. (1999) Pigment recovery from encapsulated-dehydrated Vaccinium pahalae (ohelo) cryopreserved cells. *Plant Cell Tiss. Org. Cult.* **55,** 119–123.
15. Bachiri, Y., Gazeau, C., Hansz, J., Morisset, C., and Dereuddre, J. (1995) Successful cryopreservation of suspension cells by encapsulation-dehydration. *Plant Cell Tiss. Org. Cult.* **43,** 241–248.
16. Jitsuyama, Y., Suzuki, T., Harada, T., and Fujikawa, S. (1997) Ultrastructural study on mechanism of increased freezing tolerance due to extracelluar glucose in cabbage leaf cells. *Cryo-Lett.* **18,** 33–44.
17. Touchell, D. H., Chiang, V. I., and Tsai, C. J. (2001) Cryopreservation of embryogenic cultures of *Picea mariana* (black spruce) using vitrification. *Plant Cell Rep.* **21,** 118–124.
18. Gazeau, C., Elleuch, H., David, A., and Morisset, C. (1998) Cryopreservation of transformed *Papaver somniferum* cells. *Cryo Letters* **19,** 147–159.
19. Ribeiro, R. C. S., Jekkel, Z., Mulligan, B. J., et al. (1996). Regeneration of fertile plants from cryopreserved cell suspensions of *Arabidopsis thaliana* (L.) Heynh. *Plant Science* **115,** 115–121.
20. Touchell, D. K., Chiang, V. L., and Tsai, C. J. (2002) Cryopreservation of embryogenic cultures of *Picea mariana* (Black spruce) using vitrification. *Plant Cell Rep.* **21,** 118–124.
21. Plessis, P., Leddet, C., and Dereuddre, J. (1991) Resistance to dehydration and to freezing in liquid nitrogen of alginate coated shoot tips of grapevine (*Vitis vinif*era L. cv. Chardonnay). *Comptes Rendus de l'Acad Sci. Paris.* Serie III. **313,** 373–380.
22. Shibli, R. A., Haagenson, D. M., Cunningham, S. M., Berg, W. K., and Volenec, J. J. (2001) Cryopreservation of alfalfa (*Medicago sativa* L.) cells by encapsulation-dehydration. *Plant Cell Rep.* **20,** 445–450.
23. Jitsuyama, Y., Suzuki, T., Harada, T., and Fujikawa, S. (2002) Sucrose incubation increases freezing tolerance of asparagus (*Asparagus officinalis* L.) embryogenic cell suspensions. *Cryo Letters* **23,** 103–112.
24. Ishikawa, M., Tandon, P., Suzuki, M., and Yamaguishi-Ciampi, A. (1996) Cryopreservation of bromegrass (Bromus inermis Leyss) suspension cultured cells using slow prefreezing and vitrification procedures. *Plant Sci.* **120,** 81–88.
25. Nishizawa, S., Sakai, A., Amano, Y., and Matsuzawa, T. (1993) Cryopreservation of asparagus (Asparagus officinalis L.) embryogenic suspension cells and subsequent plant regeneration by vitrification. *Plant Sci.* **91,** 67–73.
26. Sakai, A. K., Kobayashi, S., and Oiyama, I. (1991) Survival by vitrification of nucelar cells of navel orange (*Citrus Sinensis* var. *brasiliensis* Tanaka) cooled to −196°C. *J. Plant Physiol.* **137,** 465–470.
27. Langis, R., Schnabel, B., Earle, E. D., and Steponkus, P. L. (1989) Cryopreservation of *Brassica campestris* L. suspensions by vitrification. *Cryo Letters* **10,** 421–428.

28. Rall, W. F. (1987) Factors affecting the survival of mouse embryos cryopreserved by vitrification. *Cryobiol.* **24**, 367–402.
29. Matsumoto, T. and Sakai, A. (2003). Cryopreservation of axillary shoot tips of in vitro-grown grape (*Vitis*) by a two-step vitrification protocol. *Euphytica.* **131**, 299–304.
30. Steponkus, P. L., Langis, R., and Fujikawa, S. (1992) Cryopreservation of plant tissue by vitrification, in *Advances in Low Temperature Biology*, vol. 1. (Steponkus, P. L., ed.), London, JAI Press, pp. 1–61.
31. Dussert, D., Mauro, M. C., and Engelmann, F. (1992) Cryopreservation of grape embryogenic cell suspensions. 2: Influence of post-culture conditions and application to different strains. *Cryo Letters* **13**, 15–22.
32. Kuriyama, A., Watanabe, K., Ueno, S., and Mitsuda, H. (1990) Effect of post-thaw treatment on the viability of cryopreserved *Lavandula vera* cells. *Cryo Letters* **11**, 171–178.
33. Shashi, G. and Vasil, I. K. 1992. Cryopreservation of immature embryos, embryogenic callus and cell suspension cultures of gramineous species. *Plant Sci.* **83**, 205–215.

8

Somatic Embryogenesis in *Picea* Suspension Cultures

Claudio Stasolla

Summary

Generation of somatic embryos in spruce is achieved through the execution of five steps designated as: (1) induction of embryogenic tissue, (2) maintenance of embryogenic tissue, (3) embryo development, (4) embryo maturation, and (5) conversion into plants. Depending on species and genotypes within the same species, each step must be optimized for obtaining maximum results. In general, embryogenic tissue is generated from immature and mature zygotic embryos and maintained in either liquid or solid conditions in the presence of plant growth regulators auxin and cytokinin. Initiation of embryo development in suspension cultured is induced by removal of plant growth regulators, whereas continuation of development and completion of maturation require applications of abscisic acid and imposition of a desiccation period. Both treatments are needed for conferring morphological and physiological maturation to the embryos. Mature somatic embryos are germinated in the absence of plant regulators and embryo conversion (i.e., formation of a functional shoot and root, occurs after a few weeks in culture).

Key Words: Conversion; drying treatment; embryogenic tissue; embryo development and maturation; germination; spruce.

1. Introduction

Somatic embryogenesis is the process in which asexual cells (haploid or diploid) can be induced to generate viable embryos in culture. This process leads to the production of bipolar structures with a well-defined root/shoot axis *(1)*. When implemented with conventional breeding programs and novel molecular biology techniques, somatic embryogenesis represents a powerful tool for increasing the pace of genetic improvement of commercial species. Compared to flowering plants, generation of somatic embryos in conifers is a relatively recent event.

Spruce is one of the most widely distributed coniferous genus' in the Northern hemisphere and represents an economically important species because of its utilization for wood and lumber production *(2)*. In white spruce (*Picea glauca*), which is the predominant spruce species in North America, development of embryo-like structures from cultured cells was observed almost 25 yr ago *(3)*. These structures, however, were not able to complete the maturation process and failed to regenerate viable plants. Morphologically mature embryos were only produced a few years later by two independent groups *(4,5)*, as a result of optimized protocols for culture media and judicious selections of the culture environment. In order to efficiently improve the yield and quality of the embryos produced in culture it is essential to understand how embryos develop and how changes in culture conditions affect this process. In spruce a "developmental map" that delineates important stages of development has been created *(6)* and it will be used here as a basis to correlate culture treatments to morphological changes of the maturing embryos. As a general rule for all spruce species so far regenerated in culture, the embryogenic process can be divided into five consecutive developmental steps, with unique culture conditions. These steps include the induction of embryogenic tissue, the maintenance of embryogenic tissue, embryo development, embryo maturation, and embryo conversion (*see* **Fig. 1**). It must be noted that successful completion of each step is dependent upon proper execution of the previous steps and that several factors, including types and levels of growth regulators, length of each step and manipulation of the tissue must be optimized for each spruce species and for each genotype within the same species. Therefore, the protocol provided in this chapter, which has been optimized for the generation of somatic embryos of white spruce, requires modifications if somatic embryogenesis has to be induced in other spruce species. Detailed references for the production of somatic embryos in red spruce, black spruce, and Norway spruce are available elsewhere *(7–9)*.

2. Materials

1. All chemicals needed for the generation of somatic embryos are listed in **Table 1** and they can be purchased from Sigma.
2. Purified agar (Becton Dickinson, cat. no. 211852) or Phytagel (Sigma, cat. no. P 8196).
3. Petri dishes (10 × 15 cm) and 24-well tissue culture plate (Falcon, cat. no. 3847).

3. Methods

In white spruce, induction of embryogenic tissue, development and maturation of somatic embryos, and conversion of embryos into viable plants can be achieved using either a modified Arnold and Erikkson (AE) medium *(10,11)* or

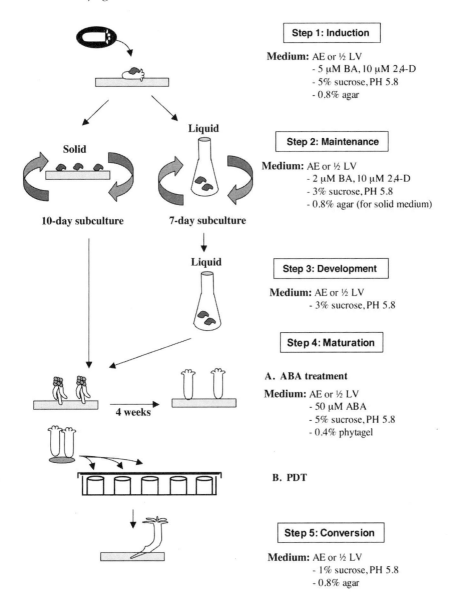

Fig. 1. Diagram showing the different steps required for the induction and maintenance of embryogenic tissue (**steps 1** and **2**), development and maturation of somatic embryos (**steps 3** and **4**), and conversion into plants (**step 5**). *See* **Subheadings 3.1.–3.5.** for detailed descriptions. Supplementations of media components for each step are also illustrated. AE, von Arnold and Eriksson medium *(10)*; LV, Litvay medium *(12)*; BA, N^6-benzyladenine; 2,4-D, 2,4-dichlorophenoxyacetic acid; ABA, abscisic acid; PDT, partial drying treatment.

Table 1
Composition of Basal Media LV and AE Basal Media

Components	AE	LV
1. Macronutrients	(mg/L)	(mg/L)
NH_4NO_3	1200	1650
KNO_3	1900	1900
$MgSO_4 \cdot 7H_2O$	370	1850
KH_2PO_4	340	340
$CaCl_2 \cdot 2H_2O$	180	22
2. Micronutrients		
KI	0.75	4.15
H_3BO_3	0.63	31
$MnSO_4 \cdot 4H_2O$	2.2	27.7
$ZnSO_4 \cdot 7H_2O$	4.1	43
$Na_2MoO_4 \cdot 2H_2O$	0.025	1.25
$CuSO_4 \cdot 5H_2O$	0.0025	0.5
$CoCl_2 \cdot 6H_2O$	0.0025	0.125
3. Iron stock		
$FeSO_4 \cdot 7H_2O$[a]	5.57	27.8
$Na_2 \cdot EDTA$[a]	12.85	37.3
4. Vitamins and organics		
Thiamine HCl	5	0.1
Pyridoxine HCl	1	0.1
Nicotinic acid	2	0.5
Myo-inositol	500	100

[a]Solutions should be prepared separately in hot sterile distilled water and combined together on a hot plate. The final solution should be yellow

Note: Individual components should be prepared as stock solutions and stored at 4°C (macronutrients, micronutrients, and iron stock) and –0°C (vitamins and organics).

the Litvay (LV) medium *(12)* (**Table 1**). Both media have been successfully used in independent studies *(11,13)*. Progression through the five steps of the embryogenic process is achieved by changing types and levels of plant growth regulators as well as sucrose concentration (**Fig. 1**).

3.1. Induction of Embryogenic Tissue

Explants of choice for the generation of embryogenic tissue in white spruce, as well as for other spruce species, are immature zygotic embryos, even though production of embryogenic tissue at high frequency (40%) can be obtained from fully mature zygotic embryos *(13)*. Attempts to utilize mature tissues as explants, including cotyledons from seedling have been reported, but with no satisfactory results *(14)*. Although the stage of development of immature zygotic embryos does not seem to be a critical factor for the induction of embryogenic tissue, it is preferable if early cotyledons embryos are used. Depending on location and environment, maturation of seeds can be asynchronous and can vary from year to year. It is therefore advisable to sample cones regularly and to check the development of the embryos. It is always better to select "healthy looking" cones devoid of brown spots, which may indicate the presence of contaminants.

3.1.1. Sterilization of Immature Seeds

Because immature cones are often covered with resin and have their scales pressed against each other, a harsh surface sterilization is needed. The cones can be sterilized in a solution of Javex bleach (25%) with a few drops of Tween-20 for 30 min under continuous agitation. After rinsing three to four times with sterilized water, the cones can be further sterilized in ethanol (70%) (*see* **Note 1**). Following additional rinsing in water (three times), the immature seeds are removed from the scales and the embryos are dissected and placed on the solid induction medium (*see* **Fig. 1**).

3.1.2. Sterilization of Mature Seeds

Mature seeds can be easily removed from the cone by loosening the scales. Prior to sterilization, the seeds are wrapped in a small piece of cheese cloth, which is then tied with a rubber band (*see* **Note 2**). The small "bag" containing the seeds is then immersed in a solution of 25% Javex bleach for 25 min under continuous agitation. It is important that all air pockets that tend to form between the cheese cloth and seeds are eliminated with sterile forceps to allow for complete sterilization. The cheese cloth bag is then rinsed three times in sterile distilled water, placed in a sealed Petri dish, and incubated at 4°C for 24 h. After this incubation period, which facilitates the dissection of the embryos as the seeds absorb water and swell, the cheese cloth bag containing the seeds is surfaced sterilized again in 20% Javex bleach and washed three times with sterile distilled water. The bag is then opened in a Petri dish containing 10 mL of sterile water. Only seeds that sink at the bottom of the sterile Petri dish should be retained, whereas those floating in water should be discarded because they have not been imbibed properly. With the help of a dissecting microscope and sterile scalpel and forceps, the embryos are dissected from the seeds and placed on solid induction medium (**Fig. 1**).

To avoid cross-contamination no more than 10 embryos (immature or mature) should be plated in one Petri dish. After being sealed with Parafilm™, the cultures should be incubated in the dark at room temperature and checked at regular intervals over the course of the first week. In the presence of contaminated embryos, the noncontaminated ones should be removed and placed in new induction medium. It is important that the forceps are flamed after each transfer to avoid further contamination among embryos.

3.1.3. Production of Embryogenic Tissue

Depending on the type of explant, the percentage of induction frequency and the time required for the production of embryogenic tissue can vary a great deal. As a general rule, immature embryos are more responsive to the process. In these embryos embryogenic tissue is produced after 2–3 wk in culture, compared to the 4–5 wk of mature embryos. Cultured embryos can produce embryogenic tissue, characterized by milky friable masses composed of early filamentous embryos, or nonembryogenic tissue, composed of nodules that are often brown in color. This latter tissue is not able to generate embryos upon further development. In a few instances, however, embryogenic tissue is initiated later in culture from the same embryos producing a primary nonembryogenic tissue. Once the embryogenic tissue has reached a size of 5–7 mm, it should be gently removed, placed on solid maintenance medium (**Fig. 1**), and cultured in the dark at room temperature.

3.2. Maintenance of Embryogenic Tissue

After a few days on solid maintenance medium, the embryogenic tissue that has reached 1 cm in diameter can be transferred in 125-mL flasks containing 50 mL of liquid maintenance medium and incubated at room temperature in the dark under continuous agitation (120 rpm) using a gyratory shaker. After several days the tissue will become loose and a suspension culture will be generated. Liquid cultures can be maintained for several years with weekly transfers into fresh medium. Responsiveness to the liquid environment varies among genotypes. Some cell lines do not proliferate in suspension and should be maintained on solid medium. In these cases, tissue (0.5 mm in diameter) is placed on fresh maintenance solid medium, incubated in the dark and subcultured every 10 d (*see* **Fig. 1**). As a general rule, it is always better to maintain the same genotype on both solid and liquid conditions, because the different environments may have some effects on yield and quality of embryos produced during the following steps.

The morphology of the embryogenic tissue in the maintenance medium varies among spruce species and also among cell lines within the same species. In several lines of white spruce the embryogenic tissue has a disorganized structure which is composed of two different types of cells: vacuolated cells and densely cytoplasmic cells. In other lines the embryogenic tissue is more organized and it consists of filamentous embryos. These immature embryos are formed by a small group of cytoplasmic apical cells (embryo proper), subtended by a suspensor region composed of elongated cells with large vacuoles. No correlation between embryogenic tissue morphology and embryogenic potential seems to exist, because both types are able to produce good quality embryos. The organization of the embryogenic tissue in Norway spruce appears to be more defined *(15)*. Analyses of this tissue in the maintenance medium have revealed the presence of proembryogenic masses (PEM I) at d 0, which are composed by cytoplasmic cell aggregates, subtended by a single vacuolated cell. By d 3, more vacuolated cells develop at the base of the cytoplasmic cells. These aggregates are referred to as PEM II. Progression of PEM II into larger aggregates of both cytoplasmic and vacuolated cells (PEM III) occurs between d 3 and 7 in culture (*see* **Fig. 3**) *(6)*. Somatic embryos will develop from PEM III following the changes in culture conditions that are described below.

3.3. Embryo Development

In both white spruce and Norway spruce, initiation of embryo development is promoted by removal of the plant growth regulators (PGRs), auxin, and cytokinins, which are present in the maintenance medium (*see* **Fig. 1**) and that promote cell proliferation and inhibit development. This is achieved by transferring the embryogenic tissue from the liquid maintenance medium to a PGR-free medium for 7 d under similar culture conditions. It is advisable to wash cells in PGR-free medium before the transfer to remove excess of auxin and cytokinin. In the medium devoid of PGRs, the cell aggregates of the embryogenic tissue increase in size and assume a more organized structure, which resembles that of early somatic embryos. These embryos have a well defined apical-basal axis and are formed by an embryo proper subtended by an elongated suspensor.

Embryo development is quite synchronous in Norway spruce, as a large number of organized somatic embryos are visible after 7 d in PGR-free medium *(15)*. In this system, transfer of the tissue in the PGR-free medium results in the degeneration of PEM III though extensive programmed cell death, followed by formation of somatic embryos *(6)*. Programmed cell death appears important for shaping the embryo body, because its inhibition via manipulation of culture medium reduces the number of somatic embryos produced *(16)*. It is important to keep in mind that, at least for white spruce, the length of culture in PGR-free medium, which is generally of 7 d, must be adjusted for each cell line. For some lines 1 wk in the medium devoid of PGRs may be too long, as the cells start becoming brown and lose viability. Other lines can be cultured for a longer amount of time (up to 10 d) in the absence of PGRs and produce a large number of somatic embryos.

Embryogenic tissue of white spruce maintained on solid medium does not need to be transferred on PGR-free medium, but can be directly placed on the abscisic acid (ABA)-containing maturation medium (*see* **Subheading 3.4.** and **Fig. 1**). It is likely that tissue on solid medium is less exposed to auxin and cytokinin, and therefore does not require the PGR-free treatment.

3.4. Embryo Maturation

If initiation of embryo development is stimulated by withdraw of PGRs, continuation of development and completion of maturation are induced by applications of ABA and the imposition of a drying period. Both treatments are required for the production of morphologically and physiologically mature embryos. Early attempts to generate somatic embryos only used ABA *(4)* and resulted in the production of embryos that were morphologically mature, but not physiologically ready to germinate. This notion is important in that generation of embryo-like bipolar structures does not guarantee successful regeneration, which necessitates both morphological and physiological maturation of the embryos.

3.4.1. Applications of Abscisic Acid

About 0.7 mL of settled embryogenic cells are removed from the PGR-free medium using a large-mouth serological pipet and spread onto a Whatman no. 2 (5 cm in diameter) sterile filter paper, which is placed onto solid maturation medium containing ABA (*see* **Note 3**). The role of the filter paper is to absorb excessive water, which has a negative effect on the maturation process. As the density of the embryogenic tissue is critical for proper maturation, it is recommended that the amount of tissue transferred on solid medium be optimized for each line. Continuation of cell proliferation and delay of embryo maturation can be observed for some lines. In these cases, adjustments in the amount of embryogenic tissue transferred can improve the maturation process. The plates are then incubated in the dark for 4 wk (*see* **Fig. 1**). After 2 wk, the tissue supported by the filter paper should be placed on fresh maturation medium, to ensure continual supply of ABA. Depletion of ABA in the medium, in fact, arrests maturation and promotes precocious germination.

Responsiveness of the embryogenic tissue to ABA has been found to vary among genotypes. Usually high levels of ABA (40–50 μM) are required to promote proper maturation *(11,17–18)*, although a similar effect was achieved in other lines with a lower concentration of ABA (12 μM) *(14)*.

Embryogenic tissue cultured on solid maintenance medium can be directly placed on the ABA-containing maturation medium without the PGR-free treatment (*see* **Fig. 1**). A small amount of tissue is removed from the edge of the embryogenic callus and it is gently teased on the maturation medium to form a thin layer. Once again, the amount of tissue plated must be optimized for each line. However, the thickness of the layer of culture should not exceed 1 mm, to allow better contact of the tissue with ABA.

3.4.2. Embryo Desiccation

Embryos treated with ABA only are not able to convert into viable plants when transferred onto germination medium. Full maturation is only achieved after the imposition of a drying period, which is needed for the termination of the developmental program and the initiation of germination *(19)*. Embryo desiccation in spruce somatic embryos has been achieved through a variety of methods, which include the applications of osmotic agents into the ABA-containing maturation medium *(20)*. Although these methods have been found to be reliable and routinely used in several laboratories, a more traditional method for desiccating embryos (i.e., the partial drying treatment [PDT] will be illustrated here). This method, which was developed by Roberts et al. *(21)*, promotes a gradual and limited water loss in the embryos and enhances their post-embryonic performance considerably.

Cotyledonary somatic embryos with fully expanded cotyledons are harvested and placed on sterile filter paper disks of 5 mm in diameter, which are kept moist on the maturation medium (*see* **Note 4**). No more than six embryos should be placed on one filter paper to avoid the sticking of the embryos to one another. The sterile filter disks supporting the embryos are then placed in the central wells of a 24-well tissue culture plate (maximum of three filter disks per well), whereas the remaining wells are filled with sterile distilled water (*see* **Fig. 1**). The plates are sealed with Parafilm™ and incubated in the dark at room temperature for 10 d. Under these conditions the moisture content of the embryos declines by 20% *(22)*.

3.4.3. Morphology of Maturing Embryos

If somatic embryogenesis needs to be used for physiological or molecular studies, it is important to know the exact stages of embryo development during maturation. Therefore, it is suggested to observe the cultures on a weekly basis so that embryos of defined stages can be collected and used for further studies. Although detailed descriptions on the morphology of maturing embryos are available *(20)*, the major phases of embryo maturation are outlined below. The reader must remember that structural differences among embryos of difference species and genotypes are often observed.

After 1 wk in the maturation medium, the mitotic activity within the embryonal mass increases and small embryos emerge from the surface of the culture. Under a dissecting microscope, the embryos appear as yellow nodules surrounded by the translucent embryogenic tissue. After 2 wk the embryos increase in size and they become more elevated from the tissue mass, as a result of increased mitotic activity. A well-defined protoderm and organized root and shoot apices are visible at this time. Soon the apical poles of the embryos become dome-shaped, because of the formation of a functional shoot apical meristem. At the root pole the root cap starts forming. It is only after 3 wk in culture that the apical meristems become fully differentiated. At this stage, embryos have developed a visible procambium and the cotyledons *(4–6)* start emerging from the apical pole. The embryo axis elongates and the embryos acquire a creamy-yellow coloration indicative of deposition of storage products, namely starch and later proteins. Cotyledonary embryos are produced after 4 wk in maturation. They have reached a length of about 2–3 mm and are characterized by the presence of fully expanded cotyledons. At this time embryos have accumulated a large quantity of storage products, including proteins and lipids *(11)*.

Depending on the cell line, the maturation of the embryos can be quite synchronous. One way to improve synchrony during embryogenesis is to spread the tissue on the maturation period uniformly, to ensure that most of the cells on the surface of the layer are at equal distance from the medium. This guarantees uniform supply of ABA.

3.5. Embryo Conversion

Conversion is the ability of the embryos to produce a functional shoot and root during germination and is the final test for estimating the quality of the embryos produced in culture. The conversion frequency of the embryos varies from genotype to genotype and, within the same genotype, and it often depends on the level of maturity reached by the embryos. Morphological abnormalities within a population of embryos are often observed.

After 10 d of the PDT, embryos are removed from the filter paper disk and gently placed on the germination medium (*see* **Fig. 1**). Some embryos may be sticking together, therefore it is advisable to exercise maximum care not to damage the partially dried embryos. No more than eight embryos should be placed in one Petri dish containing germination medium (*see* **Note 5**). Germination should be carried out under a 16-h photoperiod (80 $\mu Em^2/s$). Conversion of shoot and root should occur after 2 wk.

For some cell lines with low conversion number, applications of ascorbic acid can aid the process *(18)*. Ascorbic acid (around 100 μM) can be applied to the autoclaved medium before the plates are poured. The medium should be used the following day, as ascorbic acid degrades spontaneously. Embryos can be kept in the germination medium for up to 1 mo. After that, plantlets can be transplanted in glass jars that contain sterile peat pellets saturated with strength LV medium without sucrose. After 4–5 wk, plantlets can be transferred to the greenhouse.

4. Notes

1. If the immature cones have a lot of resin or if the scales are tightly pressed against one another, the ethanol and bleach solutions will not reach the seeds. It is therefore advisable to loosen the scales by rotating the ends of the cone in opposite directions. If preferable, scales with the attached seeds can be removed from the cone and sterilized.
2. During this step no more than 100 seeds should be wrapped in the cheese cloth bag. Also, the bag should be loose, so that the bleach solution can reach the surface of the seeds. If the number of seeds is too high, or if the bag is wrapped tightly, contamination may occur.
3. Differently from 6-benzylaminopurine and 2,4-dichlorophenoxyacetic acid, which can be autoclaved with the culture medium, ABA cannot be autoclaved. Fresh ABA stock solution is prepared by dissolving ABA in a minimum amount of sodium hydroxide solution. After adjusting the pH to 5.8, the solution is filter-sterilized and added to warm maturation medium prior to the pouring of plates.
4. Embryos should not dry during this step, as this will preclude their post-embryonic growth. If dealing with many embryos, the filter paper disks should be placed on solid maturation medium while embryos are harvested. Once done, all the filter papers supporting the embryos should be placed in the wells of the 24-well tissue culture plate that is sealed immediately.
5. For better conversion, embryos should be placed vertically on the germination medium with their root in contact with the medium. Embryo conversion can also be achieved by transferring partial dried embryos directly into glass jars containing germination medium. Culture in large volume vessels (i.e., glass jars, may prevent accumulation of ethylene, which precludes growth) *(17)*.

Acknowledgments

The author acknowledges with gratitude the support received from the Natural Sciences and Engineering Research Council of Canada in the form of a Discovery Grant.

References

1. Thorpe, T.A. and Stasolla, C. (2001) Somatic embryogenesis, in *Current Trends in the Embryology of Angiosperms* (Bhojwani, S.S., ed.), Kluwer Acad. Publisher, Dordrecht. pp. 35–64.
2. Hosie, R.C. (ed.) (1979) *Native Trees of Canada*. 8th ed. Fitzhenry and Whiteside, Don Mills.
3. Durzan, D.J. (1980) Progress and promise in forest genetics. In: proceedings of the 50th Anniversary Conference, Paper Science and Technology, The Cutting Edge, May 8–10, pp 31–60. The Institute of Paper Chemistry, Appleton, WI.

4. Lu, C. Y. and Thorpe, T. A. (1987) Somatic embryogenesis and plantlet regeneration in cultured immature embryos of *Picea glauca*. *J. Plant Physiol.* **128**, 297–302.
5. Hakman, I., Rennie, P., and Fowke, L.C. (1987) A light and electron microscopy study of *Picea glauca* (white spruce) somatic embryos. *Protoplasma* **140**, 100–109.
6. Von Arnold, S., Sabala, I., Bozhkov, P., Dyachok, J., and Filonova, L. (2002) Developmental pathways of somatic embryogenesis. *Plant Cell Tiss. Org. Cult.* **69**, 233–249.
7. Harry, I.S. and Thorpe, T.A. (1991), Somatic embryogenesis and plant regeneration from mature zygotic embryos of red spruce. *Bot. Gaz.* **152**, 446–445.
8. El Meskaoui, A. and Tremblay, F. (2001) Involvement of ethylene in the maturation of black spruce embryogenic cell lines with different maturation capacities. *J. Exp. Bot.* **52**, 761–769.
9. Bozhkov, P.V. and von Arnold, S. (1998) Polyethylene glycol promotes maturation but inhibits further development of *Picea abies* somatic embryos. *Physiol. Plant.* **104**, 211–224.
10. von Arnold, S. and Eriksson, T. (1981) In vitro studies on adventitious shoot formation in *Pinus contorta*. *Can. J. Bot.* **59**, 870–874.
11. Joy IV, R.W., Yeung, E.C., Kong, L. and Thorpe, T.A. (1991) Development of white spruce somatic embryos: I. Storage product deposition. *In Vitro Cell. Dev. Biol. Plant* **27P**, 32–41.
12. Litvay, J.D., Johnson, M.A., Verma, D., Einspahr, D., and Weyrauch, K. (1985) Conifer suspension culture medium development using analytical data from developing seeds. Institute Paper Chem. Tech. Paper Series 155, Appleton, Wisconsin.
13. Tremblay, F.M. (1990) Somatic embryogenesis and plantlet regeneration from embryos isolated from stored seeds of *Picea glauca*. *Can. J. Bot.* **68**, 236–242.
14. Attree, S.M., Budimir, S., and Fowke, L.C. (1990) Somatic embryogenesis and plantlet regeneration from cultured shoots and cotyledons of seedlings from stored seeds of black and white spruce (*Picea mariana* and *Picea glauca*). *Can. J. Bot.* **68**, 30–34.
15. Filonova, L. H, Bozhkov, P. V., and von Arnold, S. (2000) Developmental pathway of somatic embryogenesis in *Picea abies* as revealed by time-lapse tracking. *J. Exp. Bot.* **51**, 249–264.
16. Bozhkov, P.V., Filonova, L.H., and von Arnold, S. (2002) A key developmental switch during Norway spruce somatic embryogenesis is induced by withdrawal of plant growth regulators and is associated with cell death and extracellular acidification. *Biotech. Bioen.* **77**, 658–667.
17. Kong, L. and Yeung, E.C. (1992) Development of white spruce somatic embryos: II. Continual shoot meristem development during germination. *In Vitro Cell. Dev. Biol. Plant* **28P**, 125–131.
18. Stasolla, C. and Yeung, E.C. (1999) Ascorbic acid improves the conversion of white spruce somatic embryos. *In Vitro Cell. Dev. Biol. Plant* **35**, 316–319.

19. Stasolla, C., Kong, L., Yeung, E.C., and Thorpe, T.A. (2002). Maturation of somatic embryos in conifers: morphogenesis, physiology, biochemistry, and molecular biology. *In Vitro Cell. Dev. Biol-Plant* **38,** 93–105.
20. Attree, S.M. and Fowke, L.C. (1993) Embryogeny of gymnosperms: advances in synthetic seed technology of conifers. *Plant Cell Tiss. Org. Cult.* **35,** 1–35.
21. Roberts, D.R., Sutton, B.C.S., and Flinn, B.S. (1990) Synchronous and high frequency germination of interior spruce somatic embryos following partial drying at high relative humidity. *Can. J. Bot.* **68,** 1086–1090.
22. Kong, L. (1994) Factors affecting white spruce somatic embryogenesis and embryo conversion. Ph.D. Dissertation, University of Calgary, Calgary.

9

Indirect Somatic Embryogenesis in Cassava for Genetic Modification Purposes

Krit Raemakers, Isolde Pereira, Herma Koehorst van Putten, and Richard Visser

Summary

In cassava both direct and indirect somatic embryogenesis is described. Direct somatic embryogenesis starts with the culture of leaf explants on Murashige and Skoog (MS) medium supplemented with auxins. Somatic embryos undergo secondary somatic embryogenesis when cultured on the same medium.

Indirect somatic embryogenesis is initiated by subculture of directly induced embryogenic tissue on auxin-supplemented medium with Gresshoff and Doy salts and vitamins. A very fine friable embryogenic callus (FEC) is formed after a few rounds of subculture and stringent selection. This FEC is maintained by subculture on auxin supplemented medium. Lowering of the auxin concentration allows the FEC to form mature somatic embryos that develop into plants when transferred to a cytokinin-supplemented medium.

Key Words: Direct; FEC; indirect; somatic embryogenesis.

1. Introduction

Cassava is an important food crop for millions of people. It is grown for its starch-containing tuberized roots, which are used for human consumption, as animal feed, and as raw material for the starch industry. Classical breeding is used to improve the crop; however, there are certain traits (e.g., resistance against viruses, cyanide content, and protein content of tuberized roots) that can only be improved by genetic modification.

An efficient regeneration system is a prerequisite for the development of a procedure for genetic modification. In cassava, only regeneration via somatic embryogenesis is well described. Somatic embryogenesis can be divided in two different types: indirect and direct. In the direct form of embryogenesis explants cultured on medium for initiation will form embryogenic units which develop directly into bipolar somatic embryos. In the indirect form of somatic embryogenesis, the initiated embryogenic units do not develop beyond the preglobular stage but instead break up into new embryogenic units. In some crops both types of embryogenesis occur simultaneously, or both types are easy to obtain by choosing the proper explant or the proper (concentration of) growth regulators. In other crops it is either the direct or indirect type of somatic embryogenesis as is the case in cassava. Direct somatic embryogenesis in cassava was described for the first time in 1982 *(1)*. Since then, the methods have been extended to a whole range of genotypes (*see* **Note 1**) using a range of different media *(2–8)*. Furthermore, methods are described to obtain continuous proliferating embryogenic cultures *(8–15)*.

Taylor et al. *(16)* showed that culture of direct induced embryogenic tissue on very specific medium combined with rigorous selection for a very specific type of embryogenic cells, described as friable embryogenic callus (FEC), leads to embryogenic cultures that have the characteristics of indirect somatic embryogenesis.

In our laboratory, FEC has been initiated successfully in 6 of the 10 tested genotypes: Adira 4 (Indonesia), R60 (Thailand), R90 (Thailand), TMS60444 (Nigeria), Thai 5 (Thailand), and M7 (Zimbabwe). In all these genotypes the FEC has been maintained for years by regular subculture.

As was shown in other crops, FEC is an excellent tissue for use in genetic modification (*see* **Notes 2 and 3**), either via particle bombardment *(17,18)* or *Agrobacterium (19,20)*. A range of different methods have been developed to select transgenic tissue and subsequently plants. The best-described method makes use of the *npt*II gene *(17,20,21)*. Less well described are methods using the *pat* gene *(22)*, *hph* gene *(20)*, *luciferase* gene *(18,23)*, and the *manose isomerase* gene *(24)*. The methods described above have been used to produce plants with altered starch *(23,25)*, increased protein content *(26)*, and reduced cyanide concentration *(27)*.

In a previous volume of this series *(15)* methods to induce direct somatic embryogenesis in cassava are described, with special emphasis on how it can be used for mass multiplication purposes. Here we describe methods to induce indirect somatic embryogenesis and how it can be used for genetic modification purposes

This chapter describes methods to initiate somatic embryogenesis in elite genotypes of cassava, the subsequent multiplication of the embryogenic tissue by secondary/cyclic somatic embryogenesis, the germination of somatic embryos into plants, and, finally, the transfer to the greenhouse.

Table 1
Composition of the Media Used in the Different Steps of FEC Initiation and Regeneration of Plants

Function	Salts and vitamins	Growth regulators	Abbreviations
Multiplication of plants	MS (**28**) + 4% sucrose	—	MS
Initiation of somatic embryos (SE)	GD (**29**) + 4% sucrose	8 mg/L 2,4-D	MS + 8 mg/L 2,4-D
Initiation of FEC	GD + 4% sucrose	10 mg/L picloram	GD + 10 picloram
Maintenance of FEC	GD + 4% sucrose	10 mg/L picloram	GD + 10 picloram
Suspension cultures	SH (**30**) + 6% sucrose	10 mg/L picloram	SH + 10 picloram
Regeneration of torpedo-shaped SE	MS + 4% sucrose	1 mg/L NAA	MS + 1 mg/L NAA
Maturation of torpedo-shaped SE	MS + 4% sucrose	0.1 mg/L BA	MS + 0.1 mg/L BA
Shoot development of mature SE	MS + 4% sucrose	1 mg/L BA	MS + 1 mg/L BA
Rooting of shoots	MS + 4% sucrose	—	MS

2. Materials
2.1. General Requirements

1. Rotary shaker (120 rpm).
2. Sterile 300-mL Erlenmeyer flasks, capped with aluminium foil.
3. Sterile tubes and Petri-plates.
4. Bottles (0.5- and 1-L) that can be autoclaved.
5. Magnetron to dissolve solid medium.
6. pH meter.
7. Sterile working surface.
8. Murashige and Skoog (MS) medium, Gresshoff and Doy medium, Schenk and Hildebrandt medium.
9. Scalpel and pincet.
10. Binocular microscope for isolation of explants.
11. Refrigerator for storage of stock solutions.
12. Growth chamber: temperature 30°C, day length of 12 h, light intensity of 40 mmol/m^2/s.
13. Patience.

3. Methods
3.1. Isolation of Leaf Explants and Culture for Primary Somatic Embryogenesis

1. Multiplication of plants in in vitro by subculturing single node cuttings on solid (8 g/L micro agar) MS. Plants can be maintained in jars, tubes, or Petri dishes (*see* **Notes 4** and **5**).
2. A binocular microscope is needed for isolation of leaf explants. Harvest shoot tips of in in vitro plants, remove leaves larger than 8 mm, isolate single leaves smaller than 4 mm, if possible divide leaves into single leaf lobes, and isolate meristems of shoot tip. Culture leaves, leaf lobes, and meristems on solid (8 g/L micro agar) MS + 8 mg/L 2,4-dichlorophenoxyacetic acid (2,4-D) for initiation of somatic embryos (*see* **Note 6**).

3.2. Isolation of Compact Embryogenic Callus and Culture for Initiation of Friable Embryogenic Callus

1. After 14–25 d: isolate high-quality embryogenic tissue (*see* **Note 7**) from the leaf explants and transfer to solid (10 g/L micro agar) GD + 10 mg/L picloram for initiation of FEC (*see* **Note 8**). Most of the explants will form a white/yellow nonfriable and nonembryogenic type of callus. Some of the explants form, as in direct embryogenesis, again CEC.
2. After 14–28 d the CEC is isolated and cultured on the same medium as in **step 1**. In this second subculture again most of the explants form white/yellow nonfriable and nonembryogenic callus. A few explants will also form small colonies of very small globular embryos, or small colonies of friable embryogenic units, or a mixture of both.

Indirect Somatic Embryogenesis

3. Subculture the colonies of globular embryos/FEC units and select for pure colonies of FEC units. FEC units contain no more than a few hundred cells and are not enclosed by an epidermis (*see* **Note 9**).

3.3. Maintenance of Friable Embryogenic Callus Cultures and Imitation of Suspension Cultures

1. FEC is maintained by a 3 wk subculture regime on solid (10 g/L micro agar) GD + 10 mg/L picloram. Usually some selection has to be applied during transfer. Part of the FEC either has become nonembryogenic whereas other FEC has regenerated into globular shaped embryos. Only fine FEC is used for multiplication.
2. Suspension cultures are initiated by transfer of 100 mg of FEC to 300 mL flaks filled with 50 mL SH + 10 mg/L picloram. The liquid medium is refreshed at least once per week, preferably twice. Every 2–3 wk the FEC is divided over new flasks. The Erlenmeyer flasks are cultured in the light on a shaker at 120 rpm.

3.4. Regeneration of Plants From Friable Embryogenic Callus

1. FEC is first cultured on solid (10 g/L micro agar) MS4 + 1 mg/L NAA regeneration of torpedo-shaped somatic embryos. Every 2–3 wk the tissue is transferred to fresh maturation medium. The first torpedo-shaped embryos appear after about 4–6 wk of culture. Torpedo-shaped embryos are translucent, have a size from 2 to 10 mm and have cotyledon primordial (*see* **Note 10**).
2. Torpedo-shaped embryos are isolated and transferred to solid (8 g/L micro agar) MS + 0.1 mg/L bovine albumin (BA), which allows further maturation of the torpedo-shaped embryos into mature somatic embryos. Mature somatic embryos have a size from 0.5 to 3 cm, a distinct hypocotyls and large green cotyledons. Every 2–3 wk the somatic embryos are transferred to fresh maturation medium. Usually it takes 2–6 wk of culture before torpedo-shaped embryos become mature (*see* **Note 11**).
3. Mature somatic embryos are than transferred to solid (8 g/L micro agar) MS4 + 1.0 mg/L BAP for shoot development. Again the somatic embryos have to be transferred to fresh medium every 2–3 wk. After about 4–8 wk the first mature somatic embryos have initiated shoots. Most of the shoots do not possess roots (*see* **Note 12**).
4. Shoots are transferred to rooting medium (MS4). Roots are formed within 2 wk (*see* **Note 13**).

3.5. Transfer of Plants to the Greenhouse

1. Two- to four-centimeter tall plantlets can be transferred to the greenhouse. Remove the plantlets carefully from the tubes or containers and wash away thoroughly the agar from the roots.
2. Grow plants in pots (1:3 sterilized mixture of soil with fine sand) and keep at high humidity in an incubator. After 1 wk the plants can be acclimatized to the greenhouse conditions by gradual reduction of the humidity (*see* **Note 14**).

4. Notes

1. Primary somatic embryogenesis has been successfully accomplished in at least 50 different genotype originating from such diverse places as Nigeria, Zimbabwe, Columbia, Venezuela, Indonesia, and Thailand *(1–8)*.
2. The FEC system is the most common and wide spread method to produce genetically modified plants in cassava. Less often used are systems based on direct somatic embryogenesis *(27,31)*. In another system, direct induced somatic embryos are used as explant for genetic modification and cultured hereafter on medium for the initiation of adventitious shoots *(21)*.
3. FEC has been shown to be an excellent source of tissue for protoplast regeneration *(32)* which can also be used to obtain transgenic plants by electroporation (Raemakers et al., unpublished results).
4. Cassava plants can be maintained in the same medium for a period up to 9 mo when cultured in jars. If the plants are used for primary somatic embryogenesis, freshly multiplied plants are preferred.
5. Leaf explants from greenhouse grown plants can also be used as the starting material for primary somatic embryogenesis. However, it has been shown that the embryogenic response of these leaf explants depends on the growth conditions in the greenhouse *(6)*.
6. Other auxins such as Dicamba and Picloram or combinations of auxins can also be used to induce primary somatic embryogenesis *(5)*. Danso et al. *(8)* add $CuSO_4$ to the auxin supplemented medium to improve primary embryogenesis.
7. In genotypes where the quality of the embryogenic tissue formed on the leaf explants is too low for FEC initiation high quality embryogenic tissue can be obtained from secondary embryogenic cultures. Secondary somatic embryogenesis is rather simple and (almost) genotype independent *(10,11,13)*.
8. A higher amount of agar is used for this step as compared to other steps. This is also the case for the maintenance of FEC and the regeneration of torpedo-shaped somatic embryos from FEC.
9. FEC initiation is strongly genotype dependent. A significant number of genotypes will never produce FEC. Amongst the genotypes which can produce FE, the time period of culture on GD + 10 mg/L Picloram needed to produce FEC, varies from 2 to 8 mo.
10. In our laboratory torpedo-shaped embryos have been obtained from FEC in all tested genotypes; however, the frequency and time period needed varies considerably between genotypes.
11. Development of torpedo-shaped embryos into mature somatic embryos is relatively easy and occurs at high frequencies in all tested genotypes. It is important to isolate well formed torpedo-shaped embryos.
12. Depending on the genotype, between 20 and 80% of the mature somatic embryos develop into shoots.
13. Usually highly branched shoots are formed or one embryo has initiated more than one shoot. The apical part of these shoots is isolated and transfer to MS4

medium for rooting. Usually a normal cassava plant with roots is obtained after one transfer. At times a second transfer is needed to obtain a nonbranched plant.
14. The main problem in using the FEC culture systems for genetic modification is the occurrence of somaclonal variation. In some genotypes it is almost absent especially if the FEC culture period is very short. In for example TMS60444, a 6-mo-old FEC regenerates normal looking plants, whereas 2-yr-old FEC regenerates abnormal plants. In other genotypes, as for example R60 even young FEC regenerates abnormal plants.

References

1. Stamp, F. A. and Henshaw, G. G. (1982) Somatic embryogenesis in cassava. *Z. Planzenphysiol.* **105**, 183–187.
2. Stamp, J. A. and Henshaw, G. G. (1987) Somatic embryogenesis from clonal leaf tissue of cassava. *Ann. Bot.* **59**, 445–450.
3. Szabados, L., Hoyos, R., and Roca, W. (1987) In vitro somatic embryogenesis and plant regeneration of cassava. *Plant Cell Rep.* **6**, 248–251.
4. Mathews, H., Schöpke, C., Carcamo, R., Chavarriaga, P., Fauquet, C., and Beachy, R. N. (1993) Improvement of somatic embryogenesis and plant regeneration in cassava. *Plant Cell Rep.* **12**, 328–333.
5. Sudarmonowati, E. and Henshaw, G. G. (1993) The induction of somatic embryogenesis of recalcitrant cassava cultivars using Picloram and Dicamba, in *Proceedings of the First International Scientific Meeting of the Cassava Biotechnology Network* (Roca, W. M. and Thro, A. M., eds.), Centro International de Agricultura Tropical, Cartagena de Indias, Columbia, pp. 128–134.
6. Raemakers, C. J. J. M., Bessembinder, J., Staritsky, G., Jacobsen, E., and Visser, R. G. F. (1993) Induction, germination and shoot development of somatic embryos in cassava. *Plant Cell Tiss. and Org. Cul.* **33**, 151–156.
7. Taylor, N. J. and Henshaw, G. G. (1993) The induction of somatic embryogenesis in 15 African and one South American cassava cultivars, in *Proceedings of the First International Scientific Meeting of the Cassava Biotechnology Network* (Roca, W. M. and Thro, A. M., eds.) Centro International de Agricultura Tropical, Cartagena de Indias, Columbia, pp. 229–240
8. Danso, K. E. and Ford-Lloyd, B. V. (2002) Induction of high frequency somatic embryogenesis in cassava for cryopreservation. *Plant Cell Rep.* **21**, 226–232.
9. Stamp, J. A. and Henshaw, G. G. (1987) Secondary somatic embryogenesis and plant regeneration in cassava. *Plant Cell Tiss. and Org. Cult.* **10**, 227–233.
10. Raemakers, C. J. J. M., Amati, M., Staritsky, G., Jacobsen, E., and Visser, R.G.F. (1993) Cyclic somatic embryogenesis in cassava. *Ann. Bot.* **71**, 289–294.
11. Raemakers, C. J. J. M., Schavemaker, C. M., Jacobsen, E., and Visser, R. G. F. (1993) Improvements of cyclic somatic embryogenesis of cassava (*Manihot esculenta* Crantz). *Plant Cell Rep.* **12**, 226–229.
12. Li, H. Q., Huang, Y. W., Liang, C. Y., and Guo, J. Y. (1995) Improvement of plant regeneration from secondary somatic embryos of cassava, in *Proceedings of second international meeting of cassava biotechnology network,* Bogor, Indone-

sia, 22–26 August, Centro Internacional de Agricultura Tropical, Cali, Columbia, pp. 289–302.
13. Sofiari, E., Raemakers, C. J. J. M., Kanju, E., et al. (1997), Comparison of NAA and 2,4-D induced somatic embryogenesis in cassava. *Plant Cell Tiss. Org. Cult.* **50,** 45–56.
14. Raemakers, C. J. J. M., Jacobsen, E., and Visser, R. G. F. (1997) Micropropagation of *Manihot esculenta* Crantz (cassava), in *Biotechnology in Agriculture and Forestry*, vol. 39 (Bajaj, Y. P. S., ed.), Springer Verlag, Berlin, pp. 77–103.
15. Raemakers, C. J. J. M., Jacobsen, E., and Visser, R. G. F. (1999) Direct, cyclic somatic embryogenesis in cassava for mass production purposes, in *Methods in Molecular Biology: Plant Cell and Tissue Culture* (Hall, R. D., ed.) Humana Press, Totowa, NJ, pp. 61–71.
16. Taylor, N. J., Edwards, M. Kiernan, R. J., Davey, C. D. M., Blakesley, D., and Henshaw, G. G. (1996) Development of friable embryogenic callus and embryogenic suspension culture systems in cassava (*Manihot esculenta* Crantz). *Nat. Biotechnol.* **14,** 726–730.
17. Schöpke, C., Taylor, N., Carcamo, R., et al. (1996) Regeneration of transgenic cassava plants (*Manihot esculenta* Crantz) from microbombarded embryogenic suspension cultures. *Nat. Biotechnol.* **14,** 731–735.
18. Raemakers, C. J. J. M., Sofiari, E., Taylor, N., Henshaw, G. G., Jacobsen, E. and Visser, R. G. F. (1996) Production of transgenic cassava (*Manihot esculenta* Crantz) plants by particle bombardment using luciferase activity as selection marker. *Mol. Breeding* **2,** 339–349.
19. Gonzalez, A. E., Schöpke, C., Taylor, N., Beachy, R. N., and Fauquet C. (1998) Regeneration of transgenic cassava plants (*Manihot esculenta* Crantz) through *Agrobacterium tumefaciens*-mediated transformation of embryogenic suspension cultures. *Plant Cell Rep* **17,** 827–831.
20. Schreuder, M. M., Raemakers, C. J. J. M., Jacobsen, E. and Visser, R. G. F. (2001) Efficient production of transgenic plants by *Agrobacterium*-mediated transformation of cassava. *Euphytica* **120,** 35–42.
21. Li, H. Q., Sautter, C., Potrykus, I., and Puonti-Kaerlas, J., (1996) Genetic transformation of cassava (*Manihot esculenta* Crantz). *Nat. Biotech.* **14,** 736–740.
22. Snepvangers, S. C. H. J., Raemakers, C. J. J. M., Jacobsen, E., and Visser, R. G. F. (1997) Optimization of chemical selection of transgenic friable embryogenic callus of cassava using the luciferase reporter gene system. *African Crop Sci. J.* **2,** 196–200.
23. Munyikwa, T. R. I., Raemakers, C. J. J. M., Schreuder, M., et al. (1998) Pinpointing towards improved regeneration and transformation of cassava. *Plant Sci.* **135,** 87–101.
24. Zhang, P., Potrykus, I., and Puonti-Kaerlas, J. (2000) Efficient production of transgenic plants using negative and positive selection. *Trans. Res.* **9,** 405–415.
25. Raemakers, C. J. J. M., Schreuder, M., Suurs, L., et al. (2005) Properties of amylose-free cassava starch, produced by antisense inhibition of granule-bound starch synthase. *Nat. Biotech.* (submitted).

26. Zhang, P., Jaynes, J. M., Potrykus, I., Gruisem, W., and Puonti-Kaerlas, J. (2003) Transfer and expression of an artificial storage protein (ASP1) gene in cassava. *Trans. Res.* **12,** 243–250.
27. Siritunga, D., Arias-Garcon, D., White, W., and Sayre, R. T. (2004) Overexpression of hydroxynitrile lyase in transgenic cassava roots accelerates cyanogenesis and food detoxification. *Plant Biotech. J.* **2,** 37–43.
28. Murashige, T. and Skoog, F. (1962) A revised medium for rapid growth and bioassay with tobacco cultures. *Physiol. Plant.* **15,** 473–497.
29. Gresshoff, P. M. and Doy, C.H. (1974) Development and differentiation of haploid *Lycopersicon esculentum* (tomato). *Planta* **107,** 161–170.
30. Schenk, R. U. and Hildebrandt, A. C. (1972) Medium and techniques for induction and growth of monocotyledonous and dicotyledonous plant cell cultures. *Can. J. Bot.* **50,** 199–204.
31. Sarria, R., Torres, E., Chavarriaga, P., and Roca, W. M. (2000) Transgenic plants of cassava with resistance to basta obtained by *Agrobacterium*-mediated transformation. *Plant Cell Rep.* **19,** 339–344.
32. Sofiari, E., Raemakers, C. J. J. M., Bergervoet, J. E. M., Jacobsen, E., and Visser, R. G. F. (1997) Plant regeneration from protoplasts isolated from friable embryogenic callus of cassava. *Plant Cell Rep.* **18,** 159–165.

10

Direct Somatic Embryogenesis in *Coffea canephora*

Francisco Quiroz-Figueroa, Miriam Monforte-González,
Rosa M. Galaz-Ávalos, and Victor M. Loyola-Vargas

Summary

Somatic embryogenesis (SE) provides a useful model to study embryo development in plants. In contrast to zygotic embryogenesis, SE can easily be observed, the culture conditions can be controlled, and large quantities of embryos can be easily obtained. In *Coffea* spp several model systems have been reported for in vitro SE induction. SE for coffee was first reported in *Coffea canephora*. Several systems have been developed since then, including SE from callus cultures derived from leaf explants; a two-phase experimental protocol for SE from leaves of *Coffea arabica*; and from leaf explants of Arabusta or *C. arabica* using a medium with cytokinins. Here we report a protocol using young leaves from in vitro seedling pre-conditioned with growth regulators. This is a simplified method to obtain a faster and more efficient protocol to produce direct somatic embryos in *C. canephora*.

Key Words: Coffee; *Coffea* spp; somatic embryogenesis; direct; leaves.

1. Introduction

The capacity to produce morphologically well-formed and normally developed embryos from somatic cells resided uniquely within the plant kingdom *(1)* up to the present time. In addition, the development of somatic and zygotic embryos is highly similar *(1,2)*. Therefore, somatic embryogenesis (SE) provides a useful model to study embryo development in plants. In contrast to zygotic embryogenesis, SE can easily be observed, the culture conditions can be controlled, and large quantities of embryos can be easily obtained *(3)*.

Two types of SE are recognized. The term "direct" is applied to explants that undergo a minimum of proliferation before forming somatic embryos, whereas "indirect" refers to explants that undergo an extensive proliferation before the development of somatic embryos *(4)*. It has been suggested that in direct embryogenesis, embryogenic cells are present and simply require favorable conditions for embryo development, whereas indirect embryogenesis requires the re-determination of differentiated cells *(5)*. However, the terms "direct" and "indirect" are still useful in describing cases in which either very little or a great deal of explant proliferation precedes embryogenesis although not necessarily indicating fundamental differences in the cells involved *(6)*.

In *Coffea* spp several model systems have been reported for in vitro SE induction. SE In *Coffea canephora* was first reported by Staritsky *(7)*, who described the induction of callus tissue from orthotropic internodes. Subsequently, Herman and Haas *(8)* obtained SE for *Coffea arabica* from callus cultures derived from leaf explants. Söndahl and Sharp *(9,10)* developed a two-phase experimental protocol for SE from leaves of *C. arabica*. Dublin *(11)* reported SE from leaf explants of Arabusta using a medium with cytokinins (6-Benzylaminopurine [BA] and kinetin). Whereas, that of Yasuda et al. *(12)* induced embryogenic calli and somatic embryos from *C. arabica* leaf explants using only BA.

The physiological and morphological maturity of the tissue, as well the components of the culture medium, determine the time of occurrence and types of possible responses in coffee SE *(13)*. Using young leaves from in vitro seedling pre-conditioned with growth regulators, we *(14)* have shorted the embryogenic response time in direct SE of *Coffea* spp and preconditioning with growth regulators was critical for embryogenic response, because, no embryos were observed when explants came from plants cultured in the absence of growth regulators. The protocol presented here is a simplified method to obtain a faster and more efficient protocol to produce direct somatic embryos in *C. canephora*.

2. Materials

1. Flow cabinet, dry-hot sterilizer, scissors, Petri plates, scalpel, forceps, sterile cotton, 70% (v/v) ethanol.
2. Seeds of *C. canephora*.
3. Sterile distilled water by autoclaving (121°C during 20 min), commercial bleach.
4. MS basal medium is based on formula of Murashige and Skoog *(15)*. The vitamins, amino acids, and plant growth regulators are prepared in independence stocks solutions. The medium is semi-solidified by addition of 0.25% (w/v) gelrite™ as gelling agent.

3. Methods

3.1. In Vitro Plantlets

1. Seeds are washed and imbibed for 24–48 h in sterile distilled water (*see* **Note 1**).
2. Seeds are disinfected with commercial bleach (1.25% v/v free sodium hypochlorite) for 20 min with agitation. Remove the solution and rinse the seed three times with sterile water.

3. Isolate the zygotic embryos (*see* **Note 2**) using a scalpel and forceps, the embryos are puts onto zygotic germination medium (*see* **Table 1**). Ten embryos are placed in each Magenta plastic box (*see* **Note 3**) containing 40 mL of medium and cultivated at $25 \pm 2°C$ under a 16/8-h (light/dark) photoperiod. Subculture with fresh medium is doing every 60 d. The boxes are sealing with strips of cling film.
4. The zygotic plantlets are growing in Magenta boxes containing 40 mL of zygotic plantlet medium (*see* **Table 1**) and cultured at $25 \pm 2°C$ under a 16/8-h (light/darkness) photoperiod and transferred to fresh medium every 60 d. One plantlet is put for each container.
5. When the plantlet has six pairs of leaves (size) it can be used as an explant.

3.2. Direct Somatic Embryogenesis

1. Plantlet leaves are cut avoiding mid-vein and edges (*see* **Fig. 1A**).
2. The foliar explants (ca. 0.25 cm2) are placed on direct embryogenesis induction medium (**Table 1**) in glass bottle, aluminum sheet or plastic are used as tap (*see* **Note 4**) and sealed with cling film strip (*see* **Fig. 1B**). Those bottles are incubated under photoperiod (16 h/8 h light/darkness) at $25 \pm 2°C$.
3. Somatic embryo can be observed after 3 wk of induction, however, on wk 6 the edge explant is cover with somatic embryos (*see* **Fig. 1C,D**).

3.3. Germination of Somatic Embryo

1. Cotiledonary somatic embryos are separate from explants using forceps and put onto developmental medium (**Table 1**) for their germination, ensure that embryos are in good contact with the medium (*see* **Fig. 1E** and **Note 5**).
2. The subculture of embryos in germination fresh medium is done every 8 wk.
3. The plantlets obtained (*see* **Fig. 1F**) can be maintenanced in the same developmental medium.
4. Direct embryogenesis induction (*see* **Notes 6** and **7**) can be done following the same protocols as in **Subheading 3.1., step 5** to **Subheadings 3.3., step 3** for somatic plantlets.

4. Notes

1. The beans are imbibed in water with aim of to take off the parchment. The coat is removed by manual manipulation.
2. The developmental stage of zygotic embryo affects response of germination. Optima embryo stage is cotyledonary.
3. Each magenta box has a thread in border (*see* **Fig. 1F**, *arrow*) that service as filter to gasses interchange.
4. When glass bottle is cover with aluminum the embryogenic yield is higher than when use plastic tap.
5. After 5 wk somatic embryos have develop their radicular system (*see* **Fig. 1E**, *arrows*) and 5 wk after can be seen the first pair leaves (*see* **Fig. 1E**, *headarrows*).
6. The position of leaves on the plant is very important, because the first two pairs of leaves do show poor embryogenic response as compared with the mature

Table 1
Composition of Mediums-MS Medium: Inorganic and Organic Salt (15), Vitamins and Others Supplements

	ZG	ZP	DEI	D
Macro salts -mg/L (mM)-				
NH_4NO_3	1650 (20.61)	1650 (20.61)	412.0 (5.15)	1650 (20.61)
KNO_3	1900 (18.79)	1900 (18.79)	475.0 (4.7)	1900 (18.79)
$CaCl_2 \cdot 2H_2O$	440 (2.99)	440 (2.99)	110.0 (0.748)	440 (2.99)
KH_2PO_4	170 (1.249)	170 (1.249)	85.0 (0.624)	170 (1.249)
$MgSO_4 \cdot 7H_2O$	370 (1.500)	370 (1.500)	92.5 (0.375)	370 (1.500)
Micro Salts -mg/L (μM)-				
KI	0.830 (5)	0.830 (5)		0.830 (5)
$COCl_2 \cdot 6H_2O$	0.025 (0.1)	0.025 (0.1)		0.025 (0.1)
$Na_2MO_4 \cdot 2H_2O$	0.250 (1)	0.250 (1)		0.250 (1)
H_3BO_3	6.200 (100.27)	6.200 (100.27)	0.125 (0.5)	6.200 (100.27)
$MnSO_4 \cdot 7H_2O$	22.300 (80.5)	22.300 (80.5)	3.100 (50)	22.300 (80.5)
$CuSO_4 \cdot 5H_2O$	0.025 (0.1)	0.025 (0.1)	6.83 (40)	0.025 (0.1)
$ZnSO_4 \cdot 7H_2O$	8.100 (28)	8.100 (28)	0.05 (0.2)	8.100 (28)
$FeSO_4 \cdot 7H_2O$	27.950 (100)	27.950 (100)	4.3 (15)	27.950 (100)
Na_2EDTA	37.230 (100)	37.230 (100)	21 (75.53)	37.230 (100)
			27.9 (74.95)	
Vitamins -mg/L (μM)-				
Piridoxine HCl			1 (4.86)	2 (9.72)
Nicotinic acid			1 (8.12)	2 (16.24)
Thiamine HCl	10 (29.6)	4 (11.86)	10 (29.6)	4 (11.85)
Myo-inositol	100 (550)	100 (550)	100 (550)	100 (550)
Biotin	0.01 (0.041)			
Amino acids -mg/L (μM)-				
Cysteine	25 (158)	25 (158)		25 (158)
Sugars g/L (μM)-				
Sucrose	30 (166.48)	30 (87.64)	30 (87.64)	30 (87.64)
Glucose				
Growth regulator -mg/L (μM)-				
Naphthaleneacetic acid	0.1 (0.54)	0.1 (0.54)	1.12 (5)	0.1 (0.54)
6-Benzyl-aminopurine				
Kinetin	0.5 (2.32)	0.5 (2.32)		0.5 (2.32)

^{a}The pH of all medium are adjusted to 5.8 before autoclaving (20 min, 110°C). Gelrite is used as gelling agent at 0.25% (w/v) and is added after the pH has been adjusted.

Abbr:ZG, zygotic germination medium; ZP, zygotic plantlet medium; DEI, direct embryogenesis medium, D, developmental medium.

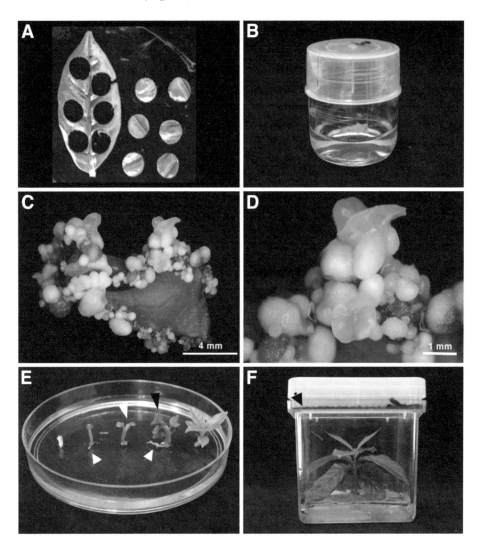

Fig. 1. Direct embryogenesis system in *Coffea* spp. (**A**) Cut usual of explant leaf avoiding midvein and edges. (**B**) Glass bottle with plastic tap where direct embryogenesis is induced. (**C**) Direct somatic embryos obtained after 6 wk embryogenesis induction. (**D**) Close up of **C**, showing a cotyledonary somatic embryo. (**E**) Different stages of embryo germination, arrows radicular system, head arrows first pair of leaves. (**F**) Plantlet from somatic embryo after 25 wk. Arrow, thread of cotton.

leaves. Also, explants coming from the distal part of the leaf are less responsive than those coming from the basal part of the leaf.

7. The preconditioning of plantlet for 2 wk in medium with 0.54 μM naphthaleneacetic acid and 2.33 μM kinetin is critical for the induction of direct somatic embryogenesis. Reduced embryogenic response is observed when the plantlet is not preconditioning.

References

1. Zimmerman, J. L. (1993) Somatic embryogenesis: a model for early development in higher plants. *The Plant Cell* **5**, 1411–1423.
2. Dodeman, V. L., Ducreux, G., and Kreis, M. (1997) Zygotic embryogenesis versus somatic embryogenesis. *J. Exp. Bot.* **48**, 1493–1509.
3. Kawahara, R. and Komamine, A. (1995) Molecular basis of somatic embryogenesis, in Biotechnology in Agriculture and Forestry. Vol. 30. Somatic Embryogenesis and Synthetic Seed I (Bajaj, Y. P. S., ed.), Springer-Verlag, Berlin, pp. 30-40.
4. Sharp, W. R., Söndahl, M. R., Caldas, L. S., and Maraffa, S. B. (1980) The physiology of in vitro asexual embryogenesis. *Hort. Rev.* **2**, 268-310.
5. Yeung, E. C. (1995) Structural and developmental patterns in somatic embryogenesis, in In Vitro Embryogenesis in Plants, (Thorpe, T. A., ed.), Kluwer Academic Publishers, Netherlands, pp. 205–247.
6. Halperin, W. (1995) In vitro embryogenesis: some historical issues and unresolved problems, in In Vitro Embriogenesis in Plants, (Thorpe, T. A., ed.), Kluwer Academic Publishers, Netherlands, pp. 1–16.
7. Staritsky, G. (1970) Embryoid formation in callus tissues of coffee. *Acta Bot. Need.* **19**, 509–514.
8. Herman, E. B. and Haas, G. J. (1975) Clonal propagation of Coffea arabica L. from callus culture. *HortScience* **10**, 588–589.
9. Söndahl, M. R. and Sharp, W. R. (1977) High frequency induction of somatic embryos in cultured leaf explants of Coffea arabica L. *Z. Pflanzenphysiol.* **81**, 395–408.
10. Söndahl, M. R. and Sharp, W. R. (1979), Research in Coffea spp. and applications of tissue culture methods, in Plant Cell and Tissue Culture Principles and Applications, (Sharp, W. R., Larsen, P. O., Paddock, E. F., and Raghavan, V., eds.), Ohio State University Press, Columbus, pp. 527–584.
11. Dublin, P. (1981) Embryogenèse somatique directe sur fragments de feuilles de caféier Arabusta. *Café Cacao Thé* **25**, 237–242.
12. Yasuda, T., Fujii, Y., and Yamaguchi, T. (1985) Embryogenic callus induction from Coffea arabica leaf explants by benzyladenine. *Plant Cell Physiol.* **26**, 595–597.
13. Loyola-Vargas, V. M., Fuentes-Cerda, C. F. J., Monforte-González, M., Méndez-Zeel, M., Rojas-Herrera, R., and Mijangos-Cortés, J. (1999) Coffee tissue culture as a new model for the study of somaclonal variation. ASIC; Helsinsky. 18e Colloque Scientifique International sur le café, pp. 302–307.
14. Quiroz-Figueroa, F. R., Fuentes-Cerda, C. F. J., Rojas-Herrera, R., and Loyola-Vargas, V. M. (2002) Histological studies on the developmental stages and differentiation of two different somatic embryogenesis systems of Coffea arabica. *Plant Cell Rep.* **20**, 1141–1149.

15. Murashige, T. and Skoog, F. (1962) A revised medium for rapid growth and bioassays with tobacco tissue cultures. *Physiol. Plant.* **15,** 473–497.

III

Plant Propagation In Vitro

11

A New Temporary Immersion Bioreactor System for Micropropagation

Manuel L. Robert, José Luis Herrera-Herrera, Gastón Herrera-Herrera, Miguel Ángel Herrera-Alamillo, and Pedro Fuentes-Carrillo

Summary

A new type of bioreactor system for plant micropropagation is described that incorporates a number of features specifically designed to simplify its operation and reduce production costs. The BioMINT unit is a mid-sized (1.2 L) reactor that operates on the principle of temporary immersion. It is built of polypropylene and is translucent, autoclavable, and reusable. It consists of two vessels, one for the plant tissues and the other one for the liquid culture media coupled together through a perforated adaptor piece that permits the flow of the liquid media from one vessel to the other. This flux is driven by gravity through a see-saw movement provided by equipment (SyB) consisting of electric motor powered platforms that change position. The structural simplicity and the modular and independent nature of the bioreactors simplify their operation and reduce the amount of hand labor required for transfers, thereby reducing the cost of the whole micropropagation process.

Key Words: Bioreactor; BioMINT; liquid culture; plant micropropagation; temporary immersion; RITA.

1. Introduction
1.1. Micropropagation in Semi-Solid Medium

Most micropropagation processes are carried out in small culture vessels containing a culture medium solidified with a gelling agent to create a substrate on which the plant tissues are cultured. In spite of its general use, this method has some disadvantages. The culture conditions are heterogeneous because not all the tissues are in contact with the nutrient medium. Different media compositions and growth regulator concentrations are required for each stage of the micropropagation process (*see* **Note 1**), which means that the tissues or plants need to be continuously transferred to new containers with fresh medium. The multiplication stage also requires frequent transfers as the biomass increases and fills the culture vessels. Consequently, micropropagation is a labor intensive method that greatly increases the production costs of the plants produced in vitro (*see* **Note 2**) and is only economically viable on a commercial scale in the case of high-value-added species *(1)*.

In order to simplify the whole process, reduce production costs, and make micropropagation available to a larger number of species it is necessary to develop simpler and cheaper methods that can decrease the amount of labor. A first step in this direction was the design of semi-automated bioreactors to culture the plants in liquid media *(2–4)*.

1.2. Culture in Liquid Media

Culture in liquid media can be scaled up and automated *(4)* and should, in theory, offer the best conditions for efficient, large scale micropropagation because it:

- Reduces manual labor.
- Enables the change of culture media instead of transferring the plants.
- Does not require jellifying substances that considerably increase costs.
- Nutrients are more readily available to the cultured tissues.
- Can be sterilized through microfiltration.
- Culture in liquid media, however, also has some disadvantages.
- The plants become vitrified.
- Microbial contamination is more difficult to control.
- It is more complicated to scale up.

1.3. Temporary Immersion

A method that combines the advantages of both semi-solid and liquid-culture media is the Temporary Immersion System designed by Teisson et al. *(5)*. This system alternates short periods of total immersion in liquid medium with longer ones of complete aeration. Satisfactory results for the propagation of various species have been reported using two bioreactors based on this principle *(5–8)*. The Automated Temporary Immersion Reactor (RITA) system that has been used mainly for the culturing of somatic embryos is driven by an air compressor that pushes the liquid medium from a lower container to an upper one in which the plant tissues are cultured *(5)*. The second one is comprised of two large and separate glass containers (10 L), one for the plant tissues and the other for the liquid medium which is also moved from one container to the other by pressurized air *(7)*.

Although the two systems work adequately, they both have disadvantages and are not suitable for the propagation of all species.

The RITA bioreactor is too small for the propagation of large vitroplants and the sponge that supports the plants at the base of the top chamber maintains very high humidity that induces a certain degree of vitrification in the plants that remain at the bottom. The large bioreactor used by Escalona and colleagues *(9)* is too large and is therefore difficult to sterilize. Both bioreactors require complicated systems of tubes and filters that are not easy to manipulate, which connect them to machines that drive the liquid medium from one vessel to another.

1.4. BioMINT

In order to eliminate these inconveniences, we designed and built a mid-sized, modular bioreactor (BioMINT™) (*see* **Note 3**) in which the movement of the liquid medium from one vessel to the other is passively driven by gravity and therefore does not require connection to pumps or air compressors. It has a larger volume capacity (1200 mL) as compared with the RITA (250 mL), but it is small and is light enough to be easily autoclaved and transported from the airflow cabinet to the culture room. Its structural simplicity permits easy opening and closing that expedites the introduction and removal of the cultured plant materials and the change of the liquid culture media according to the objectives and needs of the micropropagation system.

BioMINT can be used for all stages of the micropropagation process, from induction to in vitro weaning, of many species. It considerably reduces the amount of labor required for transfers and the period of time required for some micropropagation stages, it also produces larger and more vigorous plants as compared with the ones cultured in semi-solid medium in magenta boxes. Its low production costs and efficiency make it an economic alternative for low-value-added species such as agaves.

2. Materials

2.1. Equipment

The BioMINT unit consists of two cylindrical vessels (*see* **Fig. 1**) closed at one end that are joined together through their open ends being screwed into an adaptor piece (*see* **Fig. 2**). One vessel is for the plant tissues and the other for the liquid culture medium. The adaptor piece is also cylindrical with two open ends that have female screw threads into which the two vessels can be tightly screwed, it is closed in the middle by a perforated plate that permits the free flow of the liquid culture medium while keeping the plants or tissues in their place when the bioreactors change position. Rubber O-rings are placed at the joints to ensure a tight fit and to prevent the liquid medium from leaking. The adaptor piece has two connector inlets that can be closed or connected to tubes

Fig. 1. BioMINT (Modular Temporay Immersion Bioreactor).

or filters to allow for passive atmosphere exchange or forced ventilation, as well as for the injection of gases such as CO_2, ethylene, or chemical substances. The bioreactors are made of polycarbonate and are fully autoclavable and translucent.

The liquid culture medium is displaced from one vessel to the other by gravity when the bioreactors change their inclination (*see* **Fig. 3**), which is carried out by means of the SyB (see-saw) (*see* **Note 4**). The electromechanical unit consists of a set of four platforms (*see* **Figs. 3** and **4**) on which nine bioreactors are placed. The platforms are joined to a supporting structure by their middle point and their free edges alternate positions, moving up and down, driven by a piston connected to an electrical motor. The entire system is controlled automatically through a programmable control panel that regulates the timing and the speed at which the platforms change position. The system of four platforms and 36 BioMINT units is equivalent to a 45 L bioreactor.

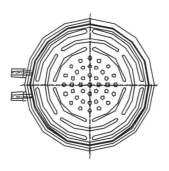

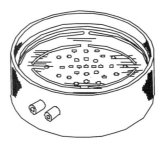

Fig. 2. Adapter piece to which the two culture vessels are assembled.

3. Methods

3.1. Sterilization

The culture medium and the assembled bioreactors (BioMINT) are autoclaved separately at 121°C for 20 min at 1 kg/cm^2 and transferred to the air flow cabinet.

3.2. Filling

In order to fill the BioMINT, the assembled bioreactor is placed vertically in the airflow cabinet and opened by unscrewing the top vessel, which is also placed vertically with its open end up and filled with the plant material and the liquid culture medium. The salt, supplementary growth regulator, and vitamin composition of the culture medium depends on the species being cultured and the micropropagation stage (i.e., induction, multiplication, rooting) (*see* **Note 1**). The proportion of culture medium to biomass also depends on the species, the type of tissue and the stage of micropropagation.

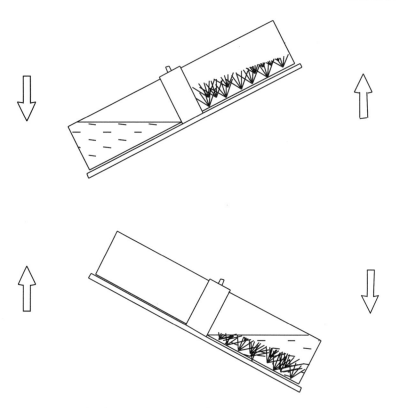

Fig. 3. The two stages of the immersion system.

The vessel containing the plants and culture medium is then closed by assembling the other vessel and the adapter on top and screwing them tightly. The BioMINT is now ready to be transferred to the growth room.

3.3. Running the SyB

The bioreactors are placed horizontally on the SyB platforms and tightened with a Velcro® strip to hold them in place. Care must be taken to ensure that the plant tissues are placed on the platform side that will remain in the "up position" during the longer aeration stage (*see* **Notes 5** and **6**). The timer must be set for the length of time of both the immersion and the aeration stages.

3.4. Incubation

The incubation period and the environmental conditions (light and temperature) will depend on the species and micropropagation stage involved.

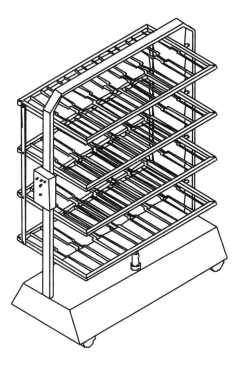

Fig. 4. Diagram showing the stand that supports the BioMINT™ system.

3.5. Changing the Culture Media

If the culture medium needs to be changed, the BioMINT can be taken back to the air-flow cabinet and opened by removing the vessel containing the plant tissues and coupling it to another one containing fresh or a different medium before returning it to the SyB.

3.6. Cutting and Grading

The cultured plants in the multiplication stage that require separation and grading can be easily taken out of the bioreactor through its wide mouth simply by draining the liquid culture medium and disassembling the vessel containing the tissues.

3.7. Control of Microbial Contamination

Microbial contamination is a more serious problem in a liquid culture than in a semi-solid culture because the bacteria or fungi from one contaminated plant will more rapidly spread to all others in the former. Microbial contamination is more easily controlled in modular bioreactors than in larger systems. This is because there is a smaller number of plants exposed to contamination, which reduces the risk, and also because the modular design allows for easy individual operation and treatment of the vessels with small amounts of antimicrobial chemicals such as plant preservative mixture PPM (*see* **Note 7**).

3.8. Addition of Chemicals During Culture

Growth regulators (or other chemicals such as antimicrobials) can be added safely, either periodically or at specific stages of the culture, by injection through one of the openings where a micro filter has been coupled.

4. Notes

1. According to Murashige *(10)*, micropropagation is divided into five stages:

   Stage 0: Selection and conditioning of explant donor plants for tissue culture.
   Stage I: Induction. Explants are placed in culture media to initiate organogenesis or embryogenesis.
   Stage II: Multiplication. Shoots multiply repeatedly increasing the biomass.
   Stage III: In vitro rooting.
   Stage IV: Transfer to soil and hardening.

2. Manual labor represents the main expense in commercial micropropagation. Although estimates vary these expenses typically account for a minimum of 60–70% of total production costs *(11)*.
3. BioMINT stands for "modular temporary immersion bioreactor" in Spanish (PA/a2004/003837).
4. SyB stands for "up" and "down" in Spanish, referring to the see-saw mechanism (PA/a2004/003837).
5. The immersion/aeration cycles can vary drastically from one species to another but, in general, only very short (5–10 min) immersion periods are sufficient for every 2–4 h of aeration.
6. To avoid vitrification it is imperative that the angle of inclination is sufficient to allow the entire liquid medium to drain to the lower vessel.
7. PPM is available from Phytotechnology Laboratories, Shawnee Mission, KS; http://phytotechlab.com.

References

1. Chu, I. (1995) Economic analysis of automated micropropagation, in automation and environmental control, in *Plant Tissue Culture* (Aitken–Christie, J., Kozai, T., Smith, M.A.L., eds.), Kluwer Academic Publischers, Dordrecht, The Netherlands, pp. 19–27.
2. Debergh, P. (1988) Improving mass propagation of in vitro plantlets, in *Horticulture in High Technology Era* (Kozai, T., ed.), International Symposium on High Technology in Protected Cultivation, Tokyo, pp. 45–57.

3. Aitken–Christie, J. (1991). Automation, in *Micropropagation: Technology and Application* (Debergh, P. C. and Zimmerman, R. J., eds.), Kluwer Academic Publishers, Dordrecht, The Netherlands, pp. 363–388.
4. Ziv, M., Ronen, G., and Raviv, M. (1998) Proliferation of meristematic clusters in disposable presterilized plastic bioreactors for the large-scale micropropagation of plants. *In Vitro Cell Dev Biol.* **34,** 152–158.
5. Teisson, C. and Alvard, D. (1995) A new concept of plant in vitro cultivation liquid medium: temporary immersion, in *Current Issues in Plant Molecular and Cellular Biology* (Terzi, M., Cella, R., and Falavigna, A., eds.) Kluwer Academic Publishers, Dordrecht, The Netherlands, pp. 105–110.
6. Etienne–Barry, D., Bertrand, B., Vásquez, N., and Etienne, H. (1999) Direct sowing of Coffea arabica somatic embryos mass-produced in a bioreactor and regeneration of plants. *Plant Cell Rep.* **19,** 111–117.
7. Escalona, M., Lorenzo, J. C., González, B., et al. (1999). Pineapple (*Ananas comosus* L. Merr) micropropagation in temporary immersion systems. *Plant Cell Rep.* **18,** 743–748.
8. Espinosa, P., Lorenzo, J. C., Iglesias, A., et al. (2002). Production of pineapple transgenic plants assisted by temporary immersion bioreactors. *Plant Cell Rep.* **21,** 136–140.
9. Escalona, M., Samson, G., Borroto, C., and Desjardins, Y. (2003). Physiology of effects of temporary immersion biorreactors on micropropagated pineapple plantlets. *In Vitro Cell. Dev. Biol. Plant.* **39,** 651–656.
10. Murashige, T. (1974). Plant propagation through tissue cultures. *Annu. Rev. Plant Physiol.* **25,** 135–166.
11. COST 843 1st International Symposium on "Liquid systems for *in vitro* mass propagation of plants." As, Norway, 2002.

12

Protocol to Achieve Photoautotrophic Coconut Plants Cultured In Vitro With Improved Performance Ex Vitro

Gabriela Fuentes, Carlos Talavera, Yves Desjardins, Jorge M. Santamaría

Summary

This chapter presents a protocol that will show ways to obtain photoautotrophic coconut in vitro plants and outlines protocol for improving photosynthesis and field performance. This protocol involves reducing sucrose concentration from the growing medium while simultaneously increasing light intensity and enriching the CO_2 concentration of growth rooms.

Key Words: Acclimatization; CO_2; field performance; photoautotrophy; light, sugars.

1. Introduction

Coconut palm (*Cocos nucifera* L.) growers require plant breeding programs to improve the trees' resistance to diseases such as lethal yellowing. This creates a necessity for the exchange of germoplasma between countries. The only accepted system for the safe exchange of germoplasma is the use of zygotic embryos cultured in vitro. Protocols are available to produce plantlets derived from zygotic embryos. At Centro de Investigación Cientifica de Yucatán's (CICY) lab, we have developed a protocol to produce in vitro plantlets derived from zygotic embryos *(1,2)*. However, these protocols show constraints, such as a very slow growth of plantlets, when transferred to the greenhouse or field as compared with their seedling counterparts (*see* **Fig. 1**). It is possible that the slow growth of in vitro plantlets is caused by their limited photosynthetic capacity. It has been suggested that the limited photosynthesis of plantlets is caused by the high-sucrose concentration normally added to the medium, the low intensity of growth rooms, and the high variability in CO_2 concentration inside the container during the photoperiod.

Fig. 1. Comparison of the growth capacity (plant height and number of leaves) of coconut plants cultured in vitro (**A**) with that of seedlings (**B**) of comparable age (ruler = 30 cm).

In other species, the major limitation for an adequate ex vitro establishment is the photosynthetic capacity of plants cultured in vitro *(3,4)*. Species such as beetroot, *Dieffenbachia*, asparagus, potato, *Spathiphyllum*, tobacco, and grapes have shown high-photosynthetic capacity when grown in vitro and they will eventually adapt to the autotrophic conditions needed when transferred ex vitro *(5–10)*. On the other hand, different species cultured in vitro—such as *Brassica*, *Calathea*, cauliflower, and strawberry—show limited photosynthetic capacity *(11–14)*. Net photosynthetic rate (Pn) of in vitro plantlets cultured is affected by factors such as light intensity, irradiance (I), radiation quality, CO_2 concentration in the vessels, and exogenous sugar concentrations.

1.1. Exogenous Sucrose

The low light intensity of conventional growth rooms (40–60 μmol photosynthetic photon flux density [PPFD]/m^2/s) is one of the reasons why sucrose has to be added to the culture medium to promote adequate growth and development of in vitro cultured plants *(15)*. Indeed, sugars play an important role in the nutrition and structure of plants *(16)*. They are also considered to act as signal molecules that affect a variety of physiological responses, regulating genes involved in photosynthesis, sink-source relationships, and defense mechanisms in plants *(17,18)*.

Sucrose is the main sugar transported in plants. Its concentration in plant tissues varies during leaf development and is affected by pathogen attack and environmental conditions. Sucrose plays three major roles in plants: (1) it is the main product of photosynthesis and represents a high proportion of the absorbed CO_2; (2) it is the main way in which carbon is transported; and (3) it is the main reserve sugar. In tissue culture, sugars are added to the culture medium as a source of carbon and energy; however, it is important to evaluate the sugar concentration, type, and stage at which it is being to ensure for the best plant growth. By analyzing the exudates from the phloem sap of the particular species it is possible to optimize exogenous sugar to add to the culture medium *(16)*. In the case of coconut, it has been reported that the liquid endosperm has a high sucrose (9.93 mg/mL) and sorbitol (15 mg/mL) content, whereas glucose (3.93 mg/mL) and fructose (4.3 mg/mL) can be found in lower concentrations. Before haustorium development, sucrose accounted for 91% of the total extracted sugars *(19)*. During the in vitro culture of coconut callus, sucrose content decreased by two-thirds after 14 d in culture, whereas the uptake of other sugars was insignificant *(20,21)*.

In general, exogenous sucrose induces changes in the carbon metabolism of tissue cultured plants. Species such as *Actinidia deliciosa*, *Chenopodium rubrum*, carnation, *Clematis*, *Cymbidium*, strawberry, gardenia, potato, *Phalaenopsis*, rose, and tobacco show higher growth and photosynthesis in absence or with only a low concentration of exogenous sucrose *(22–33)*. However, species such as strawberry, potato, rose, and tobacco also respond positively to sucrose addition by increasing their biomass accumulation, leaf area, photosynthesis, and survival time *(34–39)*. Other sugars, exogenous glucose in particular, have had a positive effect on Rubisco activity and chlorophyll content in spinach *(40)*.

Plantlets suffer sudden and extreme shifts in their physiology when transferred from in vitro to ex vitro conditions. Plantlets are forced to modify from either a heterotrophic or mixotrophic metabolism, where sucrose is provided in the culture medium within the in vitro environment, to an autotrophic metabolism ex vitro where they have to produce their own sugars via photosynthesis to achieve normal growth. In some species the negative photosynthesis in vitro take two weeks or more after they have been transferred to ex vitro, to achieve positive photosynthesis *(11)*.

1.2. Light Intensity

The potential productivity of a plant is a function of the intercepted light *(41)*. Similarly, the photosynthetic rate per unit leaf area in C_3 plants is a function of the PPFD, being negative in the absence of light resulting from the CO_2 lost in respiration. Photosynthetic rates increases with light to a certain limit as long as CO_2 fixation is kept coupled to electron transport associated to the capture of photons. Light also plays an important role in plant development including chloroplast differentiation, leaf expansion, flowering, and plant morphology *(42)*.

In vitro cultured plantlets grow with low light intensity of 40–60 μmol PPFD/m^2/s in the standard growth rooms and under an artificial light spectrum, as compared with natural sunlight intensity of 1500–2000 μmol PPFD/m^2/s. It has been shown that plants grown with low light have less carotenoids, show irregular thylacoids arrangement, have reduced electron transport rates, and have a low carboxylation efficiency (low Rubisco quantity and activity). This therefore leads to low photosynthetic rates, low sugar content, and high starch content. In vitro plants are thus not adapted to high-light conditions found in the natural environment and may incur damages from photoinhibition or even photobleaching. Increasing the light intensity of growth rooms does not always have positive effects. For example in *Daphne* plantlets, increasing light from 40 to 80 μmol PPFD/m^2/s caused an increase in photosynthesis and growth *(42)*. However, in *Hosta* no difference in photosynthesis occurred when grown with high- or low-light intensities; however, *Clematis* photosynthesis and underground growth was inhibited with high-light intensities *(15)*.

1.3. CO_2

The plant growth cultured in vitro normally occurs in closed containers that allows very little gas exchange. Therefore the environment container is normally low in O_2, rich in ethylene, and contains high variations in CO_2 concentrations that correspondence to the photoperiod and, thus, to photosynthesis *(43)*.

The low-photosynthetic rate of in vitro cultured plantlets has been limited to low CO_2 concentration inside the container. The CO_2 concentration during the first hours of the photoperiod can decrease to one-third of that in the outside air (100 µmol/mol) *(26)*. In some species, CO_2 enrichment may promote photosynthesis and growth and is part of the acclimatization process to the final transfer ex vitro *(43)*.

One way to favor the gas exchange inside the culture container is the fitting of vents in their lids *(44)*. This vent allows for the diffusion of ethylene outside the container, thus allowing better diffusion of O_2 and CO_2 to and from the container atmosphere. This technique in itself has proven to be beneficial to the development of an improved capacity of plantlets to control excessive water loss commonly present in plantlets, as their stomata are nonfunctional *(45)*. In addition, the fact that CO_2 can diffuse inside the container through this vent can be used for CO_2 enrichment, allowing for the improvement of photosynthesis and growth of in vitro cultured plantlets

1.4. Photosynthetic Capacity of Coconut Plantlets

It has been reported that plantlets derived from the in vitro culture of coconut embryos show low photosynthetic rates *(46)*. Those plants also show low photosynthesis associated with low chlorophyll content and low Rubisco activity *(47–49)*. In some species, it has been shown that Calvin cycle enzymes as well as Rubisco are inhibited by high concentration of sugars *(17)*.

The in vitro culture of coconut zygotic embryos needs the addition of sugars for growth and energy input *(50)*. However, the high sugar concentration used in the growth medium negatively affects the photosynthetic capacity of coconut plantlets *(48)* and other species *(38)*. Fuentes et al. *(51,52)* also showed that reducing the concentration of sucrose in the medium resulted in an improvement of the photosynthetic capacity of coconut plantlets but that those plantlets showed limited growth because they had no source of carbon and photosynthesis as a result of the light intensity of standard growth-rooms was insufficient to maintain growth.

The principal storage products of CO_2 fixation are sucrose and starch *(53)*. The content of sucrose is estimated to be 10–20 times higher than other nonstructural carbohydrates *(54)*. However, it has been shown that exogenous sucrose inhibits photosynthesis of in vitro plantlets *(38)*.

High biomass and carbohydrate reserves obtained during the in vitro stage, are likely to be critical factors for a subsequent successful acclimatization, allowing plantlets to rapidly produce new organs (leaves and roots) ex vitro *(55)*. In vitro leaves act as nutrient storage structures during acclimatization and sustain the growth of new adapted leaves ex vitro. Hence, any attempt to increase the reserves of in vitro plantlets should improve acclimatization performance. Increased light was found to be beneficial for plant growth in the later stages of micropropagation *(56)*. Plantlets under low light at normal growth-rooms limited their Pn in vitro; under high light, however, the carbon availability becomes the limiting factor for photosynthesis during several hours in the light period. The low light of standard growth-rooms and the high exogenous sugar concentrations in the medium may be responsible for the low Pn in some plantlets.

Continuous efforts have been made to improve protocols for the in vitro culture of zygotic embryos of coconut *(1,2,57)*. The novel protocol shown here (*see* **Fig. 2**), through use of low sucrose, high light intensity, and CO_2 enrichment, generates plantlets displaying higher photosynthetic response to light and CO_2, higher plant biomass (both in terms of dry weight and plant height), as well as a higher number of leaves, and higher survival rates when transferred to the nursery *(51,52)*.

2. Materials

1. Commercial bleach, ethanol, distilled and sterile water.
2. Seeds and nuts (*see* **Note 1**).
3. Zygotic embryos (*see* **Note 2**).
4. Flow cabinet, forceps, and scalpel.
5. Glass containers containing growth medium (*see* **Note 3**).
6. Transparent plastic, isotactic polypropilene (IPP), extensions with Whatman no. 1 vent.
7. Growth room with standard light and controlled temperature and photoperiod (*see* **Note 4**).
8. Growth room fitted with high light intensities (*see* **Note 5**) and CO_2 enrichment facilities.
9. Pots (black bags) and transparent plastic covers.
10. Fogging glasshouse and shade space at the nursery.

3. Methods

3.1. Embryo In Vitro Germination

1. Cut the nut in half and remove the endosperm section containing the zygotic embryo with a corkscrew. Soak the endosperm disks in 70% ethanol for 1 min. Remove the ethanol and rinse thoroughly with distilled water. Sterilize by soaking the endosperm in 10% commercial bleach (6% [v/v] sodium hypo-chlorite) for 20 min with agitation. Remove the bleach solution, and rinse the endosperm with zygotic embryos with plenty of sterile distilled water.
2. Extract the zygotic embryos using a forceps and scalpel. Soak the embryos in ethanol for 1 min and rinse thoroughly with distilled water. Sterilize the embryos by soaking them in 10% commercial bleach (6% [v/v] sodium hypo-chlorite) for 15 min with agitation. Remove the bleach solution, and rinse thoroughly with sterile water.

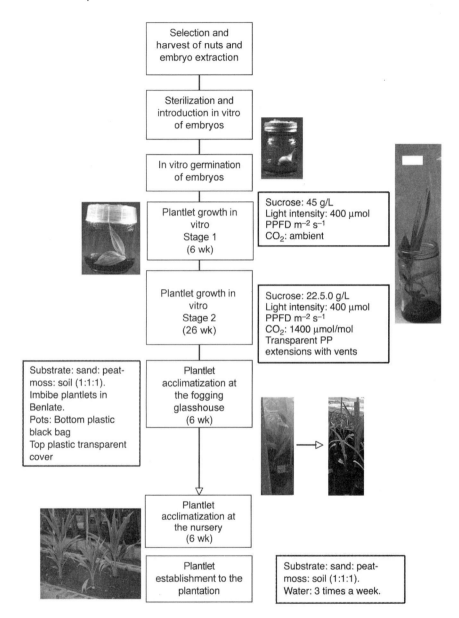

Fig. 2. Protocol to promote the photosynthetic capacity of in vitro cultured plants. For details of each stage *see* **Subheadings 3.2.–3.4.** Emphasis is made on the concentration of exogenous sugar, light intensity and CO_2 enrichment of growth rooms required to promote photoautotrophy, and enhanced ex vitro performance. The requirement for high light can be met even better with natural light by using modified glasshouse facilities as growth rooms *(60)*.

3. Gently place the embryos onto the medium containing 45 g/L of sucrose, making sure that they are in good contact with the Y3 solid medium. Seal glass containers with strips of parafilm.
4. Leave the glass containers containing the embryos in the dark at 27 ± 1°C for 8 wk.

3.2. Establishing Autotrophic Plantlets (see Fig. 2)

1. Eight weeks after last subculturing, when the embryos have been germinated, carefully transfer the germinated embryos to another glass container. Place one embryo in a glass container with 20 mL of fresh liquid medium for 6 wk. The medium should contain 45 g/L of sucrose.
2. Leave the containers in a 16/8 (light/dark) photoperiod at 27 ± 1°C for 6 wk.
3. Remove the container when one bifid leaf forms and carefully place one plantlet in a 500-mL glass flask containing 50 mL of fresh liquid medium. The medium should contain 22.5 g/L of sucrose. Cover the container with transparent plastic IPP extension with a vent. Subculture onto fresh medium every 8 mo for 32 wk until three leaves and two roots have formed.
4. Leave the containers in a 16/8 (light/dark) photoperiod at 27 ± 1°C.

3.3. Growth-Room Conditions (see Fig. 2)

1. Subculture a growth room adapted to fit lamps that allow for an increased light intensity (10 times that of standard lamps) without increasing the temperature. Halogen lamps (Osram, 50 W, 12 V, 36JC) can be used to keep light intensity at 400 µmol/m^2/s. Standard growth rooms provide from 40 to 60 µmol/m^2/s.
2. The extra heat should be avoided by passing bellow the light source a film of running water to dissipate the heat.
3. The CO_2 concentration of the growth room should be kept at 1400 µmol/mol. CO_2 enrichment can be controlled through use of a computerized system program using a PP Systems unit (EGM-1, Environmental Gas Monitor for CO_2, UK) as a CO_2 monitor (*see* **Note 6**).

3.4. Acclimatization Ex Vitro (see Fig. 2)

1. Thirty-two weeks after subculturing in vitro, when three leaves and two roots have formed, carefully move the plantlets to fogging conditions (*see* **Note 7**). Clean the plantlets by rinsing them thoroughly with sterile water.
2. Sterilize the roots by soaking them in 1 g/L, benlate for 5 min. Remove the fungicide solution and rinse the roots thoroughly with sterilized water.
3. Put plantlets in 0.5 kg polyethylene black bags with a sand:peat moss:soil (1:1:1 proportions) sterilized substrate. Carefully put a clear polyethylene bag over the plantlets and seal with no. 18 rubber bands.
4. Rinse the plantlets with 250 mL of distilled water every 2 d. Subculture pots every 2 wk.
5. Put plantlets in 1 kg clear polyethylene bags and carefully make two perforations over the transparent top bag once every week. Subculture for 6 wk for a total of 12 perforations.

6. Remove the polyethylene bag and subculture for 1 wk.
7. Remove plantlets from 1 kg black polyethylene bag and place them in 2 kg clear polyethylene bags with a sand:peat moss:soil (1:1:1 proportions) sterilized substrate. Place the pots in the nursery and maintain 65% shade with polyethylene black mesh (see **Note 8**).
8. Rinse thoroughly with water every 2 d. The plantlets can be cleaned with 1 g/L Benlate and 0.75 g/L Avamectine every month. Subculture for 8 wk.
9. Remove plantlets from the nursery and carefully place them in field conditions with a sand:peat moss:soil (1:1:1 proportions) sterilized substrate. Thoroughly rinse every 2 d. Subculture for 8 wk (see **Note 9**).

4. Notes

1. Coconut seeds should be selected from healthy, 10- to 15-yr-old coconut palms (Malayan Green Dwarf) at the stage when the liquid endosperm is still noticeable upon shacking.
2. Mature embryos are generally taken from nuts 12 mo after flowering.
3. The growth medium used is Y3 *(58,59)*. Embryos are germinated in solid medium *(2)*.
4. The light of a standard growth-room is provided by cool, white, fluorescent lamps (Phillips 39 W).
5. The requirement for high light intensity can be better met with natural light by using modified glasshouse facilities as growth rooms.
6. The CO_2 concentration is provided with a CO_2 tank and compressor. A solenoid valve controlled by a computer maintains them.
7. The fogging conditions are 0–120 µmol $PPFD/m^2/s$, temperatures of 26 ± 2°C, and 80–95% relative humidity.
8. The shade conditions are 0–600 µmol $PPFD/m^2/s$, temperatures of 18–38°C, and 50–95% relative humidity.
9. The field conditions are full-sun, 0–2000 µmol $PPFD/m^2/s$, temperatures of 20–40°C, and 40–95% relative humidity.

References

1. Talavera, C., Oropeza, C., Cahue, A., Coello, J., and Santamaría J. (1998) Status of research on coconut zygotic embryo culture and acclimatization techniques in Mexico, in *Proceedings of the First Workshop on Embryo Culture* 27-31 October 1997 (Batugal, P.A. and Engelmann, F., eds.), Banao, Guinobatan, Albay, Philippines, pp. 43–54.
2. Pech, A., Santamaría, J., Souza, R., Talavera, C., Maust, B., and Oropeza C. (2002) Changes in culture conditions and medium formulation to improve efficiency of *in vitro* culture of coconut embryos in *Mexico, in Coconut Embryo In vitro Culture Part II* (Engelmann, F., Batugal, P., and Oliver, J., eds.), IPGRI-APO, Serdang, Malaysia, pp. 122–137.
3. Le, V. Q., Samson, G., and Desjardins, Y. (2001) Opposite effects of exogenous sucrose on growth, photosynthesis and carbon metabolism of in vitro plantlets of

tomato (*L. esculentum* Mill.) grown under two levels of irradiances and CO_2 concentration. *J. Plant Physiol.* **158**, 599–605.
4. Arigita, L., González, A., and Sánchez, R. (2002) Influence of CO_2 and sucrose on photosynthesis and transpiration of Actinidia deliciosa explants cultured *in vitro*. *Physiol. Plant.* **115**, 166–173.
5. Cournac, L., Dimon, B., Carrier, P., Lohou, A., and Chagvadieff, P. (1991) Growth and photosynthetic characteristics of *Solanum tuberosum* plantlets cultivated *in vitro* in different conditions of aeration, sucrose supply and CO_2 enrichment. *Plant Physiol.* **97**, 112–117.
6. Galzy, R. and Compan, D. (1992) Remarks on mixotrophic and autotrophic carbon nutrition of *Vitis* plantlets cultured *in vitro*. *Plant Cell Tiss. Org. Cult.* **31**, 239–244.
7. Yue, D., Desjardins, Y., Lamarre, M., and Gosselin, A. (1992) Photosynthesis and transpiration of *in vitro* cultured asparagus plantlets. *Sci. Hort.* **49**, 9–16.
8. Paul, M. and Stitt, M. (1993) Effects of nitrogen and phosphate deficiencies on levels of carbohydrates, respiratory enzymes and metabolites in seedlings of tobacco, and their response of exogenous sucrose. *Plant Cell Environ.* **16**, 1047–1057.
9. Kovtun, Y. and Daie, J. (1995) End-product control of carbon metabolism in culture culture-grown sugar beet plants. *Plant Physiol.* **108**, 1647–1656.
10. Van Huylenbroeck, J. M., Piqueras, A., and Debergh, P. C. (1998) Photosynthesis and carbon metabolism in leaves formed prior and during *ex vitro* acclimatization of micropropagated plants. *Plant Sci.* **134**, 21–30.
11. Grout, B. W. W. and Aston, M. (1978) Transplanting of cauliflower plants regenerated from meristem culture. II. Carbon dioxide fixation and the development of photosynthetic ability. *Hortic. Research* **78**, 65–71.
12. Wardle, K., Quinlan, A., and Simpkins I. (1979) ABA and regulation of water loss in plantlets of *Brassica oleracia* L. var. Botrytis regenerated through apical meristem culture. *Ann. Bot.* **43**, 745–752.
13. Grout, B. W. W. and Millam, S. (1985) Photosynthetic development of micropropagated strawberry plantlets following transplanting. *Ann. Bot.* **55**, 129–131.
14. Grout, B. W. W. and Donkin, M. E. (1987) Photosynthetic activity of cauliflower meristem cultures in vitro and at transplanting time. *Acta Hortic.* **212**, 323–327.
15. Lees, R. P. (1994) Effects of the light environment on photosynthesis and growth *in vitro*, in *Physiology, Growth and Development of Plants in Culture* (Lumsden, P.J., Nicholas, J. R., and Davies, W. J., eds.), Kluwer Academic Publishers, Dordrecht, pp. 31–46.
16. Welander, M. and Pawlicki, N. (1994) Carbon compounds and their influence on in vitro growth and organogenesis, in *Physiology, Growth and Development of Plants in Culture* (Lumsden, P. J., Nicholas, J. R., and Davies, W. J., eds.), Kluwer Academic Publishers, Dordrecht, pp. 83–93.
17. Koch, K. E. (1996) Carbohydrate-modulated gene expression in plants. *Ann. Rev. Plant Physiol.* **47**, 509–540.

18. Sheen, J., Zhou, L., and Jang, J. C. (1999) Sugars as signaling molecules. *Curr. Opi. Plant Biol.* **2,** 410–418.
19. Sugimura, Y. and Murakami, T. (1990) Structure and function of the haustorium in germinating coconut palm seed. *Jap. Agric. Research Quart.* **24,** 1–14.
20. Dussert, S., Verdeil, J. L., and Buffard-Morel, J. (1995) Specific nutrient uptake during initiation of somatic embryogenesis in coconut calluses. *Plant Sci.* **111,** 229–236.
21. Dussert, S., Verdeil, J. L., Rival, A., Noirot, M., and Buffard-Morel, J. (1995) Nutrient uptake and growth of in vitro coconut (*Cocos nucifera* L.) calluses. *Plant Sci.* **106,** 185–193.
22. Mousseau, M. (1986) CO_2 enrichment *in vitro*: effect on autotrophic and heterotrophic cultures of *Nicotiana tabacum* (var. Samsun). *Photosynth. Res.* **8,** 187–191.
23. Kozai, T., Oki, H., and Fujiwara, K. (1987) Effects of CO_2 enrichment and sucrose concentration under high photosynthetic photon fluxes on growth of tissue cultured Cymbidium plantlets during the preparation stage, in *Symposium on Plant Micropropagation in Horticultural Industries* (Ducat, G., Jacob, M., and Simeon, A., eds.), BPTC Group, Arlon Belgium, pp. 135–141.
24. Langford, P. J. and Wainwrigth, H. (1987) Effects of sucrose concentration on the photosynthetic ability of rose shoots *in vitro*. *Ann. Bot.* **60,** 633–640.
25. Fujiwara, K., Kozai, T., and Watanabe, I. (1988) Development of a photoautotrophic tissue culture system for shoots and/or plantlets at rooting and acclimatization stages. *Acta Hort.* **230,** 153–158.
26. Kozai, T., Koyama, Y., and Watanabe, I. (1988) Multiplication of potato plantlets in vitro with sugar free medium under high photosynthetic photon flux. *Acta Hortic.* **230,** 121–126.
27. Kozai, T. and Iwanami, Y. (1988) Effects of CO_2 enrichment and sucrose concentration under high photon fluxes on plantlet growth of carnation (*Dianthus caryophyllus* L.) in tissue culture during preparation stage. *J. Jap. Soc. Hortic. Sci.* **57,** 279–288.
28. Kozai, T. and Kubota, C. (1988) The growth of carnation plantlets *in vitro* cultured photoauto- and photomixotrophically on different media. *Environ. Cont. Biol.* **28,** 21–27.
29. Kozai, T. and Sekimoto, K. (1988) Effects of the number of air changes per hour of the closed vessels and the photosynthetic photon flux on the carbon dioxide concentration inside the vessel and the growth of strawberry plantlets *in vitro*. *Environ. Cont. Biol.* **26,** 21–29.
30. Doi, M., Oda, H., and Asahira, T. (1989) *In vitro* atmosphere of cultured C_3 and CAM plants in relation to day-length. *Environ. Cont. Biol.* **27,** 9–13.
31. Mezzeti, B., Lafranco, S., and Pasquel, R. (1991) Actinidia deliciosa in vitro II. Growth and exogenous carbohydrates utilization by explants. *Plant Cell Tiss. Org. Cult.* **26,** 153–160.
32. Schäfer, C., Simper, H., and Hofmann, B. (1992) Glucose feeding results in coordinated changes of chlorophyll content, ribulose-1,5-bisphosphate carboxylase-

oxygenase activity and photosynthetic potential in photoautotrophic suspension cultured cells of *Chenopodium rubrum*. *Plant Cell Environ.* **15**, 343–350.

33. Serret, M.D., Trillas, M.I., Matas, J., and Araus, J.L. (1997) The effect of different closures types, light, and sucrose concentrations on carbon isotope composition and growth of *Gardenia jasminoides* plantlets during micropropagation and subsequent acclimation *ex vitro*. *Plant Cell Tiss. Org. Cult.* **47**, 217–230.
34. Grout, B. W. W. and Price, F. (1987) The establishment of photosynthetic independence in strawberry cultures prior to transplanting, in *Plant Micropropagation in Horticultural Industries*, Symposium Florizel, Arlon. Belgium, pp. 56–60.
35. Solarova, J., Posposilova, J., Catski, J., and Santrucek J. (1989). Photosynthesis and growth of tabacco plantlets in dependence on carbon supply. *Photosynthetica* **23**, 629–637.
36. Capellades, M., Fontarnau, R., Carulla, C., and Debergh, P. (1990) Environment influences anatomy of stomata and epidermal cells in tissue-cultured *Rosa multiflora*. *J. Am. Soc. Hort. Sci.* **115**, 141–145.
37. Fujiwara, K., Kira, S., and Kozai, T. (1992) Time course of CO_2 exchange of potato cultures in vitro with different sucrose concentrations in the culture medium. *J. Agr. Met.* **48**, 49–56.
38. Hdider, C. and Desjardins, Y. (1994) Effects of sucrose on photosynthesis and phosphoenolpyruvate carboxylase activity of in vitro cultured strawberry plantlets. *Plant Cell Tiss. Org. Cult.* **36**, 27–33.
39. Tichá, I., Cap, F., Pacovska, D., et al. (1998) Culture on sugar medium enhances photosynthetic capacity and high light resistance grown *in vitro*. *Physiol. Plant.* **102**, 155–162.
40. Krapp, A., Quick, W. P., and Stitt, M. (1991) Ribulose-1,5-bisphosphate carboxylase-oxygenase, other Calvin cycle enzymes and chlorophyll decrease when glucose is supplied to mature spinach leaves via the transpiration stream. *Planta* **186**, 58–69.
41. Ohler, J. G. (1999) Historical background, in *Modern Coconut Management Palm Cultivation and Products* (Ohler, J.G., ed.), Intermediate Technology Publications The food and Agriculture Organization of the United Nations Universiteit Leide, The Netherlands, pp. 3–4
42. Vince-Prue D. (1994) Photomorphogenesis and plant development, in *Physiology, Growth and Development of Plants in Culture* (Lumsden, P. J., Nicholas, J. R., and Davies, W. J., eds.), Kluwer Academic Publishers, Dordrecht, pp. 19–30.
43. Buddendorf-Joosten, J. M. C. and Woltering E. J. (1994) Components of the gaseous environmental and their effects on plant growth and development *in vitro*, in *Physiology, Growth and Development of Plants in Culture* (Lumsden P. J., Nicholas J. R., and Davies, W. J., eds.), Kluwer Academic Publishers, Dordrecht, pp. 165–190.
44. Talavera, C., Espadas, F., Aguilar, M., Maust, B., Oropeza, C., and Santamaría, J. (2001) The control of leaf water loss by coconut plants cultured in vitro depends on type of membranes used for ventilation. *J. Hort. Sci. Biotech.* **76**, 569–574.

45. Santamaría, J., Davies, W. J., and Atkinson, C. J. (1993) Stomata of micropropagated Delphinium plants respond to ABA, CO2, light and water potential, but fail to close fully. *J. Exp. Bot.* **44,** 99–107.
46. Karun, A., Sajini, K., and Parthasarathy, V. (2002) Increasing the efficiency of embryo culture to promote germplasm collecting in India, in *Coconut Embryo In vitro Culture Part II*, (Engelmann, F., Batugal, P., and Oliver, J., eds.), IPGRI-APO, Serdang, Malaysia, pp. 7–26.
47. Triques, K., Rival, A., Beule, T., et al. (1997) Developmental changes in carboxylases activities *in vitro* cultured coconut zygotic embryos. Comparison with corresponding activities in seedlings. *Plant Cell Tiss. Org. Cult.* **49,** 227–231.
48. Del Rosario, A. G. (1998) Status of research on coconut embryo culture and acclimatization techniques in UPLB, in *Coconut Embryo In Vitro Culture* (Batugal, P. A., and Engelmann, F., eds.), IPGR-APO, Serdang, Banao, Guinobatan, Albay, Philippines pp. 12–16.
49. Santamaría, J. M., Talavera, C., Lavergne, D., et al. (1999) Effect of medium sucrose on the photosynthetic capacity of coconut vitroplants formed from zygotic embryos, in *Current advances in Coconut Biotechnology* (Oropeza, C., Verdeil, J. L., Ashburner, G. R., Cardeña, R., and Santamaría, J. M., eds.), Kluwer Academic Publishers, Dordrecht, pp. 371–381.
50. Ashburner, G. R., Thompson, W. K., and Burch, J. M. (1993) Effect of a-naphthalene acetic acid and sucrose levels on the development of cultured embryos of coconut. *Plant Cell Tiss. Org. Cult.* **15,** 157–163.
51. Fuentes, G., Talavera, C., Oropeza, C., Desjardins, Y., and Santamaría, J. M. (2004) Exogenous sucrose can decrease *in vitro* photosynthesis but improve field survival and growth of coconut (*Cocos nucifera* L.) *in vitro* plantlets. *In Vitro Cell Dev. Biol.–Plant* **41,** 69–76.
52. Fuentes, G., Talavera, C., Desjardins, Y., and Santamaría, J. M. (2004) High irradiance can minimize the negative effect of exogenous sucrose on the photosynthetic capacity of in vitro grown plantlets. *Biol. Plant* **49,** 7–15.
53. Goldschmidt, E. and Huber, S. (1992) Regulation on photosynthesis by end-product accumulation in leaves of plants storing starch, sucrose, and hexose sugars. *Plant Physiol.* **99,** 1443–1448.
54. Farrar, J. and Gunn, S. (1996) Effects of temperature and atmospheric carbon dioxide on source-sink relations in the context of climate change, in *Photoassimilate Distribution and Crops: Source-Sink Relationships* (Zamski, E. and Schaffer, A., ed.), Marcel Dekker, New York, pp. 389–406.
55. Fila, G., Ghashghaie, J., Hoarau, J., and Cornic, G. (1998) Photosynthesis, leaf conductance and water relations of *in vitro* cultured grapevine rootstock in relation to acclimatisation. *Physiol. Plant.* **102,** 411–418.
56. Donnelly, D., Vidaver, W., and Lee, K. (1985) The anatomy of tissue cultured red raspberry prior to and after transfer to soil. *Plant Cell Tiss. Org. Cult.* **4,** 43–50.
57. Chan, L. J., Sáenz, L., Talavera, C., Hornung, R., Robert, M., and Oropeza, C. (1998) Regeneration of coconut (*Cocos nucifera* L.) from plumule explants through somatic embryogenesis. *Plant Cell Rep.* **17,** 515–521.

58. Eeuwens, C. J. (1976) Mineral requirements for growth and callus initiation of tissue explants excised from mature coconut palms (*Cocos nucifera* L.) and cultured *in vitro*. *Physiol. Plant.* **36,** 23–28.
59. Rillo, E.P. and Paloma, M.B. (1992) *In vitro* culture of Macapuno coconut embryos. *Coconuts Today* **9,** 90–101.
60. Talavera, C., Espadas, F., Fuentes, G., and Santamaría, J. M. (2005) Cultivating in vitro coconut palms (Cocos nucifera L.) under glasshouse conditions with natural light, improves in vitro photosynthesis and nursery survival and growth. *Plant Cell Tiss. Org. Cult.* (In Press).

13

Use of Statistics in Plant Biotechnology

Michael E. Compton

Summary
Statistics and experimental design are important tools for the plant biotechnologist and should be used when planning and conducting experiments as well as during the analysis and interpretation of results. This chapter provides some basic concepts important to the statistical analysis of data obtained from plant tissue culture or biotechnological experiments, and illustrates the application of common statistical procedures to analyze binomial, count, and continuous data for experiments with different treatment factors as well as identifying trends of dosage treatment factors.

Key Words: Analysis of variance; binomial data; continuous data; concentration treatment factors; count data; data analysis; logistic regression; mean separation tests; plant tissue culture; Poisson regression; regression analysis.

1. Introduction

Most research projects are initiated to solve a problem. In response, we plan and design experiments in the hope that some treatment will be found that solves the problem. Researchers use a logical and stepwise approach for problem solving. We examine the published literature to gain knowledge from prior experiences and apply what we have learned to formulate a hypothesis. Experimental objectives are developed that can be tested objectively and data is gathered from observations made during experimentation. Finally, data are analyzed and evaluated to ascertain the effectiveness of each treatment at correcting the problem.

We can easily make measurements of plant cells and tissues, and mathematically calculate and rank the averages of the tested treatment variables. However, simply calculating averages does not consider the variation present among specimens receiving the treatments and does not accurately examine treatment effectiveness. For this reason statistics should be employed when evaluating data.

From: *Methods in Molecular Biology, vol. 318: Plant Cell Culture Protocols, Second Edition*
Edited by: V. M. Loyola-Vargas and F. Vázquez-Flota © Humana Press Inc., Totowa, NJ

It is important that we avoid personal bias during all phases of experimentation. Personal bias, either occurring intentionally or by accident, will produce unreliable results that will inhibit the researcher's attempts to solve problems. Various experimental designs such as the completely randomized (CR), randomized complete block (RCB), incomplete block (IB), and split plot (SP) have been developed to help researchers reduce personal bias during experimentation and data collection *(1–3)*. For more information on how these designs are used in plant biotechnology the readers should consult prior publications *(4–7)*.

Use of the most appropriate and efficient statistical procedures can aid researchers in planning, conducting, and interpreting the results of experiments. This chapter is intended to serve as an introduction to the use of statistics in plant biotechnological research. After reading this chapter the reader should be able to select the appropriate statistical methods to analyze, interpret, and present plant biotechnology data.

2. Materials

2.1. Computer Software Used to Perform Statistical Analyses

Few plant biotechnologists calculate statistical analyses by hand or use a calculator. Instead, most scientists chose to use a statistical software package that is compatible with their personal computer (PC). Several Windows©-compatible software packages are available for statistical analysis procedures using a PC and include those available from SAS *(8)*, SPSS *(9)*, SYSTAT *(10)*, MINITAB *(11)*, and STATISTX *(12)*. All software programs can be used to efficiently perform data analysis procedures used for plant cell and tissue culture and plant molecular biology data.

To illustrate the use of statistical procedures commonly used in plant biotechnology two hypothetical experiments will be used. The first evaluates the effectiveness of six cytokinins (zeatin, zeatin riboside, N6-[Δ] 2-Isopentenyl adenine [2iP], benzyladenine [BA], kinetin, or thidiazuron [TDZ]) on stimulating adventitious shoots from petunia leaf explants while the second was designed to identify the optimum BA dose required for shoot organogenesis in petunia. The hypothetical data collected included the number and percentage of explants that produce shoots, the number of shoots per responding explant, and the length of adventitious shoots.

2.2. Plant Materials

1. Plants of *Petunia x hybrida* 'Blue Picotee' were grown from seed in the greenhouse for 12 wk before excising leaves for explant preparation. Only the youngest, fully expanded leaves were harvested.

2. Plants were grown at 20°C and natural light intensities and photoperiod at Platteville, WI from September to March.
3. The freshest quiescent seeds should be obtained. Seeds can be stored at 4°C for about 5 yr and remain regenerable.

2.3. Shoot Regeneration Medium

1. Petunia shoot regeneration medium: Murashige and Skoog (MS) salts and vitamins with (per 1 L) 0.1 g myo-inositol, 30 g sucrose, and 5 g Agar-Gel (Phytotechnology Laboratories, LLC, Overland Park, KS) at pH 5.8 *(13)*.
2. The plant growth regulators tested included zeatin, zeatin riboside, 2iP, BA, kinetin, or TDZ. All growth regulators were obtained from Phytotechnology Laboratories, LLC, Overland Park,KS and supplied at 0, 2, 4, 6, 8, or 10 M concentrations.

3. Methods
3.1. Explant Preparation and Culture Conditions

The methods used for explant preparation were outlined previously by Preece *(13)*:

1. The youngest, fully expanded leaves were removed from plants actively growing in the greenhouse. Leaves were washed with soapy water and immersed in a 10% bleach solution (0.6% NaOCl with 1 mL/L antibacterial soap) for 15 min before three rinses with sterile reverse osmosis water.
2. To prepare the explants, the margins were trimmed from disinfested leaves and the lamina cut into 5 × 5 mm explants, each containing a portion of the mid rib.
3. Explants were placed abaxial side down in test tubes containing 20 mL of shoot induction medium supplemented with test concentrations of the selected cytokinins. One leaf explant was cultured per vessel.
4. Test tubes containing explants were incubated in a growth chamber providing a 16 h photoperiod (30–50 mol/m^2/s) at 25°C. Test tubes containing explants were selected at random and placed in racks that held 32 vessels. All racks were placed on the same shelf in the growth room and arranged in a completely randomized design (*see* **Note 1**).
5. Explants were transferred to vessels containing 20 mL of fresh shoot induction medium every 4 wk.

3.2. Statistical Analysis Procedures

Statistix for Windows was used to analyze the experimental data. Like other Windows-based versions of statistical software, Statistix for Windows uses drop down dialog boxes to guide the user through data analysis *(12)*.

The following paragraphs will be divided into sections demonstrating the use of statistical analysis software to analyse binomial, count, and continuous data for experiments with different treatment factors as well as identifying trends of dosage (concentration) treatment factors.

Explants were placed in the treatment media previously described. Data for all experiments were recorded at 12 wk and included the number and percentage of explants with shoots, the number of shoots per responding explant, and the length of each shoot recorded. Data analysis procedures were performed based on a completely randomized design with one replicate per vessel, making each explant a replicate (*see* **Note 2**). Vessels were randomly placed into racks positioned on the same shelf in the growth chamber. Therefore, the model statements used for each analysis were constructed to reflect a completely randomized design.

3.2.1. Analysis of Binomial Data

Logistic regression is the statistical procedure suggested for analysing response data *(14)*. This is because, unlike analysis of variance (ANOVA), logistic regression does not produce separate estimates of experimental error. Because of this feature, logistic regression produces more accurate results than ANOVA when analyzing response data. The logistic regression procedure can be found in Statistix for Windows in the linear models selection of the statistics drop-down menu. The following steps should be followed when using this software to conduct logistic regression analysis.

1. While working in the appropriate data file, select the **Statistics** drop down menu in the tool bar and click on **Linear Models**. This action creates a new box listing several linear models-based procedures, one of which is logistic regression.
2. Clicking on **Logistic Regression** will produce a new window containing boxes in which the dependent and independent variables must be entered. In this example the dependent variable is **Response** and the independent variable is **Cytokinin**. Click **OK** after entering names of the requested variables.
3. Clicking the **OK** button initiates the analysis. Test results will appear in a new window that can be printed or saved.

When reading the analysis output, significance of the independent variable, cytokinin type in this case, is indicated by the p-value (*see* **Note 3**).

3.2.2. Analysis of Count Data

Data such as the number of shoots produced per regenerating explant or number of somatic embryos per explant are considered counts *(14)*. Count data are not normally distributed because the variance of each treatment is equal to the average response of the treatment *(15)*. Because of their distribution, Poisson regression (also known as discrete regression) is suggested for analyzing count data *(14)*.

The advantage of Poisson regression is that the procedure uses a logarithmic value of the mean counts, which normalizes the data during analysis *(14)*. The following steps should be followed when using Statistix to conduct Poisson regression analysis.

1. While working in the appropriate data file, select the **Statistics** drop down menu and click on **Linear Models**. This will create a new box listing several linear models-based procedures, one of which is **Poisson regression**.
2. Clicking on **Poisson regression** will produce a new window in which the dependent and independent variables must be entered. In this example the dependent variable is the number of shoots (**NoShoots**) and the independent variable is **Cytokinin**. Click on the **OK** button after entering the information.
3. Clicking the **OK** button initiates the analysis. Test results will appear in a new window that can be printed or saved.
4. When reading the analysis output, significance of the independent variable, cytokinin type in this case, is indicated by the *p*-value (*see* **Note 4**).

3.2.3. Analyzing Continuous Data

The shoot length variable is considered continuous because the value of data observations is unrestricted. Because of this fact, continuous data tend to be normally distributed with treatments having similar variances *(14)*. ANOVA is well suited for analyzing continuous data with a normal distribution *(4,6,14)*.

During ANOVA, a model statement is created that identifies the treatment variables and the observations recorded (dependent variables) as well as treatment interactions (independent variable), if present. The model statement must be written based on the experimental design used *(4)*. The procedure generates a random error value by subtracting the value of each datum from the overall mean *(14)*. The influence of independent variables on the dependent variables is tested according to the model statement, generating a summary table that provides the results of the model tested *(2)*. Information in the summary table includes the degrees of freedom (DF), sources of variation (SS), mean square error (MSE), F-statistic (F) and estimates the probability (P) value that determines the level of significance of the F *(2)*. The following steps should be followed when conducting a general ANOVA using Statistix.

1. While working in the appropriate data file, move the cursor to the **Statistics** drop down menu. Clicking on **Linear Models** will create a new box listing several linear models-based procedures, one of which is the **General ANOVA**.
2. Click on **General ANOVA** to display a new window appears in which the dependent variable must be entered. In this example the dependent variable is shoot length (**ShootsLth**).
3. The model statement should be entered in the box identified as **AOV Model Statement**. In this case the model statement is **Cytokinin**. No other variables are entered into the model statement in this situation because cytokinin is the only treatment factor. If there were multiple treatment factors each would be written

in the model statement with the interactions of interest (*see* **ref. 4** for instructions on writing model statements for experimental designs).
4. Clicking the **OK** button initiates the analysis and the results appear in a new window that can be printed or saved.
5. In ANOVA the sources of variation are listed with their DF, SS, MS, F, and p-values. In this example there are two main sources of variation identified as the treatment factor (cytokinin type) and the residual (experimental error). When reading the analysis output, significance of the treatment factor (cytokinin type) is indicated by the p-value (*see* **Note 5**).

3.2.4. Analyzing Treatment Means

Treatment means can be analyzed to determine treatment differences once a significant value is obtained in the preliminary statistical test (ANOVA, Poisson regression, or Logistic regression). A separate mean comparison evaluation test is not necessary when there are only two treatments because the general test (ANOVA, Poisson regression, or logistic regression) alone determines the statistical difference *(4)*. However, most researchers evaluate multiple treatments simultaneously. The easiest way to compare treatment means is to rank them in ascending or descending order and pick the best one. However, this method does not measure variation within the treatments and may not, on its own, accurately represent how the explants responded to the treatments *(14,16)*. For this reason it is suggested that a mathematical procedure that calculates within treatment variation be used to compare treatment means.

There are many mean separation procedures that account for variation within treatment. The procedures most commonly used in plant biotechnological research to compare the means of unrelated or related treatments are multiple comparison and multiple range tests, standard error of the mean (SEM), and orthogonal contrasts.

3.2.5. Multiple Comparison and Multiple Range Tests

Multiple comparison and multiple range tests are statistical procedures that use the population variance to calculate a numerical value for comparing treatment means *(4,6,16)*. Means are ranked in ascending or descending order and the difference between adjacent means calculated and compared to a value computed by the statistical test *(14)*. The two treatment means are considered different if their difference exceeds the computed statistical value. However, if the difference between treatment means is equal to or less than the computed statistical value the treatments are considered similar *(14)*.

Multiple comparison tests calculate one statistical value and use it to compare adjacent and nonadjacent means. Examples of multiple comparison tests are Bonferoni, Fisher's least significant difference (LSD), Scheffe's, Tukey's Honestly Significant Difference Test (Tukey's HSD), and Waller-Duncan K-ratio T test (Waller-Duncan). One problem with these procedures is that they may over estimate treatment differences among treatments ranked far apart. This occurs because treatment variances tend to increase with increasing data values *(14)*. One should use caution if using one of the above procedures in experiments with a large number of treatments.

Multiple range tests differ from multiple comparison tests because the former employs different critical values to compare adjacent and nonadjacent means *(4)*. This modification improves the accuracy of the test making it less likely that errors will be made when comparing distantly ranked means. Examples of multiple range tests include Duncan's New Multiple Range Test (DNMRT), Ryan-Einot-Gabriel-Welsh Multiple F-test (REGWF), Ryan-Einot-Gabriel-Welsh Multiple Range Test (REGWQ), and Student-Newman-Kuels (SNK). The Ryan-Einot-Gabriel-Welsh Multiple Range Test is recommended because of its moderate level of conservancy. Unfortunately, multiple range tests are not always available in PC statistical software packages.

Mean comparisons are conducted as part of an ANOVA. The following steps will demonstrate how to conduct the Tukey HSD test using Statistix.

1. While working in the appropriate data file, select the **Statistics** drop down menu and click on **Linear Models**. This will create a new box listing several linear models-based procedures, one of which is the general **ANOVA**.
2. Clicking on **General ANOVA** will produce a new window in which the dependent variable must be entered. In this example the dependent variable is shoot length (**ShootsLth**). **Cytokinin** is written in the model statement box.
3. Clicking the **OK** button will initiate the test. The results are displayed in the general ANOVA coefficient table window.
4. Clicking **Results** in the tool bar will display a drop box listing procedures that can be used to evaluate treatment means or variances.
5. Clicking on **Comparison of Means** will display a window listing various mean comparison tests. Enter the desired main effect, or interaction, to be tested (**Cytokinin** in the case) and click on one of the available tests. In this case the desired test is Tukey HSD.
6. Click **OK** to initiate the test. An output is generated in a new window that lists the treatments and their means, the homogeneous groups and the number of significant groups as well as the critical Q value and the critical value for comparisons.
7. The means with their homogeneous group identification may then be displayed in a table or graph (*see* **Note 6**).

3.2.6. Orthogonal Contrasts

Orthogonal contrasts are used to make comparisons between treatments with similar characteristics *(4,14)*. In plant tissue culture and biotechnology studies these may be plant growth regulators (PGRS) with similar activity (e.g., natural vs synthetic cytokinins) or DNA constructs with the same promoter.

Orthogonal contrasts differ from mean comparison tests in that more than two treatments can be compared in one test. However, the number of comparisons made must be restricted to the number of degrees of freedom for the treatment variable *(2)*.

Orthogonal contrasts are performed as part of the ANOVA procedure *(4)*. A contrast statement is written that specifies the treatments or group of treatments to be compared and ANOVA calculate the DF, SS, and MS for the comparison. An *F* statistic is calculated for the contrasts and the level of significance (*p*-value) determined. One DF is used for each comparison. The following steps will demonstrate how to perform orthogonal contrasts using Statistix.

1. While working in the appropriate data file, select the **Statistics** drop menu. Click on **Linear Models** to create a new box listing several linear models-based procedures, one of which is the general ANOVA. Click on general **ANOVA**.
2. In the ANOVA window, indicate the dependent variable and type in the model statement. Click **OK** to initiate the test.
3. Click on **Results** located in the tool bar of the general ANOVA coefficient window. This will generate a list of treatment evaluation procedures.
4. Clicking on **General Contrasts** produces a new window in which the contrast main effect, or interaction, and appropriate contrast coefficients for the desired treatment, or treatment group, comparisons are entered (*see* **Note 7**).
5. Clicking **OK** will initiate the test and display the results in a new window.
6. The results are interpreted by identifying the level of significance (*p*-value) for each contrast (*see* **Note 8**).

3.2.7. Standard Error of the Mean

The standard error (SE) is frequently used for mean comparison purposes in plant tissue culture and biotechnology research. SEs are obtained by dividing the sample standard deviation by the square root of the number observations for that treatment *(15)*. Many researchers use SE values like a mean comparison test, generating an individual value for each treatment and comparing the difference between the means of paired treatments with their calculated SE. The researcher often declares two treatments similar or different if the collective values (treatments mean ± its SE) for the paired treatments do not overlap. This use of SE can be used to compare means ranked adjacently. However, problems occur when using SE to compare means that are ranked far apart. When remembering how SE values are calculated, one would realize that SE values increase with the numerical value of the data, causing the researcher to over estimates treatment differences and violate the assumption of ANOVA that treatment variances are equal *(14)*. Therefore, the use of SE to compare means ranked far apart fails to produce useful results and does not accurately reflect population variance.

The following steps can be used to generate SE values for the accurate comparison of treatment means. This test is most valuable for treatments that are unrelated *(6,14)*.

1. While working in the appropriate data file, select the **Statistics** drop down menu and click on **Linear Models**. This will create a new box listing several linear models-based procedures, one of which is the **General ANOVA**. Click on **General ANOVA**.
2. In the **ANOVA** window, indicate the dependent variable and type in the model statement. Click **OK** to initiate the test.
3. Click on **Results** located in the tool bar of the general ANOVA coefficient window. This will generate a list of treatment evaluation procedures. Select **Means and Standard Errors**. Enter the main effect, or interaction, and click **OK** to initiate the test and generate the results.
4. The results will be displayed in a window. The identity of treatment with their means and SS will be listed as well as the number of observations per cell, the SE of an average and the SE of the difference of two averages (*see* **Note 9**).

3.2.8. Dosage or Concentration Treatments

Researchers that design experiments with treatments consisting of various doses, or concentrations, of a single treatment factor are usually interested in identifying a single dose, or concentration, that produces an optimal explant response. In other words, the researcher is interested in identifying a trend present among the levels of the treatment factors. The best statistical procedure for these types of experiments is usually trend analysis *(6,14)*. In trend analysis models identifying specific trends (linear, quadratic, cubic) are tested in a stepwise fashion from simplest (linear) to most complex (cubic in most cases) until a nonsignificant trend is identified *(17)*. The last significant trend is considered to best describe the response to the treatments. Trend analysis uses SS, T, and R2 values to indicate significant trends *(4)*. Trends may be tested through regression analysis or polynomial contrasts statements in ANOVA. The most effective method of conducting trend analysis depends on the experimental objectives and the statistical software package used.

To illustrated trend analysis the treatments for the petunia example can be changed from six cytokinin types to six concentrations of BA (0 [control], 2, 4, 6, 8, and 10 *M*). The objective of this experiment would be to identify a trend in explant response to BA that would identify an optimum concentration of the growth regulator for adventitious shoot organogenesis. Data recorded included the percentage of explants that produced shoots, number of shoots per explant, and length of regenerated shoots. The following steps can be used to conduct trend analysis using Statistix.

3.2.8.1. For Simple Linear Regression

1. While working in the appropriate data file, select the **Statistics** drop down menu. Clicking on **Linear Models** will create a new box listing several linear models-based procedures, one of which is linear regression.
2. Clicking on linear regression will produce a window in which the dependent and independent variables must in entered in their respective boxes. Click **OK** after entering these variables.
3. Clicking **OK** will initiate the procedure and produce a window displaying the linear regression coefficient table (*see* **Note 10**).

3.2.8.2. For Trend Analysis Using Polynomial Contrasts

1. While working in the appropriate data file, select the **Statistics** drop down menu and click on **Linear Models**. This will create a new box listing several linear models-based procedures, one of which is the general ANOVA.
2. Clicking on general **ANOVA** window will display a box in which the dependent variable and type in the model statement must be entered. Enter the appropriate statements.
3. Clicking on **OK** will initiate the test and produce the ANOVA results with a new tool bar.
4. Clicking on **Results** located in the tool bar of the general ANOVA coefficient window will generate a list of treatment evaluation procedures.
5. Selecting **Polynomial Contrasts** will produce a new window in which the level of polynomial contrasts and the contrast main effect, or interaction must be entered (*see* **Note 11**). Click on **OK**.
6. Clicking on **OK** will produce a window displaying the results of the polynomial contrasts for the desired dependent and independent variables indicated in the ANOVA.
7. The last polynomial degree with a significant *p*-value is considered to be the most significant trend (*see* **Note 12**).

4. Notes

1. Because one explant was cultured in a vessel and the vessel is randomly placed in the test tube rack, the experimental design was a CR design *(2)*. The CR design is commonly used for plant biotechnological experiments because it is easy to use and allows researchers to maximize the number of replicates examined as well as utilize either equal or unequal replication for the treatments *(1)*. The CR design is also amenable to statistical procedures for binomial, count, continuous, and concentration data *(6)*. These advantages are important as maximizing replicate numbers can lead to a more powerful test in determining treatment differences. Flexibility in the number of replicates for each treatment is important because unequal replication often occurs in due to contamination or death of explants.

    In a CR design the sources of variation are the applied treatment(s), in this case cytokinin type, and experimental error, the latter representing variation not

Table 1
Results of Analysis of Variance of Petunia Shoot Length Data Analyzed Using a Completely Randomized Design

Source	DF	SS	MS	F	P
Cytokinin	5	1803.35	360.67	12.46	<0.0001
Experimental error	54	1563.50	28.954		
Total	59	3366.85			

Note: Explants were prepared from leaves of petunia plants grown in the greenhouse. The influence of six cytokinins (zeatin, zeatin riboside, 2iP, BA, kinetin, and TDZ) on the length of regenerated shoots was examined. Ten replicate test tubes were culture per treatment. Data are hypothetical.

associated with the applied treatments. Interpretation of computer outputs from the CR design is considered simple and straightforward (**Table 1**).

2. Whereas the petunia example uses a CR design with one explant per vessel, the randomization scheme can be modified to include multiple explants per vessel. Such a randomization scheme is referred to as subsampling *(18)* and is often used in plant biotechnology to conserve resources, space, and time. Subsampling can also be used to minimize the variation associated with explants by "averaging out" the response among culture vessels *(7)*. When using subsampling, it is important that the researcher write the analysis statements in a manner that the variation present among explants within each culture vessel is calculated, otherwise the experimental error would be overestimated leading to erroneous results.

   Other experimental designs such as the randomized complete block, incomplete block, and split plot can be used in situations in which isolation of specific nontreatment or treatment factors are desired. The use of experimental designs that employ blocking has been discussed thoroughly by others *(4,6,7,19)*.

3. A *p*-value of 0.7969 was obtained for cytokinin when analyzing the petunia data using logistic regression (**Table 2**). This indicates that the explants similarly to cytokinin treatments.

   Following logistic regression, a mean comparison test can be used to determine which cytokinin elicited the optimum explant response. When performing a mean comparison test, it is usually best to perform an appropriate data transformation before comparing means of response data if an ANOVA-based mean comparison or multiple range tests will be used *(4)*.

   Before analyzing response data, it is important to determine if there are treatments in which the response did not vary. This usually occurs when all or none of the explants responded. These values should either be changed or deleted before analyzing data *(14)*. If a decision is made to change the values, zeros should be changed to a slightly higher value (0.000001) and 100% values reduced slightly (0.999999). It is important to decide during planning if values are going to be

Table 2
Results of Logistic Regression Analysis of the Percentage of Leaf Explants That Produced Adventitious Shoots

Predictor variables	Coefficients (COEF)	Standard error (SE)	COEF/SE	p
Constant	1.296	0.743	1.74	0.0813
Cytokinin	0.057	0.196	0.29	0.7696
Deviance	57.08			
p-value	0.509			
DF	58			

Table 3
Results of Poisson Regression Analysis of the Number of Adventitious Shoots Produced per Responding Leaf Explants

Predictor variables	Coefficients (COEF)	Standard error (SE)	COEF/SE	p
Constant	1.120	0.1504	7.45	< 0.0001
Cytokinin	0.101	0.0362	2.79	0.0052
Deviance	205.94			
p-value	< 0.0001			
DF	58			

altered or deleted. You should also indicate if treatments were dropped or values altered when writing reports and manuscripts.

4. In the petunia example, Poisson regression revealed that cytokinin type influenced the number of shoots per responding explant as determined by reading the p-value of the **Cytokinin** variable **(Table 3)**. In this case a p-value of 0.0052 was obtained indicating that there was a significant regression associated with cytokinin type. It is important to delete observations from non-responding explants before analyzing the data, as retention of zeros will skew the data set and lead to inaccurate results and interpretation of treatment effectiveness *(14)*. As always, it is important to indicate that values from nonresponding explants were deleted before analysis in all reports and manuscripts.

5. In the petunia example, the results of ANOVA indicated that shoot length was influenced by the cytokinin used **(Table 1)**. An F of 12.46 was highly significant ($p < 0.0001$). The cytokinin MS reflects the amount of variation caused by the cytokinin treatments whereas the MSE measures variation due to culture vessels and other non-treatment sources.

Statistics in Plant Biotechnology

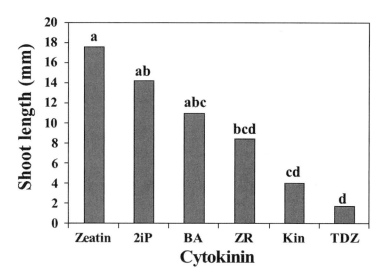

Fig. 1. Bar graph demonstrating the use of Tukey's Honestly Significant Difference (HSD) mean comparison test. The hypothetical experiment tested the influence of cytokinin type on the elongation of adventitious shoots regenerated from leaf explants of *Petunia x hybrida* incubated in Murashige and Skoog medium containing 4 µM of zeatin, 2-(Δ) Isopentenyl adenine (2iP), benzadenine (BA), zeatin riboside (ZR), kinetin (Kin), or thidiazuron (TDZ). Bars with the different letters are significantly different according to Tukey's HSD. (From Compton *[6]* with permission.)

6. In the petunia example, Tukey's HSD was used to compare the effect of different cytokinin on elongation of regenerated shoots. The test divided the treatment means into four homogeneous groups as represented by the letters next to each treatment mean. When presenting treatment means in a table or graph, the results of multiple comparison and multiple range tests are placed with the treatment means in a table or graph. A letter is assigned to each treatment mean and treatment means assigned different letters are considered statistically different. Means assigned the same letter are considered statistically similar (*see* **Fig. 1**).
   Based on these results, shoot length was greater among explants cultured in medium containing zeatin compared to zeatin riboside, kinetin, and TDZ. However, there was no difference in the length of shoots regenerated from explants cultured in medium supplemented with 2iP or BA.
   As you can see, results of mean comparison tests can be complicated and confusing. For this reason researchers should seriously examine their options for evaluating treatment observations when planning experiments and consider using procedures that produce the most meaningful results. Often procedures such as orthogonal contrasts can produce the most meaningful comparisons *(4–6,14)*.
7. In the petunia example, five contrasts of interest were planned before the experiment: (1) natural cytokinins (zeatin, zeatin riboside, and 2iP) vs synthetic types

Table 4
Use of Orthogonal Contrasts to Demonstrate the Influence of Cytokinin Type on the Elongation of Adventitious Shoots Regenerated From Petunia Leaf Explants

Cytokinin type	Number of explants	Shoot length (mm)
Zeatin (Z)	9	17.6
2iP	9	14.3
Benzyladenine (BA)	10	11
Zeatin Riboside (ZR)	5	8.5
Kinetin (K)	9	4.1
Thidiazuron (TDZ)	7	1.8

Orthogonal Contrasts	F-value	p-value
Z and ZR vs 2iP, BA, K, and TDZ	2.54	0.039
Z, ZR and 2iP vs BA, K, and TDZ	6.36	0.0001
BA vs Z, ZR, 2iP, K, and TDZ	0.17	0.9711
BA vs TDZ	2.92	0.0209
TDZ vs Z, ZR, 2iP, BA, and K	4.98	0.0008

(BA, kinetin, and TDZ); (2) cytokinins 5 containing zeatin (zeatin and zeatin riboside) vs non-zeatin containing cytokinins; (3) BA (as a control) vs all other cytokinins; (4) BA vs TDZ; and (5) TDZ vs all other cytokinins. Contrasts were conducted for all dependent variables as part of ANOVA. The software instruction manual will explain how orthogonal coefficients are written.

8. Results of orthogonal contrasts are usually presented in a table containing ANOVA computations (**Table 4**). Means of each contrast may be placed in a table or presented in the text with the means and p-value given for each contrast. Examining results for shoot length. Shoot elongation was promoted when pgrs containing zeatin (Z and ZR) were used ($p = 0.039$). Mean shoot length was 13 mm when explants were cultured in medium containing a zeatin-based cytokinin whereas non-zeatin-based cytokinins inhibited shoot length (7.8 mm). Likewise, natural cytokinins (Z, ZR, and 2iP) promoted shoot length more than synthetic formulations of BA, kinetin, and TDZ ($p = 0.0001$; 13.5 mm vs 5.3 mm, respectively). Shoot elongation was inhibited ($p = 0.0008$) when TDZ (1.8 mm) was used compared to any of the other cytokinin formulations (11.1 mm).

Orthogonal contrasts can be used to analyze binomial or count data given that the data are transformed before analysis. Information gained from orthogonal contrasts is often much more useful than outcomes of multiple comparison or multiple range tests. Orthogonal contrasts are under utilized in plant biotechnological

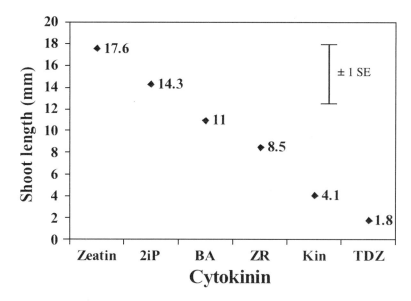

Fig. 2. Graph demonstrating the use of standard error for comparing treatment means. The hypothetical experiment tested the influence of cytokinin type on the elongation of adventitious shoots regenerated from leaf explants of *Petunia x hybrida* incubated in Murashige and Skoog medium containing 4 μM of zeatin, 2-(Δ) Isopentenyl adenine (2iP), benzyadenine (BA), zeatin riboside (ZR), kinetin (Kin), or thidiazuron (TDZ).

studies possibly because researchers have not been exposed to the concept or because their usefulness is not completely understood. *See* Little and Hills *(1)*, Lentner and Bishop *(2)*, Compton *(4–6)*, or Zar *(15)* for more information regarding the use of orthogonal contrasts.

9. When using SE values to compare the means of several treatments, one SE value should be calculated from the ANOVA MSE or SE values obtained from Poisson or Logistic regression *(14)*. This use of SE is more likely to yield realistic results and reveal true treatment differences compared to the calculation of individual SEs each treatment, which violates the equal variances assumption of ANOVA *(14)*. The correct use of SE is demonstrated in **Fig 2**.

10. In the petunia example, trend analysis was used to examine the effect of cytokinin concentration on shoot length. Because no shoots were produced by explants cultured in medium without BA, observations recorded for 0 BA were deleted before analysis. In addition, zero values from nonresponding explants were deleted because only shoot length values from responding explants were of interest.

  A regression coefficient table is generated by trend analysis. The table identifies the predictor variables with their coefficients, standard errors, student's T, and *p*-values along with the R2, adjusted R2, MSE and standard deviation are also displayed (**Table 5**). Based on information from responding explants, shoot

Table 5
Results of Regression Analysis of Evaluating the Influence on Benzyladenine Concentration on the Elongation of Adventitious Shoots Regenerated From Petunia Leaf Explants

Predictor variables	Coefficients (COEF)	Standard error (SE)	COEF/SE	p
Constant	22.9535	0.9164	25.05	< 0.0001
BA concentration	−2.1014	0.1366	−15.38	< 0.0001
R^2	0.8616	EMS	4.8968	
Adjusted R^2	0.858	SD	2.2129	

Regression Summary Table

Source	DF	MS	F	p
Regression	1	1158.70	236.62	< 0.0001
Residual	38	4.8968		
Total	39			

Note: Explants were prepared from leaves of petunia plants grown in the greenhouse. The influence of various concentrations of benzyladenine (0, 2, 4, 6, 8, and 10 μM) on the length of regenerated shoots was examined. Data from 0 μM BA were deleted before analysis because no explants produced shoots. Likewise, data from nonresponding explants incubated in media containing 2 to 10 μM were deleted before analysis. Data are hypothetical.

length was dependent on BA concentration with the linear model (X = 22.954 − 2.1014*BA) best describing the effect of BA on shoot length (p < 0.00001; R^2 = 0.86).

Means of dosage treatments should be presented in graphs displaying the regression equation and fitted line, R^2, individual data values, and confidence intervals (*see* **Fig. 3**). Similar to mean comparison tests, confidence intervals of treatments that overlap are considered similar. In the petunia example, confidence intervals for 2 and 4 μM overlap. However, confidence intervals for 6, 8, and 10 μM do not overlap with 2 and 4 μM, and are considered significantly different from 2 and 4 μM. The conclusion would be that shoot elongation was inhibited with increasing concentration of BA above 4 μM.

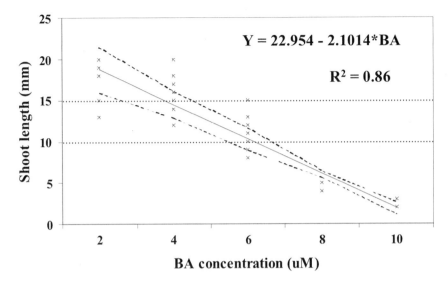

Fig. 3. Graphical presentation obtained from regression analysis testing the influence of benzyadenine (BA) concentration on elongation of adventitious shoots regenerated from leaf explants of Petunia x hybrida incubated in Murashige and Skoog medium containing 0, 2, 4, 6, 8 or 10 µM BA. (Graph from Compton [6] with permission.)

Trend analysis and polynomial contrasts may be used for evaluating optimum treatment levels, even if those levels were not directly tested but are within treatment boundaries (4). Predicted values can be obtained using information generated from the analyses. These values allow researchers to estimate the effects of treatments that were not evaluated as long as they are within the tested boundaries (17).
Trend analysis can be used to evaluate binomial and count data provided that the data are transformed prior to analysis. As with mean comparison procedures, data are converted back to the original scale for presentation.

11. For the petunia described previously, the level of polynomial contrasts is three and the main effect is BA. These values are entered in the appropriate boxes and the computer will substitute the correct contrast coefficients.
12. Polynomial contrasts often provide results similar to simple regression. As with regression analysis, the last significant contrast is considered to provide the best description of the trend. Polynomial contrasts can be used when treatments are unequally replicated or spaced as long as specific contrast coefficients are calculated (4,20).

Table 6
Use of Polynomial Contrasts to Demonstrate the Influence of Benyzladenine Concentration on the Elongation of Adventitious Shoots Regenerated From Petunia Leaf Explants

Benzyladenine concentration (μM)	Number of explants	Shoot length (mm)
0	–	–
2	5	17.0
4	9	15.9
6	10	11.0
8	9	4.56
10	7	2.6

Polynomial Contrasts	SS	F-value	p-value
Linear	1615.3	452.91	<0.0001
Quadratic	7.7861	2.18	0.1485

Note: Explants were prepared from leaves of petunia plants grown in the greenhouse. The influence of various concentrations of benzyladenine (0, 2, 4, 6, 8, and 10 μM) on the length of regenerated shoots was examined. Data from 0 μM BA were deleted before analysis because no explants produced shoots. Likewise, data from nonresponding explants incubated in media containing 2 to 10 μM were deleted before analysis. Data are hypothetical.

In the petunia example, results of polynomial contrasts indicated that the linear regression best describes the trend in shoot length (**Table 6**). This is evident by the fact that the linear contrast was the last significant trend.

Results of polynomial contrasts are presented in a table with ranked means (**Table 6**). For instructions on how to generate a regression equation from polynomial contrasts *see* Carmer and Seif *(20)*, Kleinbaum et al. *(17)*, or Zar *(15)*.

References

1. Little, T. M. and Hills, F. J. (eds.) (1978) *Agricultural Experimentation: Design and Analysis.* John Wiley and Sons, Inc., New York.
2. Lentner, M. and Bishop, T. (eds.) (1986) *Experimental Design and Analysis.* Valley Book Company, Blacksburg, VA.
3. Kempthorne, O. (ed.) (1973) *The Design and Analysis of Experiments.* Robert E. Krieger Publishing Co., Malabar, FL.
4. Compton, M. E. (1994) Statistical methods suitable for the analysis of plant tissue culture data. *Plant Cell Tiss. Org. Cult.* **37,** 217–242.

5. Compton, M.E. (2000) Statistical analysis of plant tissue culture data, in *Plant Tissue Culture Concepts and Laboratory Exercises* (Trigiano, R.N. and Gray, D.J., eds.), 2nd edition. CRC Press, Boca Raton, FL, pp. 61–74.
6. Compton, M. E. (2004) Elements of *in vitro* research, in *Plant Development and Biotechnology* (Trigiano, R. N. and Gray, D. J., eds.), CRC Press, Boca Raton, FL pp. 55–71.
7. Compton, M.E. and Mize, C.W. (1999) Statistical considerations for in vitro research: I—Birth of an idea to collecting data. *In Vitro Cell. Dev. Biol., Plant* **35**, 115–121.
8. Anonymous (2004) SAS/STAT Software. SAS Institute. Accessed June 14, 2004; available at http://support.sas.com/rnd/app/da/stat.html.
9. Anonymous (2004) SPSS. Accessed June 14 2004; available at http://www.spss.com/.
10. Anonymous (2004) Systat Software, Inc. Software, Services, Solutions for the Statistics, Scientific Community. Accessed June 14 2004; available at http://www.systat.com/.
11. Anonymous (2004) Minitab. Accessed June 14 2004; available at http://www.minitab.com.
12. Anonymous (2004) STATISTIX. Accessed June 14 2004; available at http://www.statistix.com.
13. Preece, J. E. (2000) Shoot organogenesis from petunia leaves, in *Plant Tissue Culture Concepts and Laboratory Exercises*, (Trigiano, R. N. and Gray, D. J., eds.), 2nd edition. CRC Press, Boca Raton, FL, pp. 167–173.
14. Mize, C. W., Koehler, K. J., and Compton, M. E. (1999) Statistical considerations for in vitro research: II data to presentation. *In Vitro Cell. Dev. Biol., Plant* **35**, 122–126.
15. Zar, J. H. (ed.) (1984) *Biostatistical Analysis, second edition.* Prentice-Hall, Inc., Upper Saddle River, NJ.
16. Mize, C. W. and Chun, Y. W. (1988) Analysing treatment means in plant tissue culture research. *Plant Cell Tissue Organ Cult.* **13**, 201–217.
17. Kleinbaum, D. G., Kupper, L. L., and Muller, K. E. (eds.) (1988) *Applied Regression Analysis and Other Multivariable Methods. Second Edition.* PWS-Kent Publishing Co., Boston.
18. Mize, C. W. and Winistorfer, P. M. (1982) Application of subsampling to improve precision. *Wood Sci.* **15**, 14–18.
19. Kuklin, A. I., Trigiano, R. N., Sanders, W. L., and Conger, B. V. (1993) Incomplete block design in plant tissue culture research. *J. Plant Tiss. Cult. Meth.* **15**, 204–209.
20. Carmer, S. G. and Seif, R. D. (1963) Calculation of orthogonal coefficients when treatments are unequally replicated and/or unequally spaced. *Agronomy J.* **55**, 387–389.

14

An Efficient Method for the Micropropagation of *Agave* Species

Manuel L. Robert, José Luis Herrera-Herrera, Eduardo Castillo, Gabriel Ojeda, and Miguel Angel Herrera-Alamillo

Summary

Despite their economic importance, the *Agave* spp. have not been genetically improved. This is probably owing to the fact that they have very long life cycles and many of them have an inefficient sexual reproduction mechanism. Micropropagation offers an alternative to this problem through the efficient cloning of selected high-yielding "elite" plants. We report here an efficient method to micropropagate agaves and a strategy for the management of large scale production that has been successfully applied to several *Agave* spp.

Key Words: *Agave*; genetic improvement; large-scale propagation; micropropagation.

1. Introduction

1.1. The Agaves

Many *Agave* spp. are of great economic importance such as henequen (*Agave fourcroydes* Lem.) and sisal (*Agave sisalana* Perr.) from which hard fibers are extracted for the fabrication of ropes, sacks, mats, as well as tequila (*Agave tequilana* Weber), and bacanora (*Agave angustifolia* Haw.), which are the source of alcoholic beverages. Some, such as *Agave victoria reginae*, are very much appreciated as ornamentals. Additional products that can be obtained from agaves include steroids for pharmaceutical products and animal feeds, high fructose syrups and inulin derivatives for the production of pre-biotics.

Despite their economic potential, the agaves, with the exception of hybrid 11648 produced by the British in Tanzania during the first half of the 20th century *(1)*, have not been genetically improved. This is probably because of their very long life cycle (8–20 yr, depending on the species) and limited sexual reproduction. Some agaves, particularly the polyploids, do not set seed or the seeds produce only aneuploids (*see* **Note 1**).

Plantation agaves are vegetatively propagated through the shoots produced by rhizomes (subterraneous shoots that grow laterally from the stem and eventually emerge to give rise to a new individual) or by means of bulbills (small plants derived from axillary buds in the branches of the inflorescences), and although strictly speaking, all the new shoots produced vegetatively from a single mother plant form a clone, a great deal of variability is observed in the size and vigour of the offspring. Under these circumstances, the only means of maintaining a high level of productivity in the plantations is through masal selection. In nature, however, a plant will only produce some 25 healthy rhizomes over a period of 5 yr, which is not sufficient to establish a continuous program of selection.

Another problem associated with the use of vegetatively propagated plants is that microbial diseases are easily transmitted to the next generation and are dispersed by man.

1.2. Applications of Tissue Culture to Agave *Species*

Through tissue culture several thousand clonal plants can be propagated in a few months from a single mother plant. The method described here permits the selection of high yielding materials, the rapid propagation of high quality (elite) selected clones with minimum unwanted genetic variability and the production of healthy stocks for plantation.

Conversely, the method can also be used to generate and introduce new genetic variability in the form of new hybrids or mutants that would take many years to become available for plantation purposes if they were to be propagated through natural means.

The basic protocol was first used for the micropropagation of henequen *(2,3)* but, with minor differences, has been successfully employed to propagate other varieties and species of agaves including *Agave fourcroydes* var. Kitam Ki, *A. fourcroydes* var. Yaax Ki; *A. tequilana, A. letonae, A. angustifolia* var. Marginata, *A. angustifolia* var. Bacanora, *A. sisalana, Agave amaniensis*; and the hybrid H11648, as well as species of the genus Yucca. Each species produces different results in terms of multiplication efficiency, rapidity of growth, and so on, but the basic procedure has proved to be an efficient micropropagation method for all the species tested.

2. Materials

2.1. Selection of Mother Plants

The first step towards the establishment of "elite clonal lines" is the adequate selection of mother plants based on their performance in the field. Outstanding healthy individuals are selected in collaboration with growers that know the age of the plantations and the conditions in which the plants have developed (*see* **Note 2**).

2.2. Sources of Explants

Apical offshoots from rhizomes from selected elite plants are generally a good source of explant (meristematic) tissue generating an average of 50 new plantlets from a single plant. Vigorous and healthy 25–40 cm tall shoots, preferably still attached to the mother plant through the rhizome, are ideal; they are big enough to yield a minimum of eight explants and are small enough to be easily transported to the laboratory (*see* **Notes 3** and **4**). It is very important, however, to make sure that the extracted shoot is the offspring of the selected mother plant and not of a neighboring one.

2.3. Pre-Conditioning of Donor Plants

1. The shoots, extracted from the rhizomes in the plantations, can be taken to the laboratory to be used immediately as a source of explants or can be placed individually in bags or pots with sterile soil and placed for a few weeks or months in a partially controlled environment in a nursery or a green house, preferably the latter.
2. Under these conditions the plants will be in a controlled environment that will make them physiologically homogeneous and more adequate as a source of explants. During this period the plants:
   a. Can be treated with fungicides or growth regulators to control infections or change physiological conditions.
   b. Can be shaded to soften the tissues by reducing the amount of sunlight and temperatures.
   c. Can be watered at will to eliminate water stress.

   This pre-conditioning has not been extensively investigated because it is has not been necessary for most of the *Agave* spp. that have been successfully cultured in vitro and, although it represents additional work and additional costs to the production system, it must be remembered that it is only for the initiation stage which is a small part of the total micropropagation process. It is also recommended as a means of stabilizing production all year round, if needed.

2.4. Culture Media and Unusual Chemicals

1. Murashige and Skoog (MS)-B. Culture media: MS modified in its nitrogen content 10 mM KNO_3 and 5 mM NH_4NO_3.
2. Plant Preservative Mixture (PPM). Phytotechnology Laboratories, Shawnee Mission, KS; http://phytotechlab.com.
3. Gelrite gellan gum (Sigma, cat. no. G1910).

3. Methods

The method described permits the efficient cloning of thousands of individuals from a single mother plant and has been successfully applied to many different agaves. Basically, it consists of the induction of shoots from meristematic tissues extracted from "elite" plants, their multiplication through direct organogenesis, and the in vitro pre-adaptation of the plants that produces a very high survival rate when transferred to soil.

3.1. Extraction of Meristematic Tissues

Before taking the plant tissues into the laboratory, all the leaves are removed and the remaining tissues thoroughly washed with soap and a brush to remove all the soil from them (*see* **Fig. 1A**).

3.2. Disinfestations of Explants and Control of Microbial Growth In Vitro Cultures

Before taking the tissues into the air flow cabinets, as much as possible of the external tissues should be removed with a sharp butcher knife, leaving a block of tissue of 6–8 cm per side (the size of the explant depends on the species used and the age and size of the mother plant). The bases of the central leaves at the top of the cube is left in place at this stage to protect the meristematic tissue below them. This meristematic tissue is the actual source of explants that will be cultured in vitro.

The method for the sterilization and cutting of explants is as follows:

1. The blocks of meristematic tissue extracted from the stems are soaked in extran (2%) or in a Tween-20 solution for 30 min and taken into an air flow cabinet.
2. The explants are then soaked in 40% commercial bleach (containing 5% active sodium hypochlorite (NaOCl) for 30 min, followed by three rinses with sterile water.
3. The blocks are cut into smaller (0.8 cm^3) cubes (*see* **Subheading 3.3.**).
4. In some cases, where the explants are particularly difficult to clean, the smaller cubes can be immersed in 2% bleach before rinsing thoroughly with sterile distilled water and placing them in the culture media.

Fig. 1. (**A**) Extraction of meristematic tissues. (**B**) Adventitious shoot formation in the induction phase. (**C**) New shoots form from the axillary buds in the multiplication phase. (**D**) In vitro pre-adaptation of clone plants. (**E**) Adaptation of in vitro rooted planted in polystyrene trays.

5. If needed, PPM 0.8–1.6 mg/L can be used in the culture media to control microbial growth in the early stages of induction of lines that are particularly difficult to clean (*see* **Notes 5** and **6**).
6. The cultures should be checked every day, removing the contaminated cultures.

3.3. Cutting the Explants

1. The large sterile blocks have the meristematic tissues at their core. To extract these meristematic tissues, located right below the primordial leaves, it is necessary to remove the latter completely because their presence will prevent the formation of new plantlets through organogenesis or embryogenesis.
2. The base of the primordial leaves can be seen as pale rosettes at the surface of the top of the explant and can be removed by cutting thin slices off the top surface with the aid of a scalpel until no traces of the rosette are left.
3. The next step is to remove the lateral tissues damaged by the sterilization treatment. Finally, the fibrous tissues at the base are eliminated in a single cut, leaving a slice of tissue about 1-cm thick.
4. This meristematic slice can be sterilized again if needed (not recommended) and then cut into smaller, 0.8 cm^3, blocks that constitute the explants to be cultured.
5. The final size of the slice and the total number of explants extracted from it depend on the size of the base of the mother plant. Small rhizomes can produce as little as 4 explants whereas larger ones can produce as many as 20 (*see* **Note 7**).

3.4. Induction: Culturing the Explants

1. The small, 0.8 cm^3, cubes are incubated in baby food jars containing MS-B media supplemented with 2,4-dichlorophenoxyacetic acid (0.11 µM) and 6 benzyadenine (BA) (44.4 µM) and solidified with 0.8% agar. The pH is adjusted to pH 5.75. The jars are incubated in a growth room at 27 ± 2°C under a 16-h photoperiod (45 µmol/m^2/s) for 8–12 wk until new plantlets are formed on the surface of the explants.
2. The first plantlets start appearing after 5 wk and, by the end of wk 12, some 3–12 new, complete, adventitious shoots with a minimum of two leaves (*see* **Fig. 1B**), varying in size from 0.5 to 2.0 cm, will have been formed on each explant.
3. In some cases, the new shoots can be separated by gentle pushing with the back of the scalpel or by cutting those apart with the blade (*see* **Note 8**). The individualized shoots or groups of two or three of them are then grouped by size and transferred to magenta boxes to grow and multiply. The production of induced shoots can be very variable, several factors accounting for the lack of reproducibility from one initiation to the next (*see* **Notes 9–12**).

3.5. Multiplication

1. This stage represents the core of the micropropagation procedure because it is at this point that the number of plants is increased to meet requirements. The plantlets are placed in magenta boxes containing 50 mL of MS-B medium supplemented with 2,4-D (0.11 µM) and 6 BAP (4.43 µM) and solidified with agar (0.175%) and gelrite (0.175%) and incubated in a growth room at 27 ± 2°C under a 16 h photoperiod (70 µmol/m^2/s) for 4 wk.
2. New adventitious shoots form at the base of the shoots, but new shoots also form from the axillary buds (*see* **Fig. 1C**). The average numbers of new shoots can

vary widely from species to species but also from one clone to another within the same species.
3. The "multiplication factor" varies depending on the species, clone and culture conditions (*see* **Note 13**) but depends mainly on the concentration of growth regulators (*see* **Note 14**). In some lines it can be as low as 1.5× or as high as 8×.
4. The new shoots are very variable in size and it is recommended that when sorted and separated they are graded and transferred separately in order to maintain as much homogeneity as possible in the culture dishes:
   a. Small: 0.5–1.0 cm.
   b. Medium: 1.0–2.0 cm.
   c. Large: >2.0 cm.
5. The first two can be transferred to multiplication medium to continue increasing the micropropagated biomass or can be transferred to growth medium (*see* **Subheading 3.6.**) to allow them to reach the right size for pre-adaptation.
6. The large ones can be transferred directly to growth and pre-adaptation media (*see* **Subheading 3.7.**).
7. The cycle can be repeated as many times as necessary to produce the required number of plants for each cultured line.

3.6. Growth

Micropropagated shoots are too small to be taken out of in vitro culture and they must be given the opportunity to develop further by keeping them in the same culture medium without growth regulators at 27°C and a 16 h photoperiod (70 µmol/m^2/s). This will not only allow them to grow but also to use up the growth regulators that they might have accumulated during the multiplication transfers, which is likely to manifest itself in the formation of some additional new shoots.

3.7. In Vitro Pre-Adaptation

The leaves of the vitroplants are not normal leaves. They have abnormal stomata that, in most cases, are not functional and lack epicuticular waxes, which makes them extremely susceptible to desiccation through rapid loss of water (*see* **Note 15**). Furthermore, because of the high concentration of sugars present in the culture media the plants do not photosynthesize efficiently and suffer when they are removed from the cosy and aseptic in vitro environment and are planted in soil. It is therefore recommended that the last growing stage takes place in pre-adaptation media without sugars and growth regulators and with higher gelling concentrations that will limit the availability of water. It is important that these plants are placed on the shelves with the highest possible light intensity (110 µmol/m^2/s) or, preferably, under natural light.

These conditions will help the new plants to start preparing for the shock of the dryer and less nutritious ex vitro environment (*see* **Fig. 1D**).

3.8. In Vitro Rooting

The shoots sometimes, but not always, produce roots in the pre-adaptation medium. If a small amount of auxin is added this will help produce roots; however, it is important to ensure that no callus is formed. The roots formed in vitro are not essential for the survival of the plants, which can be transplanted without roots providing that adequate care is taken to protect them from desiccation once they are in soil.

3.9. Transplanting to Soil and Weaning

1. The survival and performance of the micropropagated plants will depend on how well they have been adapted before they are transplanted to soil. It is of the utmost importance that only fully grown plants are taken out of the in vitro culture containers. If the plants are smaller than 3 cm they should be transferred to medium without growth regulators for the length of time needed for them to reach the right size.
2. Plants transferred to soil in polystyrene trays must be of the same size (a minimum of 5-cm high but preferably larger) and have a minimum of three leaves (*see* **Fig. 1E**); they must be kept under shade and in high humidity (under the plastic canopies) for as long as necessary (about 1 mo) before they are placed in the open (under shade) for another 1–2 mo. By then the plants should have a minimum of five normal leaves and have reached a robust size (that depends on the species being propagated).

3.10. Field Testing of Micropropagated Materials

Vigorous plants transplanted to soil in the nurseries will grow well and will show the full potential of micropropagated plants when compared with their naturally propagated counterparts, whether bulbills or off-shoots from rhizomes. The following points should be taken into consideration when designing experimental field plots:

1. Experimental nurseries and plantations must be established in suitable locations where they can be supervised and attended at all times.
2. The work must be done in close collaboration with growers who should also be evaluating the materials and comparing them with their own.
3. The plots must be planted under the same conditions, in which they will be cultivated by the growers, though they must not be stressed unnecessarily.
4. A sample of the same lots of micropropagated plants must be planted at the research stations for quality control.
5. Controls. To test the performance of micropropagated plants in the field it is recommended that they are compared with adequate controls. Plants of equiva-

lent age or size, whether from bulbills or shoots from rhizomes, must be planted at the same time, though this is not always possible because field plants with the right characteristics are not always available.
6. Measurements. A statistically significant number of randomly distributed plants should be measured at the moment of planting (time 0) and then every 2 mo. Plant height and total number of leaves are sufficient at the nursery level. Diameter of the stem and the length and width of a middle leaf (3rd rosette) must be measured at the plantation stage.

3.11. Large-Scale Production

For the propagation of large numbers of plants, we recommend building up and then maintaining a constant working biomass in accordance with the production targets and the capacity of the laboratory. This means that the same number of plants is planted for multiplication in each new transfer so that the total number of manipulations required at any stage of the process does not increase and outstrip the operational capacity of the laboratory, as would happen if the biomass were to increase without control (*see* **Fig. 2A**).

This scheme is initiated and repeated as many times as necessary to reach the desired level of production while maintaining a constant multiplying biomass (*see* **Fig. 2B**). This should avoid bottlenecks when production mounts up and several different types of transfers have to be performed on a single day. The basic scheme is shown in **Fig. 3**.

3.12. Final Considerations

This method allows the production of clonal lines of several thousand plants from a single mother agave plant in 12 mo vs a few tens of plants that the same plant will produce in the first 6 yr of its life cycle. If adequate quality control is maintained throughout the procedure (*see* **Notes 16** and **17**), the result will be high-quality, high-yielding materials that will produce a significant increase in yield at the plantation stage.

4. Notes

1. All the seedlings formed from the few henequen seeds that do germinate have variable DNA amounts (measured by flow cytometry) that are higher than the 37.5 pg corresponding to the vegetatively propagated pentaploids.
2. It is important to know if and when the plantations have been fertilized or treated with pesticides, if there has been a severe draught and how many times the plants have been cut, and so on. The selected plants must be labeled and measured for future reference. The main parameters for selection are overall size, total number of leaves, length and width of the middle leaves, thickness and rigidity of the leaves (indicative of fiber content), diameter of the stem, presence of marginal spines, and basal metabolism.

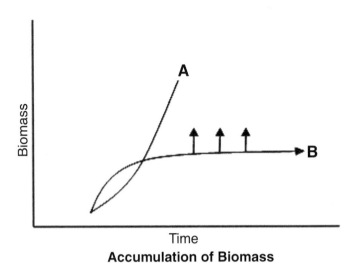

Accumulation of Biomass

Fig. 2. Biomass.

The measurement of physiological characteristics such as malate accumulation in the leaves or basic metabolic rates (isothermal calorimetry) in young plants from nurseries or plantations will speed up the selection and introduction of high-yielding lines.
3. Stems of old plants are not a suitable source of explants (meristematic tissue) for the initiation of agave tissue culture. They are hard, fibrous, have a lignified meristem and are most likely infected with one type of microorganism or another. However, selection of elite materials can only be carried out from old plants that have shown their full range of advantageous (superior) characteristics.
4. Other sources of explants. If available, other "younger" biological materials can be used for induction: plants from in vitro germinated seeds, plants from selected bulbills grown in the nursery, previously micropropagated plants maintained in the green house or the nursery, axillary buds from leaves, rhizomes, and inflorescences.

 In all cases they should preferably be from selected mother plants. The first two options, however, will not represent direct selections and will introduce genetic variability that might or might not be desirable.
5. To prevent contamination, we recommend placing cotton strings at the edge of the tops of the culture dishes and wrapping the tops of the culture dishes with commercial wrap. These also help to prevent desiccation of the gelling gum that supports the shoots.
6. Antibiotics. The use of antibiotics to control contamination in vitro is not recommended because: (a) the high costs implied, (b) limited availability, (c) most of them are toxic to plant tissues, and (d) they are not effective against fungal contamination. Furthermore, before any antibiotic is applied to a tissue culture sys-

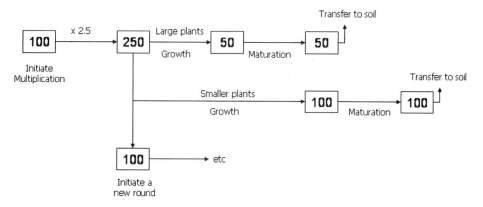

Fig. 3. The basic scheme of the propagation material is the following: 100 plants, at an assumed multiplication rate of 2.5 times, produce 250 new plants in 4 wk (depending on the multiplication rate, the numbers could be 300 or 400). Assuming that 100 plants are the basic biomass unit and that the multiplication rate is 2.5 times, only 100 shoots will be transferred back to multiplication medium while the other 150 will be transferred to growth and/or pre-adaptation media.

tem, bacterial isolations and antibiograms should be carried out to be sure of their effectiveness. The degree of toxicity on the plant tissues must also be determined prior to use.

7. After so much work one is tempted to extract as many explants as possible and many times the second layer of tissue below the first one might be worth culturing However, most of the meristematic tissue is in the top block just under the base of the primordial leaves that were removed and this region will produce most of the induced shoots. The knowledge of how much tissue is worth culturing comes with observing how the different plants respond.

8. Although it is beyond the objective of this text, it is relevant to mention that the new shoots seem to be formed through two different developmental pathways: organogenesis and somatic embryogenesis *(4)*.

9. Browning. The harsh handling of tissues during extraction and cutting and the effect of high concentrations of chemicals during disinfestations or culture will produce browning because of the production and accumulation of phenolic compounds.

10. In some cases explants from the offshoots from rhizomes do not respond to the induction treatments and turn brown and dry, producing only a few plantlets or none at all. The reasons for this can be varied:

    a. Microbial infections, not necessarily obvious or pathogenic, that produce toxins that affect the tissues.
    b. Environmental stress (very high temperatures or water scarcity) that induces the synthesis of natural inhibitors such as abscisic acid.

c. Rhizomes that are "physiologically old tissues" as a result of chemical influences that diffuse from the old mother plant to the offspring.
d. Very small meristematic regions in the tissues used as explants.
e. Seasonal variations.
f. Extreme susceptibility of the extracted tissues to the chemicals used for disinfection.
g. Very low levels of endogenous cytokinins or very high levels of endogenous auxins that are unsuitable for organogenesis.

11. Some explants do not produce any new shoots; this could be a result of the way they are cut. If explants develop only one large central shoot it is because of inadequate removal of the primordial leaves that just continue growing. What is needed for induction is the undifferentiated meristematic tissue that lies below.
Figure 4 shows that if the meristem region is too small, or the block is cut badly leaving the meristem to one side, only a few explants will give off new shoots. This is not caused by dormancy but by the lack of viable meristematic tissue in some of the explants as illustrated in **Fig. 4.**
12. Seasonal variation. The efficiency of the induction stage will vary throughout the year because of the climatic changes and the physiological status of the mother plants. It has been reported that induction is more efficient during the winter months when there is less light. Conversely, contamination is more recurrent during the rainy season.
13. Intra- and inter-specific variation. Different species will show some degree of variation in their culture requirements and their propagation efficiency. In some cases adjustments will have to be made to make the system as efficient as possible. Very big differences are, however, observed between clones of the same species that are probably a result of somatic variation within the cell populations. These variations are not always observed because the slow growing tissues or individuals are rapidly overcome by their more robust and faster growing counterparts, selected against and eliminated.

The results varied, not only according to the species in question, but also depending on the genetic properties of the different clonal lines and the seasonal or environmental conditions in which they were grown:

a. The micropropagation efficiency of *A. tequilana* varied in a cyclical manner according to the month of the year in which the mother plants were collected *(3)*.
b. Selected clonal lines of henequen showed a decrease in their micropropagation efficiency after some months in culture probably as a result of habituation of the tissues.
c. Clones of *Yucca valida* generated from seed collected from the same mother plant can vary in their multiplication efficiency from zero to five new shoots produced at every transfer.
d. *A. tequilana* clones grow very differently, depending on the type of culture container employed.

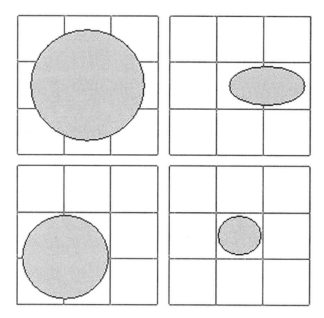

Fig. 4. Diagrammatic representation of the cutting of the explants and hypothetical sizes and positions of the meristematic tissue.

   e. Sisalana and several hybrids of the family of H11648 are being adequately induced and multiplied in Kenya but not in Tanzania. However, at ARI Mlingano using the same methodology, both *A. sisalana* and *A. amaniensis* can be effectively propagated but not H11648.
14. The high concentrations of growth regulators used or the water potential of the culture medium can induce abnormal hyperhydricity that produces the vitreous appearance of the tissues known as vitrification.
15. The leaves formed in vitro are likely to dry out and their function will be to help the plant to survive to the new environmental conditions. The new leaves, formed ex vitro will be normal leaves, with a glaucous appearance, indicative of wax deposition on their surface that will go all the way to the nursery and even the early plantation stages.
16. Quality control. Continuous observation of cultures is very important in order to be able to detect the smallest changes that might start to take place, such as vitrification, or the appearance of microbial contaminants.
17. Microbial contamination is the main enemy of efficient tissue culture and it is a problem that might emerge at any moment during the life of a micropropagation laboratory, causing serious losses.
    Contamination of in vitro cultures can arise from microorganisms present in the explants that are introduced into culture through inadequate disinfestations of the

tissues or can be caused by airborne contaminants that get into the culture dishes during the manipulations or incubation at all stages of the process.

One very important aspect to consider is the presence of latent, nonpathogenic, endophytic microorganisms that might appear after long periods of culture. On the other hand, the production of disease-free plants is important to maintain areas that are free of specific diseases (such as Korogwe Leaf Spot disease in the coastal plantations of East Africa).

References

1. Lock, G. W. (1985) On the scientific and practical aspects of sisal (*Agave sisalana*) cultivation, in *Biología y Aprovechamiento Integral del Henequen y Otros Agaves* (Cruz, C., Del Castillo, L., Robert, M. L., and Ordanza, R. N., eds.), Centro de Investigacióm Científica de Yucatán A. C., México, pp 99–119.
2. Robert, M. L., Herrera, J. L., Contreras, F., and Scorer, K. N. (1987) *In vitro* propagation of *Agave fourcroydes* Lem. (Henequen). *Plant Cell Tiss. Org. Cult.* **8,** 37–48.
3. Robert, M. L., Herrera, J. L., Chan, J. L., and Contreras, F. (1992) Micropropagation of *Agave* ssp. (Henequen), in *Biotechnology in Agriculture and Forestry*,19 (Bajaj, Y. P. S., ed.), Springer-Verlag, Berlin Heidelberg, pp 305–329.
4. Barredo-Pool, F. A., Piven, M., Borges-Argaez, I, C., Herrera, J. L., Herrera, M. A. and Robert, M. L. (2004) Formación de Brotes Adventicios y Embriones Som·ticos en Explantes de Tallo de Henequén (*Agave fourcroydes* Lem.), Cultivados *in vitro*. IV Simposio Internacional sobre Agavaceae y Nolinaceae. Mérida, Yucatán, México.

15

Micropropagation of Endangered Plant Species

Zhihua Liao, Min Chen, Xiaofen Sun, and Kexuan Tang

Summary

This chapter describes the multiple-shoot-based methods of micropropagation for endangered plant species. *Taxus* and aloe are used here as examples. For *Taxus*, the process of micropropagation includes initiating multiple shoots, elongating shoots, rooting shoots, and transplanting plantlets. For aloe, the process of micropropagation includes initiating multiple shoots, rooting shoots, and transplanting plantlets.

Key Words: Chinese aloe; endangered; micropropagation; multiple shoot; plant; *Taxus* x *media*.

1. Introduction

As more and more plant species become rare and endangered, the need to protect and save them from extinction becomes more urgent. Micropropagation is now available as an efficient method of saving these endangered plant species. Micropropagation is a term used to describe a process of plant tissue culture, which is widely used for in vitro vegetative propagation of plants. Micropropagation is generally carried out as follows: a small piece of germ-free plant tissue or explant is taken from the donor plant and cultured on designed nutrient medium in sterile containers. By altering the composition of the medium (carbon source, nitrogen source, inorganic salts, plant growth regulators, vitamins, amino acids, and so on) and the environmental conditions (temperature, light intensity and period, humidity, and others), the development of cultured plant tissue or explant can be directed along different patterns and the whole and intact plant can finally be regenerated. The offspring all come from a single plant and have identical genetic make-ups to each other and to the mother plant. These offspring are called clones.

Micropropagation as classical biotechnology has many advantages. It is a fast method of propagation by which thousands of plants can be generated within one month; it is a human-controlled method by which healthy plant material is ensured as a result of plant tissue sterilization and sterile culture procedures during the propagation cycle; it is an economical method of propagation, by which a large amount of cost of productivity can be saved by using an environment that requires no weeding, uses less water, necessitates a smaller space, and is free of pesticides, among others. Finally, it is an efficient and convenient method of propagation by which only a small piece of plant tissue is needed as the initial material (this is very important for the endangered plant species), and by which the cultured plants can meet specific targets of time and quantity because it is independent of seasonal change and weather.

The protocols for micropropagation of the endangered plant species are presented in this chapter. In the protocols, *Taxus* x *media* is used as the model woody plant species, and Chinese aloe as the model herbal plant species. The methods described here are based on widely used direct initiation of multiple shoots of endangered plant species.

1.1. Micropropagation of Taxus

Taxol®, a member of the taxanoid family of diterpenes, is one of the most potent antitumor agents. Wani et al. *(1)* first defined the structure and biological function of taxol from a bark extract of pacific yew, *Taxus brevifolia*. In the past three decades, taxol has been the first-line drug widely used in the treatment of a variety of cancers, including carcinomas of ovary, breast, lung, head and neck, bladder and cervix, melanomas, and AIDS-related Karposi's sarcoma *(2)*. All *Taxus* spp. can produce taxol, but in extremely low content. For example, the traditional isolation of 1 kg of taxol requires about 6.7 tons of *T. brevifolia* bark, equivalent between 2000 and 3000 trees *(3)*. At present, taxol is mainly extracted from the bark and needles of slow-growing taxus (yew) species, but because these species are rare and, in some cases endangered, renewable yew needles, as opposed to non-renewable bark, are receiving consideration as an alternative source of taxol for clinical use *(4)*.

Taxus cell culture is an alternative source for taxol production but taxol production through cultured *Taxus* cells is low yielding and cannot meet commercial demand. Breeding of elite *Taxus* clones containing high concentrations of taxol or taxol precursors in the needles, followed by intensive mass production of the selected clones, makes a good case for increasing taxol production. *Taxus* x *media* is a hybrid yew and the content of taxol in its perennial mature needles is relatively high. Therefore, the needles of *Taxus* x *media* are the elite and successive source for taxol production. Presented here are the protocols in micropropagation of *Taxus* x *media* *(4,5)*.

1.2. Micropropagation of Chinese Aloe

Aloe vera L. var. chinensis (Haw.) Berger, Chinese aloe, is a native aloe of South China, and has medicinal or cosmetic properties *(8)*. Chinese aloe is one of a few medicinal aloes among more than 500 aloe species, and has antiinflammation, anticancer, antivirus, antibacteria, immune-enhancing, and parasite-killing activities *(9)*. With the rapid expansion of the aloe industry in the world, the need for Chinese aloe is dramatically increasing. Without protection, Chinese aloe could be extinct in the near future. Until now there have been few reports on successful micropropagation of Chinese aloe, partly because of the difficulties in establishing primary explants in culture. In this protocol, we describe how to successfully propagate Chinese aloe by the method of micropropagation *(10)*.

2. Materials

2.1. Micropropogation of Taxus

1. Plant tissue culture equipment.
2. Artificial climate incubator.
3. 3.2 × 15-cm test tubes, 60 × 80-mm plastic pots and bottles (Sigma).
4. Petri dish.
5. Agar (Sigma).
6. Murashige and Skoog (MS) powder (Sigma).
7. Indole-3-butyric acid (IBA), benzylaminopurine (BA).
8. Woody plant medium (WPM).
9. Silver nitrate, Tween-20, 70% ethanol, and 1% sodium hypochlorite.
10. Activated charcoal.
11. Sucrose.
12. Potting soil.
13. Vermiculite.
14. River sand.
15. Clear plastic bags.
16. Filter paper.
17. *Taxus* x *media*.

2.2. Micropropogation of Chinese Aloe

1. Plant tissue culture equipment.
2. Artificial climate incubator.
3. Flask (150-mL).
4. Petri dish.
5. Agar (Sigma).
6. MS powder (Sigma).
7. BA.
8. 2,4-Dichlorophenoxyacetic acid (2,4-D).

9. α-Naphthaleneacetic acid (NAA).
10. Polyvinylpyrrolidone-60 (PVP).
11. Sucrose.
12. Potting soil.
13. Vermiculite.
14. River sand.
15. Culture pot.
16. Chinese aloe.

3. Methods
3.1. Micropropogation of Taxus

The method described in **Subheadings 3.1.1.–3.1.4.** outlines: (1) the establishment of bacteria-free explants of *Taxus x media*; (2) the induction of multiple shoots from explants and shoot elongation; (3) the rooting from multiple shoots; and (4) the transplantation of rooted plantlets.

3.1.1. Establishment of Bacteria-Free Explants of Taxus x Media

1. Two-month-old stems of *Taxus x media* are used as initial explants for in vitro micropropagation. The stems were cut off from the plants and washed under running tap water for 24 h. Then the stems were surface disinfected with 70% ethanol for 1 min, followed by emersion into 1% sodium hypochlorite solution with 0.01% Tween-20 for 15 min, and subsequently by four rinses with sterile distilled water.
2. The stems are then aseptically cut into 1- to 3-cm long segments, with one to three nodes, and erectly inserted into 3.2 × 15-cm test tubes containing 15-mL sterile culture medium, which is half-strength macro-salts of MS medium *(6)* supplemented with 30 g/L sucrose, 5 g/L activated charcoal, 4.43 μM BA, and 100 mg/L silver nitrate with pH 5.8 solidified with 8 g/L agar.
3. The segments in tubes are then incubated at 25 ± 0.5°C with 10-h daylight photoperiod at an intensity of 55 μmol/m^2/s ("daylight" fluorescent tubes). This section is modified from Chang et al. *(4)*.

3.1.2. Induction of Multiple Shoots From Bacteria-Free Taxus x Media Explants and Shoot Elongation

1. After 2 mo of culture, new shoots (approx 1.0 cm long) develop from the stem segments and the stems with shoot tips are transferred into shoot induction medium in bottles. WPM is used *(7)* with 22.19 μM BA and supplemented with 5 g/L activated charcoal and 100 mg/L silver nitrate. The culture conditions are the same as described previously. After 45 d of culture, multiple shoots are developed.
2. Subsequently, the stems with multiple shoots can be transferred into shoot elongation medium (50% MS medium free of plant grow regulators, supplemented with 30 g/L sucrose, 5 g/L activated charcoal, and 100 mg/L silver nitrate with

pH 5.8 solidified with 8 g/L agar) for subculture under the same culture condition. The shoots of more than 2 cm long can be excised and transferred to rooting medium for rooting *(4,5)*.

3.1.3. Rooting From Multiple Shoots of Taxus × Media

1. Young shoots can root on 50% MS medium supplemented with 30 g/L sucrose, 5 g/L activated charcoal, 12.30 μM IBA, and 100 mg/L silver nitrate with pH 5.8 solidified with 8 g/L agar in bottles under the same culture condition mentioned before *(5)*.

3.1.4. Transplantation of Rooted Plantlets

1. When the roots of the shoots reached more than 2 cm long in bottles, the covers of the bottles can be removed, and after another 5 d, the rooted plantlets can be transferred for acclimatization to 60 × 80 mm plastic pots containing a mixture of potting soil, vermiculite, and river sand (1:1:1), and cultured at 25 ± 0.5°C in the growth chamber under a 10-h daylight photoperiod of 55 µmol/m²/s supplied by "Daylight" fluorescent tubes.
2. To maintain cultures at high humidity, pots are covered with clear plastic bags for 3 wk.
3. Subsequently, the plastic cover is gradually removed to reduce the humidity of the pots and to adapt the plantlets to greenhouse conditions. Finally, the well-acclimated plantlets of *Taxus* × *media* can transplanted successfully into the field.

3.2. Microporopogation of Chinese Aloe

The method described in **Suheadings 3.2.1.–3.2.4.** outlines (1) the establishment of bacteria-free explants of Chinese aloe, (2) the induction of multiple shoots from explants, (3) the rooting from multiple shoots, and (4) the transplantation of rooted plantlets.

3.2.1. Establishment of Bacteria-Free Explants of Chinese Aloe

1. Young and strong underground stems of Chinese aloe are used as primary explants for rapid in vitro micropropagation. After cutting the stems with one to two shoots, the explants are washed under running tap water for 24 h.
2. Stems with buds are surface disinfected with 70% (v/v) ethanol for 1 min and 0.1% (w/v) $HgCl_2$ for 10 min, followed by five rinses with sterile distilled water.
3. The surface disinfected stems are cut into 1-cm segments each with a shoot, and the latter is transferred into 150-mL flasks containing 40 mL of medium (MS basal medium + 1.07 mM NAA + 8.87 µM 6-BA + sucrose 30 g/L + agar 8.0 g/L + PVP 0.6 g/L with pH 5.8, autoclaved at 121°C for 15 min) and cultured at 25 ± 0.5°C with 12-h daylight photoperiod at an intensity of 55 µmol/m²/s ("Daylight" fluorescent tubes).
4. Four-week-old sterile aloe shoots are used as explants for micropropagation study.

3.2.2. Induction of Multiple Shoots From Bacteria-Free Chinese Aloe

1. Bacteria-free Chinese aloe shoots obtained previously are separated from stems and transferred to the optimal medium, MS-based medium supplemented with 8.87 µM BA, 1.6 µM NAA, 30 g/L sucrose, 0.6 g/L PVP, and 8.0 g/L agar with pH 5.8, for further growth.
2. The 1-cm-long shoots were then transferred to the optimal medium in 150-mL flasks (one bud/flask). The culture condition is the same as mentioned before.
3. Four weeks later, many multiple shoots of Chinese aloe can be obtained.
4. On the optimal medium, Chinese aloe shoots can grow very well and a single shoot can propagate and multiply 15 times in 4-wk-period; that is, a single shoot can multiply theoretically about 1512 times per year under the proper conditions.

3.2.3. Rooting From Multiple Shoots of Chinese Aloe

1. Young shoots can root easily and fast on half-strength MS medium with 1.07 µM NAA, 30 g/L sucrose, 2.0 g/L PVP, and 8.0 g/L agar with pH 5.8 under the same culture condition mentioned before.
2. After 25 d, the cover of the culture vessel (flask) can be removed, and after another 5 d, the rooted plantlets can be transferred for acclimatization.

3.2.4. Transplantation of Rooted Plantlets

1. Rooted plantlets of Chinese aloe are washed off the attached medium and put in a clean, cool, dry place to evaporate the water of the surface.
2. After 2 d, they are potted in a mixture of potting soil, vermiculite, and river sand (1:1:1) and acclimatized in culture room with relative humidity of 75% at 27°C.
3. Finally, young Chinese aloes are planted in the field successfully.

Acknowledgments

The research is funded by China National High-Tech "863" Program and Shanghai Science and China Ministry of Education.

References

1. Wani, M. C., Taylor, H. L., Wall, M. E., Coggon, P., and McPhail, A. T. (1971) Plant antitumor agents. VI. The isolation and structure of taxol, a novel antileukemic and antitumor agent from Taxus brevifolia. *J. Am. Chem. Soc.* **93,** 2325–2327.
2. Miller, H. I. (2001) The story of taxol: nature and politics in the pursuit of an anticancer drug. *Nat. Med.* **7,** 148.
3. Jennewein, S. and Croteau, R. (2001) Taxol: biosynthesis, molecular genetics, and biotechnological applications. *Appl. Microbiol. Biotechnol.* **57,** 13–19.
4. Chang, S.H., Ho, C.K., Chen, Z.Z., and Tsay, J.Y. (2001) Micropropagation of *Taxus mairei* from mature trees. *Plant Cell Rep.* **20,** 496–502.
5. Majada, J. P., Sierra, M. I., and Sánchez-Tamés, R. (2001) One step more towards taxane production through enhanced *Taxus* propagation. *Plant Cell Rep.* **19,** 825–830.

6. Murashige, T. and Skoog, F. (1962) A revised medium for rapid growth and bioassays with tobacco tissue cultures. *Physiol. Plant.* **15,** 473–479.
7. Lloyd, D. G. and McCown, B. H. (1980) Commercially feasible micropropagation of mountain laurel (Kalmia latifolia) by use of shoot tip culture. *Comb. Proc. Inter. Plant Propagat. Soc.* **30,** 421–427.
8. Gui, Y. L., Xu, T. Y., Gu, S. R., et al. (1990) Studies on stem tissue culture and organogenesis of *Aloe vera*. *Acta Bot. Sin.* **32,** 606–610.
9. Reynolds, T. and Dweck, A.C. (1999) *Aloe vera* leaf gel: a review update. *J. Ethnopharmacol.* **68,** 3–37.
10. Liao, Z, Chen, M., Tan, F., Sun, X., and Tang, K. (2004) Microprogagation of endangered Chinese aloe. *Plant Cell Tiss. Org. Cult.* **76,** 83–86.

16

Clonal Propagation of Softwoods

Trevor A. Thorpe, Indra S. Harry, and Edward C. Yeung

Summary

Softwoods or gymnosperms, which make up 60% of the forested areas of the world, are economically important as a source of lumber, pulp, and paper. Reforestation is a major activity worldwide and the potential benefits of using clonal planting stock have long been recognized. Tissue culture clonal methods or micropropagation is a newer approach that can be achieved by enhancing axillary bud breaking, production of adventitious buds (organogenesis), and somatic embryogenesis. Plantlet production via organogenesis requires at least four stages: (1) bud induction on the explant, (2) shoot development and multiplication, (3) rooting of developed shoots, and (4) hardening of plantlets. Similarly, the production of plantlets via somatic embryogenesis, which has the potential to produce a larger number of plantlets, and in a shorter period of time, also requires several stages. These include (1) induction, maintenance, and proliferation of embryogenic tissue; (2) maturation (both morphological and physiological) of somatic embryos; and (3) germination and conversion of the somatic embryos. In this chapter, plantlet production via organogenesis from seedling and adolescent/mature explants and somatic embryogenesis from immature and mature seeds of white spruce (*Picea glauca*) are outlined.

Key Words: Conifer micropropagation; dormant buds; organogenesis; plantlet regeneration; seedling explants; somatic embryogenesis; white spruce (*Picea glauca*); zygotic embryos.

1. Introduction

Softwoods or gymnosperms are cone-bearing plants, of which conifers are the best known and most economically important, as they are the source of lumber, pulp, and paper *(1)*. They make up approx 60% of the forested areas of the world and include pines (*Pinus* spp.), spruces (*Picea* spp.), hemlock (*Tsuga* spp.), bald cypress (*Taxodium* spp.), redwood (*Sequoia* spp.), arbor vitae (*Thuja* spp.), and juniper (*Juniperus* spp.) *(2)*. Softwoods are a very old group of plants from the Permian period (i.e., 180–205 million yr ago), and grow best in temperate zones, where they form huge forests in North America, Europe, and

Asia. Very few are strictly tropical in distribution and usually occur at high altitudes in these regions. Although not native to Australia and New Zealand, extensive stands of certain conifers (e.g., *Pinus radiata)*, can be found there.

It is generally recognized that the forests are being harvested at a faster rate than they are being regenerated, either naturally or artificially, thus interest in clonal regeneration is high *(3)*. In addition, certain diseases and pests, as well as fires, threaten the very existence of some species. There is therefore an urgent need for large numbers of improved fast-growing trees *(4)*. The potential benefits of using clonal planting stock have long been recognized, as at least a 10% increase in gain can be expected *(5)*. However, for the maximum potential genetic gain for forest improvement, both sexual reproduction and vegetative multiplication must be used *(6)*. The former is important for both introducing new genes to prevent inbreeding and for achieving genetic gain for those characteristics controlled by additive gene effects; the latter allows for the multiplication of elite full-sib families or individuals in a family that show significant genetic gain because of nonadditive gene effects. As well, the use of recombinant gene technology also requires a clonal propagation strategy.

The traditional methods for vegetative propagation of forest species are rooted cuttings or rooted needle fascicles for pine species, and grafting. However, these methods cannot be used successfully for the majority of softwoods *(4,7)*, hence the need for tissue culture clonal methods or micropropagation. Micropropagation can be achieved by using three approaches: (1) enhancing axillary bud breaking, (2) production of adventitious buds, and (3) somatic embryogenesis. The first two approaches lead to plantlet formation via organogenesis through the production of unipolar shoots, which must then be rooted in a multi-staged process. In contrast, somatic embryogenesis leads to the formation of a bipolar structure, through steps that are often similar to zygotic embryogenesis. The potential for forming large numbers of plantlets in vitro increases in the above order, but unfortunately so does the difficulty in producing plantlets. Furthermore, a major problem in the micropropagation of woody species is that most success is achieved with juvenile tissue and not from proven mature trees *(8,9)*.

To illustrate the approaches used for plantlet regeneration with softwoods, work carried out with spruces *(Picea* spp.) in general *(10)* and white spruce *(Picea glauca)* in particular will be presented. White spruce is one of the most widely distributed conifers in North America. It represents an ecologically valuable species that is extensively utilized by the forestry industry for lumber and pulpwood. Several labs have contributed to the white spruce story, in which plantlets were obtained first by organogenesis starting in 1976 *(11,12)*, and later by somatic embryogenesis in 1987 *(13,14)*. Reference will also be made to studies carried out with red spruce *(Picea rubens) (15,16)* to point out differences and similarities to white spruce. Red spruce is a prominent species in the northeastern United States and the Maritime provinces of Canada. It is used for pulpwood and in general construction.

Production of plantlets via organogenesis requires at least four stages: (1) bud induction on the explant, (2) shoot development and multiplication, (3) rooting of developed shoots, and (4) hardening of plantlets *(3,17)*. Similarly, the production of plantlets via somatic embryogenesis also requires several stages: (1) induction, maintenance, and proliferation of embryogenic tissue, (2) maturation (both morphological and physiological) of somatic embryos, and (3) germination and conversion of the somatic embryos *(10)*.

2. Materials
2.1. Organogenesis

Stock salt formulations for bud induction and elongation, namely Schenk and Hildebrant (SH) *(18)* and Mohammed et al. (GMD) *(19)* and organic additives are outlined in **Table 1**. In addition, phytohormones including N6-benzyladenine (BA), 6-γ(-dimethylallyl-amino)-purine (2iP), and indolebutyric acid (IBA), rooting powder (Stim-root No. 1 with 0.1% IBA) (Scientific Garden Aids, Plant Products Co. Ltd., Bramalea, ON, Canada), sucrose, Difco bacto-agar, and conifer-derived activated charcoal (AC, Difco), and fine vermiculite are all used at various stages. Adjust the pH of the culture media to between 5.7 and 5.8 (for SH) and 5.6 for GMD) before adding agar (8 and 5 g/L, respectively). The SH medium is placed in 125-mL Erlenmeyer flasks (50 mL) and the GMD medium in 150 × 25-mm test tubes (20 mL/tube) or 125-mL Erlenmeyer flasks (25 mL each) (*see* **Note 1**).

2.2. Somatic Embryogenesis

Stock media are von Arnold and Eriksson (AE) and *(22)* Litvay et al. LV *(23)* (**Table 2**) as reported by Joy et al. *(24)* and Tremblay *(25)*, respectively.

3. Methods
3.1. Organogenesis From Seedling Explants (20)
3.1.1. Explant Preparation and Culture Initiation

1. Wash white spruce seed (stored at 5°C) for 12 h in tap water. Sterilize for 15 min in 20% commercial bleach (6% NaOCl) containing two drops Tween-20 per 100 mL as surfactant. Wash six times with sterile distilled water.
2. Germinate seeds on 50% strength SH mineral salts with 2% sucrose and solidified with 0.8% (w/v) Difco Noble agar in 125-mL Erlenmeyer flasks.
3. Place in cold (5°C) for 5 d and then in a culture room at 27 ± 1.5°C under a 16 h photoperiod at 80 µmol/m^2/s (*see* **Note 2**).
4. Select epicotyl explants 27–29 d following germination (i.e., radicle emergence), in seedlings in which some swelling at the base of the apex had occurred. Trim cotyledons (~1 mm removed) and cut just below swollen base (*see* **Note 3**).

Table 1
Composition of Basal SH and GMD Media

Component	SH medium	GMD medium
Macronutrients	mg/L	mg/L
KNO$_3$	2500	1820
NH$_4$NO$_3$	–	405
(NH$_4$)2SO$_4$	–	131
MgSO$_4$.7H$_2$O	400	394
NH$_4$H$_2$PO$_4$.2H$_2$O	300	293
CaCl$_2$.2H$_2$O	200	184
Micronutrients		
NaH$_2$PO$_4$.2H$_2$O	–	7.8
MnSO$_4$.4H$_2$O	10.0	11.2
ZnSO$_4$.7H$_2$O	1.0	2.0
CuSO$_4$.5H$_2$O	0.2	0.03
KI	1.0	0.9
CoCl$_2$.6H$_2$O	0.1	0.2
H$_3$BO$_3$	5.0	3.1
Na$_2$MoO$_4$.2H$_2$O	0.1	0.02
FeSO4.7H$_2$O	15.0	13.9
Na$_2$.EDTA	20.0	18.6
Organics		
myo-inositol	1000	100
Nicotinic acid	5.0	2.0
Pyridoxine·HCl	0.5	1.0
Thiamine·HCl	5.0	5.1

From **refs.** *18* and *19*.
Abbr: EDTA, ethylene diamine tetraacetic acid.

5. Place trimmed epicotyls on SH medium (**Table 1**) with sucrose (3%) benzyladenine (BA) and 2iP (5 m*M* each) for 28 d in light, followed by 20 d on cytokinin-free medium (*see* **Notes 4** and **5**).

3.1.2. Shoot Development and Multiplication

1. Cut shoot clumps vertically into four pieces and place on full-strength SH medium without cytokinins, but with sucrose (2%) and activated charcoal (0.2%) for 28 d for shoot elongation.
2. Harvest elongated shoots (>1 cm) and subdivide and subculture clumps on above medium without AC at 28–30 d for secondary shoot formation.

Table 2
Composition of the Basal AE and the Full-Strength LV Media

Component	AE medium	LV medium[a]
Macronutrients	mg/L	mg/L
NH_4NO_3	1200	1650
KNO_3	1900	1900
$MgSO_4.7H_2O$	370	1850
KH_2PO	340	340
$CaCl_2.2H_2O$	180	22
Micronutrients		
KI	0.75	4.15
H_3BO_3	0.63	31
$MnSO_4.4H_2O$	2.2	27.7
$ZnSO_4.7H_2O$	4.05	43
$Na_2MoO_4.2H_2O$	0.025	1.25
$CuSO_4.5H_2O$	0.0025	0.5
$CoCl_2.6H_2O$	0.0025	0.13
$FeSO_4.7H_2O$	5.57	27.8
$Na_2.EDTA$	12.85	37.3
Organics		
myo-inositol	500	100
Nicotinic acid	2	0.5
Pyridoxine.HCl	1	0.1
Thiamine.HCl	5	0.1

[a]For many conifers, 50% LV medium with full organics, or L-glutamine (250 mg/L) and casein hydrolysate (500 mg/L) has proven suitable *(36)*. From **refs.** *22* and *23*.

3. Repeat the above sequences to enhance shoot elongation and secondary shoot formation for a total of 128 d (*see* **Note 6**).

3.1.3. Rooting

1. Prepare trays of fine vermiculite, saturate with 50% strength SH salts, full organics, sucrose (2%), and activated charcoal (0.2%), autoclave at 121°C for 20 min.
2. Select shoots 10 mm or longer, dip lower 1–2 mm in sterile water and then into the rooting powder (sterilized by autoclaving at 121°C for 20 min) and place in trays of vermiculite.
3. Culture under a day:night regime of 20:18°C, 12 h each (*see* **Note 7**).
4. Harden rooted shoots at 20°C for 1–2 wk before transfer to greenhouse.

3.2. Organogenesis From Adolescent/Mature Explants (19,21)

3.2.1. Sterilization and Explant Preparation

1. Select dormant buds with about 3 mm subtending stem tissue from first- and second-order branches of 15- to 18-yr-old white spruce trees.
2. Surface sterilize for 15 min in 1.2% chlorine water (from commercial bleach, 5.25% NaOCl). Rinse buds three times with sterile double-distilled water.
3. Remove bud scales by a circumferential cut at the widest part of the bud. Excise the terminal portion of the exposed embryonal shoot axis comprising the meristem dome and several needle primordia and discard.
4. Select the remaining bud base, with approx 50% embryonal shoot axis and 50% subtending crown and stem tissue.
5. Place five bud base explants in 125-mL Erlenmeyer flasks, each containing 25 mL GMD medium (**Table 1**) with BA (13.3 µM) and solidified with 5 g/L agar, for 6 wk under a photon flux density of 110 µmol/m^2/s cool white fluorescent light during a 16 h photoperiod at a temperature of 25 ± 2°C (*see* **Note 8**).
6. Rooting of shoots as in **Subheading 3.1.3.**

3.3. Somatic Embryogenesis (4,13,26)

3.3.1. Embryogenic Culture Initiation

1. Immature and mature seeds can be used as explants.
2. Harvest immature cones after cotyledon initiation in embryos. Harsh surface sterilization of cones in 20% bleach (Javex) with three to four drops of Tween-20/100 mL for 30 min. Stir cones slowly using magnetic stirrer. Rinse three times with sterile distilled water for 10 min each. Soak cones in 70% ethanol for 5 min and rinse in sterile distilled water.
3. Remove seeds from cones and then dissect out embryos under aseptic conditions using a stereomicroscope on a clean bench.
4. Place embryos directly on induction medium, consisting of AE salts with organics (**Table 1**), supplemented with 10 µM 2,4-dichlorophenoxyacetic acid (2,4-D) or picloram, 5 µM BA, and 5% sucrose, solidified with 0.7–0.8% Difco Noble agar. Culture in the dark at 25°C.
5. Place mature seeds on cheesecloth and make into small bag with string. Sterilize bags in 10% bleach with Tween-20, as in **step 2**, in a beaker for 30 min, stirring gently on magnetic stirrer. Rinse 3× with sterile distilled water with continuous stirring.
6. Place bags with a few mL of distilled water inside sterile Petri dish to imbibe overnight at 4°C.
7. Re-sterilize seeds as in **step 4** for 15 min, followed by three rinses with sterile water.
8. Cut open bags and transfer seeds to sterile Petri dishes with sterile water (10 mL in a 15-cm Petri dish). Discard floating seeds (*see* **Note 9**).
9. Dissect out embryos as in **step 2** above and place directly on induction medium, as in **step 3**.

Clonal Propagation of Softwoods

10. Place embryos on induction medium in the dark. Each Petri dish contains between 10 and 15 embryos and the dishes are sealed with Parafilm®.
11. Examine Petri dishes carefully for contamination after 2–3 d. Transfer noncontaminated embryos to new Petri dishes with induction medium.
12. Remove and subculture white mucilaginous tissue, which contains proembryos, after 2–4 wk on induction medium (as in **step 9**) (*see* **Notes 10** and **11**).

3.3.2. Culture Maintenance

1. Transfer embryonic tissue to maintenance medium consisting of AE salts and organics and containing 10 µM 2,4-D, 2 µM BA, and 3% sucrose, gelled with 0.4% phytagel or 0.8% Difco Noble agar. Place in dark and subculture at 3–4 wk intervals.
2. To make cell suspension cultures, transfer maintenance tissue to liquid maintenance medium (AE or LV without gelling agent). Shake on a gyrotory shaker at approx 100 rpm and subculture at 7-d intervals in the dark. (For greater detail, *see* Chapter 8.)

3.3.3. Embryo Development and Maturation

1. Transfer embryogenic masses from solid culture to solid AE maturation medium devoid of phytohormones for 1 wk, followed by transfer to maturation medium containing 50 µM abscisic acid (ABA) and 5% sucrose. Transfer to fresh ABA medium after 20 d.
2. Transfer embryogenic masses from liquid culture to liquid AE maturation medium devoid of phytohormones for 1 wk, followed by transfer to maturation medium with ABA for 1 wk. Next spread the embryogenic tissue on sterile filter paper (Whatman no. 1) and place on solidified ABA medium. All cultures are kept in the dark (*see* **Note 12**).
3. After 40 d, transfer developed somatic embryos on filter paper for partial drying to the center wells of a 24-well tissue culture plate (Falcon 3847, Frankling Lakes, NJ), containing sterile water in the outer wells for 10 d in the dark. The filter paper (6-mm diameter) is moistened with maturation medium, with six to seven embryos per disk and three disks per well (*see* **Note 13**).

3.3.4. Embryo Germination and Conversion

1. Place partially dried embryos on germination medium (50% strength AE medium with 1% sucrose, 100 µM ascorbic acid [filter-sterilized and added after medium autoclaved]), solidified with 0.8% agar. Place in dark or dim light for 1–2 d.
2. Incubate in light (90–95 µmol/m^2/s photon flux density) under a 16 h photoperiod for approx 4 wk.
3. Place plantlets inside sterile glass jars (Pyrex 100 × 80, No. 3250) containing sterile peat pellets saturated with 50% strength LV medium without sucrose. Keep plantlets in closed, but unsealed, vessels under the **step 2** light and photoperiod conditions for 3–4 wk.
4. Finally, transfer to greenhouse for further growth under nonsterile conditions (*see* **Note 14**).

4. Notes

1. For preparing stock solutions, use of implements, equipment, and so on, and requirements for a tissue culture facility (*see* **refs.** *3,7,17,27–29*).
2. Some softwood seeds require much more elaborate pre-treatment; including stratification before excision and explantation *(3)*.
3. Other seedling explants of white spruce have been used for organogenesis, namely hypocotyls *(11,12)* and cotyledons *(30)*.
4. For red spruce, excised mature embryos, cotyledons, and hypocotyl explants produce shoot buds *(15)*.
5. Examination of white spruce epicotyls revealed that a specific ring of meristematic tissue was induced during primary shoot formation and that the secondary shoots were initiated on this tissue *(17,31)*. Thus, a key to high rates of adventitious budding was dependent on the morphogenic activity of this zone.
6. No studies were carried out on the long-term capacity of this morphogenic tissue to continue to produce shoots, capable of producing normal plantlets. Under these conditions outlined, an average of about 40 plantlets per seed or approx 50 plantlets per explant were obtained in about 170 d from seed germination *(20)*.
7. Rooting of white spruce shoots (and some other conifers) can also be carried out under nonsterile conditions. Seed trays with fine vermiculite, prepared in a similar manner as for sterile rooting, were left partly uncovered after planting. The vermiculite was allowed to almost dry out between watering with 50% strength SH salts, plus sucrose, and a fungicide (Captan®). White spruce shoots began to root after 8 d under these conditions, and the rooting success was equivalent to sterile rooting *(20)*. However, the shoots became dormant, with the apical buds becoming encapsulated in bud scales *(20)*.
8. The bud base explants increased in size from approx 2 mm to approx 4–5 mm by 6–9 wk, by which time enlargement of the needle primordia had occurred. Between 9 and 12 wk, small shoot meristems were observed on the modified needles, and these developed asynchronously into distinct shoot primordia and then 1-cm-long shoots from the wk 15 *(19)*.
9. Seeds that float should be discarded, as imbibition was incomplete and air trapped inside could harbor contaminants.
10. Although immature embryos respond faster than mature embryos, tissue proliferation begins after 2 wk in culture, ranging from compact nodular masses to friable mucilaginous tissue with filamentous structures *(26)*.
11. In addition to immature and mature embryos, excised cotyledons from young white spruce seedlings have served as explants for somatic embryogenesis *(32)*. For red spruce, mature embryos from stored seeds served as explants *(16)*.
12. Somatic embryo development takes place over several weeks, leading ultimately to an elongated embryo with four to six cotyledons. Necrosis occurs in the area between the embryo proper and the suspensor system, allowing for easy separation of the mature embryo *(26)*.
13. As an alternative to the partial drying treatment, somatic embryos can be dried using nonpermeating nonplasmolysing osmotica such as polyethylene glycol 4000 or dextran 6000 *(33)*.

14. Acclimatization conditions are important to maximize survival and provide vigorous somatic seedlings or emblings *(34)*. White spruce emblings acclimatized under high humidity and low light conditions survived transfer to ex vitro conditions and overwintering *(35,36)*.

References

1. Wenger, K. F. (1984) *Forestry Handbook*. John Wiley, NY.
2. Haynes, J. D. (1975) *Botany: An Introductory Survey of the Plant Kingdom*, John Wiley and Sons, NY, pp. 365–421.
3. Thorpe, T. A. and Harry, I. S. (1991) Clonal propagation of conifers, in *Plant Tissue Culture Manual* (Lindsey, K., ed.), Kluwer Academic Publishers, Dordrecht, pp. C3, 1–16.
4. Thorpe, T. A. and Biondi, S. (1984) Conifers, in *Handbook of Plant Cell Culture, vol. 2* (Sharp, W. R., Evans, D. A., Ammirato, P.V., Yamada, Y., eds.), Macmillan, NY, pp. 435–470.
5. Kleinschmit, J. (1974) A programme for large-scale cutting propagation of Norway spruce. *New Zealand J. For. Sci.* **4,** 359–366.
6. Hasnain, S. and Cheliak, W. (1986) Tissue culture in forestry: economic and genetic potential. *For. Chron.* **62,** 219–225.
7. Thorpe, T. A., Harry, I. S., and Kumar, P. P. (1990) Application of micropropagation to forestry, in *Micropropagation: Technology and Application* (Debergh, P. and Zimmerman, R. H., eds.), Kluwer Academic Publishers, Dordrecht, The Netherlands, pp. 311–336.
8. Dunstan, D. I. and Thorpe, T. A. (1986) Regeneration in forest trees, in *Cell Culture and Somatic Cell Genetics in Plants, vol. 3* (Vasil, I. K., ed.), Academic Press, NY, pp. 233–241.
9. Harry, I. S. and Thorpe, T. A. (1990) Special problems and prospects in the propagation of woody species, in *Plant Aging: Basic and Applied Approaches* (Rodriguez, R., Tamés, R. S., and Durzan, D. J., eds.), Plenum, NY, pp. 67–74.
10. Thorpe, T. A. and Harry, I. S. (2000) Micropropagation of Canadian spruces (*Picea* spp.), in *Transplant Production in the 21st Century* (Kubota, C. and Chun, C., eds.), Kluwer Academic Publishers, Dordrecht, The Netherlands, pp. 197–206.
11. Campbell, R. A. and Durzan, D. J. (1975) Induction of multiple buds and needles in tissue cultures of *Picea glauca*. *Can. J. Bot.* **53,** 1652–1657.
12. Campbell, R. A. and Durzan, D. J. (1976) Vegetative propagation of *Picea glauca* by tissue culture. *Can. J. For. Res.* **6,** 240–243.
13. Lu, C. -Y. and Thorpe, T. A. (1987) Somatic embryogenesis and plantlet regeneration in cultured immature embryos of *Picea glauca*. *J. Plant Physiol.* **128,** 297–302.
14. Hakman, I. and Fowke, L. C. (1987) Somatic embryogenesis in *Picea glauca* (white spruce) and *Picea mariana* (black spruce). *Can. J. Bot.* **65,** 656–659.
15. Lu, C.-Y., Harry, I. S., Thompson, M. R., and Thorpe, T. A. (1991) Plantlet regeneration from cultured embryos and seedling parts of red spruce (*Picea rubens* Sang). *Bot. Gaz.* **152,** 42–50.
16. Harry, I. S. and Thorpe, T. A. (1991) Somatic embryogenesis and plant regeneration from mature zygotic embryos of red spruce. *Bot. Gaz.* **152,** 446–452.

17. Thorpe, T. A. and Patel, K. R. (1984) Clonal propagation: adventitious buds, in *Cell Culture and Somatic Cell Genetics in Plants, vol. 1* (Vasil, I. K., ed.), Academic Press, N. Y., pp. 49–60.
18. Schenk, R. V. and Hildebrandt, A. C. (1972) Medium and techniques for induction and growth of monocotyledonous and dicotyledonous plant cell cultures. *Can. J. Bot.* **50,** 199–204.
19. Mohammed, G. H., Dunstan, D. I., and Thorpe, T. A. (1986) Influence of nutrient medium upon shoot initiation on vegetative explants excised from 15- to 18-year-old *Picea glauca*. *New Zealand J. For. Sci.* **16,** 297–305.
20. Rumary, C. and Thorpe, T. A. (1984) Plantlet formation in black and white spruce. I. *In vitro* techniques. *Can. J. For. Res.* **14,** 10–16.
21. Dunstan, D. I., Mohammed, G. H., and Thorpe, T. A. (1987) Morphogenetic response of vegetative bud explants of adolescent and mature *Picea glauca* (Moench) Voss *in vitro*. *New Phytol.* **106,** 225–236.
22. von Arnold, S. and Eriksson, T. (1981) *In vitro* studies of adventitious shoot formation in *Pinus contorta*. *Can. J. Bot.* **59,** 870–874.
23. Litvay, J. D., Johnson, M. A., Verma, D., Einspahr, E., and Weyrauch, K. (1981) Conifer suspension culture medium development using analytical data from developing seeds. IPC Technical Paper Series, No. 115, pp. 1–17.
24. Joy, R. W., IV, Yeung, E. C., Kong, L., and Thorpe, T. A. (1991) Development of white spruce somatic embryos: I. Storage product deposition. *In Vitro Cell. Devel. Biol.* **27P,** 32–41.
25. Tremblay, F. M. (1990) Somatic embryogenesis and plantlet regeneration from embryos isolated from stored seeds of *Picea glauca*. *Can. J. Bot.* **68,** 236–242.
26. Yeung, E. C. and Thorpe, T. A. (2004) Somatic embryogenesis in *Picea glauca*, Chap. 29, in *Protocols of Somatic Embryogenesis in Woody Plants* (Jain, S. M., and Gupta, P., eds.), Kluwer Academic Publishers, Dordrecht, The Netherlands (in press).
27. Biondi, S. and Thorpe, T. A. (1981) Requirements for a tissue culture facility, in *Plant Tissue Culture—Methods and Application in Agriculture* (Thorpe, T. A., ed.), Academic Press, NY, pp. 1–20.
28. Brown, D. C. W. and Thorpe, T. A. (1984) Organization of a plant tissue culture laboratory, in *Cell Culture and Somatic Cell Genetics of Plants, vol. 1, Laboratory Procedures and Their Applications* (Vasil, I. K., ed.), Academic Press, NY, pp. 1–12.
29. Harry, I. S. and Thorpe, T. A. (1994) In vitro methods for forest trees, in *Plant Cell and Tissue Culture* (Vasil, I. K., and Thorpe, T. A., eds.), Kluwer Academic Publishers, Dordrecht, The Netherlands, pp. 539–560.
30. Toivonen, P. M. A. and Kartha, K. K. (1988) Regeneration of plantlets from in vitro cultured cotyledons of white spruce (*Picea glauca* (Moench) Voss). *Plant Cell Rep.* **7,** 318–321.
31. Rumary, C., Patel, K. R., and Thorpe, T. A. (1986) Plantlet formation in black and white spruce. II. Histological analysis of adventitious bud formation *in vitro*. *Can. J. Bot.* **64,** 997–1002.

32. Lelu, M.-A. and Bornman, C. H. (1990) Induction of somatic embryogenesis in excised cotyledons of *Picea glauca* and *Picea mariana*. *Plant Physiol. Biochem.* **28**, 785–791.
33. Attree, S. M., Moore, D., Sawhney, V. K., and Fowke, L. C. (1991) Enhanced maturation and desiccation tolerance of white spruce (*Picea glauca* (Moench) Voss) somatic embryos. Effects of a non-plasmolysing water stress and abscisic acid. *Ann. Bot.* **68**, 519–525.
34. Roberts, D. R., Webster, F. B., Flinn, B. S., Lazaroff, W. R., and Cyr, D. R. (1993) Somatic embryogenesis of spruce, in *Synseeds* (Redenbaugh, K., ed.), CRC Press, Boca Raton, FL, pp. 428–450.
35. Attree, S. M., Pomeroy, M. K., and Fowke, L. C. (1995) Development of white spruce (Picea glauca (Moench) Voss) somatic embryos during culture with abscisic acid and osmoticum and their tolerance to drying and frozen storage. *J. Exp. Bot.* **46**, 433–439.
36. Attree, S. and Fowke, L. C. (1995) Conifer somatic embryogenesis, embryo development, maturation drying, and plant formation, in *Plant Cell Tissue and Organ Culture—Fundamental Methods* (Gamborg, O. L. and Phillips, G. C., eds.), Springer-Verlag, Berlin, pp. 103–113.

IV

APPLICATIONS FOR PLANT PROTOPLASTS

17

Isolation, Culture, and Plant Regeneration From Leaf Protoplasts of *Passiflora*

Michael R. Davey, Paul Anthony, J. Brian Power, and Kenneth C. Lowe

Summary

The family Passifloraceae contains many species exploited in the food, pharmaceutical, and ornamental plant industries. The routine culture of isolated protoplasts (naked cells) followed by reproducible plant regeneration, is crucial to the genetic improvement of *Passiflora* spp. by somatic cell technologies. Such procedures include somatic hybridization by protoplast fusion to generate novel hybrid plants, and gene introduction by transformation. Seedling leaves are a convenient source of totipotent protoplasts. The protoplast-to-plant system developed for *Passiflora edulis* fv. flavicarpa is summarized in this chapter. The procedure involves enzymatic degradation of leaf tissue using commercially-available Macerozyme R10, Cellulase R10, and Driselase. Isolated protoplasts are cultured in Kao and Michayluk medium, semi-solidified with agarose. The medium containing the suspended protoplasts is dispensed as droplets or thin layers and bathed in liquid medium of the same composition. Shoot regeneration involves transfer of protoplast-derived tissues to Murashige and Skoog-based medium. The protocols developed for *P. edulis* are applicable to other *Passiflora* spp. and will underpin the future biotechnological exploitation of a range of species in this important plant family.

Key Words: Agarose culture media; leaf protoplasts; morphological, cytological, and molecular analyses; *Passiflora edulis*; passionfruit; protoplast-to-plant systems.

1. Introduction

More than 580 woody and herbaceous species constitute the family Passifloraceae *(1)*, most of which are endemic to tropical South America. *Passiflora edulis* fv. flavicarpa is the most important commercial species *(2)*, because of the use of its fruit for juice. This species is also resistant to the soil-borne pathogen *Fusarium oxysporum*. Consequently, it is often used as a rootstock onto which *P. edulis* Sims is grafted. Fertile interspecific hybrids have been difficult to generate in Passiflora breeding programs *(3)*. However, somatic hybridization provides a means of overcoming sexual barriers to introduce disease resistance and other desirable characteristics from the wild into cultivated species. For example, novel fertile somatic hybrids have been produced by protoplast fusion between *P. edulis* fv. flavicarpa and wild *Passiflora incarnata (4)* for transfer of cold tolerance to the commercial crop *(5)*. Somatic hybrid plants have also been generated between *P. edulis* fv. flavicarpa and *Passiflora alata, Passiflora amethystina, Passiflora cincinnata, Passiflora coccinea*, and *Passiflora giberti*, respectively *(6)*.

A prerequisite to any somatic hybridization program is the development of a reproducible protoplast-to-plant system that contains at least one of the parental species. Protoplast isolation is influenced by several parameters, particularly the plant genotype, the source tissue and its physiological status, together with environmental factors *(7–9)*. To date, plant regeneration from *Passiflora* protoplasts has been reported for *P. edulis* fv. flavicarpa *(10,11)*, *P. amethystina*, and *P. cincinnata (11)*. Most of these studies have used similar enzyme mixtures, media based on the formulation of Kao and Michayluk *(12)*, and comparable culture techniques, including embedding protoplasts in thin layers or droplets of agarose-solidified media. Novel approaches to promote division of cultured protoplast-derived cells include supplementation of media with the commercially available haemoglobin solution Erythrogen™ *(13)*, and oxygen-gassed perfluorodecalin (Flutec® PP5) *(14)*.

2. Materials

2.1. Glasshouse-Grown Seedlings of Passiflora edulis

1. Seeds of *P. edulis* fv. flavicarpa (National Collection of Passiflora, Kingston Seymour, Clevedon, UK).
2. Levington M3 soilless compost (Fisons, Ipswich, UK) and John Innes No. 3 compost (J. Bentley Ltd., Barrow-on-Humber, UK).
3. "Vac-trays" (H. Smith Plastics Ltd., Wickford, UK).

2.2. Protoplast Isolation From Leaves of Passiflora edulis

1. "Domestos" bleach (Johnson Diversey Ltd., Northampton, UK) or any similar commercial bleach solution containing about 5% available chlorine.
2. Enzyme solution *(4)*: 2g/L Macerozyme R10 (Yakult Honsha Co. Ltd., Nishinomiya Hyogo, Japan), 10 g/L Cellulase R10 (Yakult Honsha Co. Ltd.), 1 g/L Driselase (Kyowa Hakko Co. Ltd., Tokyo, Japan), 1.1 g/L 2-[N-morpholino]ethanesulfonic acid (MES; Sigma, Poole, UK), 250 mg/L polyvinylpyrrolidone (PVP-10; Sigma), 250 mg/L cefotaxime ("Claforan"; Roussel Laboratories, Uxbridge, UK) with modified CPW salts *(15)*, pH 5.8. Filter and sterilize.

Isolation, Culture, and Plant Regeneration

3. Modified CPW salts solution (15): 27.2 mg/L KH_2PO_4, 101 mg/L KNO_3, 246 mg/L $MgSO_4 \cdot 7H_2O$, 0.16 mg/L KI, 0.025 mg/L $CuSO_4 \cdot 5H_2O$, 1480 mg/L $CaCl_2 \cdot 2H_2O$, pH 5.8.
4. CPW13M solution: CPW salts solution containing 130 g/L mannitol (see **Subheading 2.2.3.**), pH 5.8. Autoclave at 121°C in saturated steam for 20 min.
5. SeaPlaque agarose (BioWhittaker UK Ltd., Wokingham, UK). Autoclave.
6. Glass (Pyrex©) casserole dishes.
7. White glazed tiles.

2.3. Culture of Protoplasts From Leaves of Passiflora edulis

1. KM8P medium: prepared to the formulation of Kao and Michayluk (12) with modifications (16) (**Table 1**) and supplemented with 250 mg/L cefotaxime. For semi-solid KM8P medium, mix double strength KM8P medium with an equal volume of molten (40–60°C) 1.6% (w:v) aqueous SeaPlaque agarose, the latter prepared (as with all media) with reverse-osmosis water. Add cefotaxime to the molten medium at 40°C to give a final concentration of 250 mg/L.
2. Fluorescein diacetate (FDA): 3 mg/mL stock solution in acetone (17). Store in the dark at 4°C.
3. KM8 medium: prepared to the formulation of Kao and Michayluk (12) with modifications (16) (**Table 1**). Filter sterilize.

2.4. Plant Regeneration From Protoplast-Derived Tissues of Passiflora edulis

1. MSR1 medium: based on the formulation of Murashige and Skoog (MS) (18) with 26.85 μM α-naphthaleneacetic acid (NAA), 1.11 μM 6-benzylaminopurine (BA), 50 mg/L cysteine, 50 mg/L glutamine, 50 mg/L glutamic acid, 0.5 mg/L biotin, 0.5 mg/L folic acid, 30 g/L sucrose, and 8 g/L agar, pH 5.8. Autoclave the medium and add filter-sterilized amino acids and vitamins after autoclaving.
2. MSR2 medium: MS-based medium with 4.43 μM BA, 30 g/L sucrose, and 8 g/L agar, pH 5.8. Autoclave.
3. MSR3 medium: 50% MS-based medium containing 14.76 μM indole-3-butyric acid (IBA), 2.68 μM NAA, 30 g/L sucrose, and 8 g/L agar, pH 5.8. Autoclave.

3. Methods
3.1. Glasshouse-Grown Seedlings of Passiflora edulis

1. Prepare a compost consisting of equal parts of Levington M3 compost and John Innes no. 3 compost. Fill the individual compartments of plastic "Vac-trays" with the mixture.
2. Place two to three seeds (about 1 cm deep) into each compartment of the "Vac-trays," water from above (see **Note 1**). Place the "Vac-trays" into propagators and cover with transparent lids.
3. Maintain the propagators at a maximum day temperature of 28 ± 2°C with a minimum night temperature of 18 ± 2°C in a glasshouse under natural daylight

Table 1
Formulation of Culture Media for the Growth of Passiflora Leaf Protoplasts

Component	Concentration, mg/L	
	KM8P	KM8
Macronutrients		
NH_4NO_3	600	600
KNO_3	1900	1900
$CaCl_2 \cdot 2H_2O$	600	600
$MgSO_4 \cdot 7H_2O$	300	300
KH_2PO_4	170	300
Sequestrene 330 Fe	28	28
Micronutrients		
KI	0.75	0.75
H_3BO_3	3.0	3.0
$MnSO_4 \cdot H_2O$	10.0	10.0
$ZnSO_4 \cdot 7H_2O$	2.0	2.0
$NaMoO_4 \cdot 2H_2O$	0.25	0.25
$CuSO_4 \cdot 5H_2O$	0.025	0.025
$CoCl_2 \cdot 6H_2O$	0.025	0.025
Vitamins		
Myo-inositol	100	100
Nicotinamide	1.0	1.0
Pyridoxine HCl	1.0	1.0
Thiamine HCl	1.0	1.0
D-Ca Pantothenate	1.0	1.0
Folic acid	0.4	0.4
Abscisic acid	0.02	0.02
Biotin	0.01	0.01
Choline chloride	1.0	1.0
Riboflavin	0.2	0.2
Ascorbic acid	2.0	2.0
Vitamin A	0.01	0.01
Vitamin D_3	0.01	0.01
Vitamin B_{12}	0.02	0.02
Na pyruvate	20	20
Citric acid	40	40
Malic acid	40	40
Fumaric acid	40	40
Other Supplements		
Fructose	250	250
Ribose	250	250

(continued)

Table 1 *(Continued)*
Formulation of Culture Media for the Growth of Passiflora Leaf Protoplasts

Component	Concentration, mg/L	
	KM8P	KM8
Xylose	250	250
Mannose	250	250
Rhamnose	250	250
Cellobiose	250	250
Sorbitol	250	250
Mannitol	250	250
Vitamin free Casamino acids	250	250
Coconut milk	20 mL/L	20 mL/L
2,4-dichlorophenoxyacetic acid	0.2	0.1
Zeatin	0.5	0.2
α-naphthaleneacetic acid	1.0	1.0
Sucrose	250	20,000
Glucose	100,000	10,000
pH	5.8	5.8

supplemented with a 16-h photoperiod provided by cool white fluorescent tubes (180 µmol/s/m^2).

3.2. Isolation of Protoplasts From Seedling Leaves of Passiflora edulis

1. Surface sterilize fully expanded, young leaves excised from 45- to 60-d-old plants in 7% (v:v) "Domestos" bleach solution for 20 min. Wash the excised leaves thoroughly with sterile, reverse osmosis water (three times) (*see* **Note 2**).
2. Place the sterilized leaves on the surface of a sterile white tile. Cut the leaves transversely into 1-mm strips (*see* **Note 3**) and incubate for 30 min about 1 g FW of leaf strips in 20-mL aliquots of CPW13M solution (*see* **Note 4**) contained in 9-cm Petri dishes.
3. Remove the CPW13M solution and replace with enzyme solution (*see* **Note 5**), incubating 1 g FW of tissue in 20 mL of enzyme solution. Seal the dishes with Nescofilm and incubate on a horizontal orbital shaker (40 rpm) at 25 ± 2°C for 16 h in the dark (*see* **Note 6**).
4. Filter the enzyme-protoplast suspension through an autoclaved nylon mesh of 64-µm pore size (*see* **Note 7**). Place the filtrate in 16-mL capacity round bottom screw-capped centrifuge tubes using a Pasteur pipet (*see* **Note 8**).
5. Centrifuge for 7 min at 80*g*. Discard the supernatants and resuspend the protoplast pellets very gently, using a Pasteur pipet, in CPW13M solution. Repeat the centrifugation twice to wash the protoplasts free of any residual enzyme.

6. Resuspend the protoplasts in a known volume (e.g., 10 mL) of CPW13M solution. Remove an aliquot of the suspension (e.g., 0.1 mL) and transfer to a haemocytometer. Count the protoplasts and calculate the yield.
7. Assess protoplast viability using FDA (*see* **Note 9**). Add 100 µL of FDA stock solution to 16 mL of CPW13M solution. Mix one drop of this resulting working solution, using a Pasteur pipet, with an equal volume of protoplast suspension on a glass microscope slide. Observe the protoplasts under ultraviolet illumination (e.g., using a Nikon Optiphot microscope fitted with a high-pressure mercury vapor lamp HBO 100 W/2, a 420–485 nm exciter filter, a DM400 dichroic mirror and a 570 eyepiece absorption filter). Viable protoplasts fluoresce yellow-green (*see* **Note 10**).

3.3. Culture of Leaf Protoplasts of Passiflora edulis

1. Centrifuge the protoplast suspension from **Subheading 3.2.6.** for 7 min at $80g$, remove the supernatant with a Pasteur pipet and discard the supernatant.
2. Resuspend the protoplast pellet in the appropriate volume of molten agarose KM8P medium (*see* **Note 11**) to give a final protoplast plating density after gelling of the medium of 1.5×10^5/mL (*see* **Note 12**).
3. Dispense 40 µL droplets of KM8P medium containing suspended protoplasts in 5-cm Petri dishes (25 droplets/dish). Alternatively, dispense the medium with protoplasts as thin layers (about 5 mL volume). Allow the droplets/thin layers to gel (*see* **Note 13**) and bathe the droplets/thin layers in each dish in 2 mL of liquid KM8P medium containing cefotaxime at 250 mg/L. Seal the dishes with Nescofilm and incubate in the dark at $25 \pm 2°C$.
4. Replace the KM8P bathing medium every 5 d, with new KM8P/KM8 medium mixed in the ratios of 3:1, 2:1, 1:1, and 0:1 (v:v), both containing cefotaxime at 250 mg/L.
5. Determine the protoplast plating efficiency (number of protoplasts undergoing mitotic division expressed as a percentage of the number of viable [FDA positive] protoplasts originally plated) after 6–10 d of culture.

3.4. Plant Regeneration From Protoplast-Derived Tissues of Passiflora edulis

1. Transfer protoplast-derived colonies (*see* **Note 14**), 20 d after isolation of protoplasts, to agar-solidified MSR1 medium (50 colonies/20-mL aliquots of medium in 9-cm Petri dishes). Seal the dishes with Nescofilm and incubate at $25 \pm 2°C$ in the light under a 16-h photoperiod (25 µmol/s/m^2; cool white fluorescent tubes).
2. After a further 25 d of culture, transfer protoplast-derived tissues to agar-solidified MSR2 medium (5 calli/45-mL aliquots of medium in 175-mL glass jars). Incubate as in **Subheading 3.4., step 1**.
3. Subculture protoplast-derived tissues every 30 d to MSR2 shoot regeneration medium as in **Subheading 3.4., step 2**. Shoots should appear progressively within a further 60 d of culture.

4. Excise developing shoots when they are approx 5–6 cm in height, from protoplast-derived tissues. Transfer the shoots to agar-solidified MSR3 rooting medium for 7 d, followed by transfer to MS-based medium, lacking growth regulators (3 shoots/50 mL of medium for both stages in 175-mL jars). Maintain the shoots under the growth conditions as in **Subheading 3.4., step 1**.
5. Remove rooted shoots from the jars and wash their roots free of agar medium. Transfer the plants to compost, water and cover the potted plants with polythene bags to maintain humidity. After 7 d, remove one corner from each bag and gradually open the tops of the bags over the course of the following 10 d. Finally, remove the bags after 31 d.

3.5. Characterization of Regenerated Plants

Regenerated plants must be assessed for their morphological characteristics within 2–3 mo of transfer to the glasshouse, because the juvenile characteristics of leaf shape and pigmentation in vitro may be different to those of seed-derived plants raised in the glasshouse. This will enable floral characteristics, fertility, and fruit production to be compared with those of seed-derived (control) plants. The somatic chromosome complements of protoplast-derived plants can be assessed by observing root tip preparations *(19)*. DNA molecular studies may also be carried out on those protoplast-derived plants that exhibit phenotypic, fertility, and cytological characters different from those of seed-derived plants. Such techniques include the use of amplified fragment length polymorphism (AFLP) *(20)*, random amplified polymorphic DNA (RAPD) *(21)*, restriction fragment length polymorphism (RFLP) *(22)*, simple sequence repeats (microsatellites; SSR) *(23)*, cleaved amplified polymorphic sequences (CAPS) *(24)*, and tubulin-based polymorphism (TBP) *(25)*.

Notes

1. Store *Passiflora* seeds at 4°C. Sow *Passiflora* seeds in excess of the number of seedlings required, as germination within the genus *Passiflora* may be erratic and slow. Do not over-firm the compost. A minimum of 10 seedlings will be needed for each protoplast isolation. The leaves should be harvested from seedlings before the latter produce tendrils.
2. Place the leaves in a suitable, autoclaved glass container (such as a casserole dish) and add the sterilant. Subsequently, pour off the bleach from the leaves and replace with sterile water. Repeat the process at least 3 times. Replace the lid on the dish, following sterilization, to prevent drying of the leaves.
3. Repeatedly use a new scalpel blade to ensure precise cutting, rather than tearing and bruising, of leaf material.
4. Preplasmolysis (approx 30–60 min) of leaves in a salts solution with a suitable osmoticum (e.g., CPW13M) is beneficial in reducing spontaneous fusion of protoplasts and/or uptake of enzyme(s) into cells during enzymatic digestion.

5. The enzyme solution should be pre-filtered, using a nitrocellulose membrane filter (47-mm diameter, 0.2-µm pore size; Whatman, Maidstone, UK) to remove insoluble impurities. This prevents blockage of the sterile microbial filter (0.2-µm pore size, 30-mm diameter; Minisart NML, Sartorius AG, Göttingen, Germany) during subsequent sterilization of the enzyme solution.
6. It is essential that the inside of the lids and the space between the outer walls of the bases and the overlapping lids of Petri dishes are kept free of enzyme solution and explants.
7. Inexpensive nylon sieves, in a range of pore sizes, may be obtained from Wilson Sieves, Common Lane, Hucknall, Nottingham, UK.
8. It is important that the protoplast suspension is drawn up gently into the Pasteur pipet and dispensed slowly down the inside of the receiving centrifuge tube, to avoid disruption of the protoplasts.
9. Prepare an FDA working solution by diluting the FDA stock with CPW13M solution, immediately prior to viability assessments, because even short-term storage results in cleavage of FDA to fluorescein, especially if the solution is exposed to strong illumination.
10. FDA taken up into protoplasts is cleaved enzymatically by esterases, to release free fluorescein. The latter is excited under ultraviolet illumination and fluoresces yellow-green. Only viable protoplasts with intact plasma membranes, that prevent fluorescein leakage into the surrounding medium, will fluoresce.
11. The agarose-solidified KM8P medium must be prepared as (1) a filter-sterilized solution of double-strength components, and (2) a double-strength (1.6% [w/v]) agarose, sterilized by autoclaving. The double-strength liquid medium component is mixed with an equal volume of the molten agarose at 50–60°C, immediately prior to use.
12. Ensure that the agarose-solidified KM8P medium is at 35–40°C, prior to gentle addition of the appropriate volume, with a Pasteur pipet, to the protoplast pellet. Resuspend the protoplasts in the medium by gentle inversion of the tube containing the protoplast-culture medium.
13. Allow the KM8P droplets/thin layers containing protoplasts, to gel for at least 1 h, at room temperature, before addition of the liquid KM8P bathing medium. This allows the droplets to adhere to the bottom of the Petri dishes. Thin layers may remain suspended in the liquid phase.
14. Transfer individual colonies using fine, jeweller's forceps; flame sterilize and cool the forceps immediately prior to use.

References

1. Oliviera, J. C. (1987) Melhoramento genético in *Cultura do maracujazeiro*, vol. 1, (Ruggiero, C., ed.), Editora Legis Summa, Ribeirão Petro, Brazil, pp. 218–246.
2. Vanderplank, J. (1991) *Passion flowers,* Cassel Publishers, London.
3. Payan, F. R. and Martin, F. W. (1975) Barriers to the hybridization of *Passiflora* species. *Euphytica* **24,** 709–716.

4. Otoni, W. C., Blackhall, N. W., d'Utra Vaz, F. B., Casali, V. W., Power, J. B., and Davey, M. R. (1995) Somatic hybridization of the Passiflora species, *Passiflora edulis* fv. flavicarpa Degener. and *P. incarnata* L. *J. Exp. Bot.* **46,** 777–785.
5. Dozier, Jr W. A., Rodriguez-Kabana, R., Caylor, A. W., Himelrick, D. G., McDaniel, N. R., and McGuire, J. A. (1991) Ethephon hastens maturity of passion fruit grown as an annual in a temperate zone. *Hort. Sci.* **26,** 146–147.
6. Dornelas, M. C., Tavares, F. C. A., Oliviera, J. C., and Vieira, M. L. C. (1995) Plant regeneration from protoplast fusion in *Passiflora* spp. *Plant Cell Rep.* **15,** 106–110.
7. Davey, M. R., Power, J. B., and Lowe, K. C. (2000) Plant protoplasts, in *Encyclopedia of Cell Technology* (Spier, R. E., ed.), Kluwer Academic Publishers, Dordrecht, The Netherlands, pp. 1090–1096.
8. Davey, M. R., Anthony, P., Power, J. B., and Lowe, K. C. (2004) Protoplast applications in biotechnology, in *Encyclopedia of Plant and Crop Science* (Goodman, R. M., ed.), Marcel Dekker Inc., New York, USA, pp. 1061–1064.
9. Power, J. B., Davey, M. R., Anthony, P., and Lowe, K. C. (2004) Protoplast culture and regeneration, in *Encyclopedia of Plant and Crop Science* (Goodman, R. M., ed.), Marcel Dekker Inc., New York, USA, pp. 1065–1068.
10. d'Utra Vaz, F. B., dos Santos, A. V. P., Manders, G., Cocking, E. C., Davey, M. R., and Power, J. B. (1993) Plant regeneration from leaf mesophyll protoplasts of the tropical woody plant, passionfruit (*Passiflora edulis* fv. flavicarpa Degener.): the importance of the antibiotic cefotaxime in the culture medium. *Plant Cell Rep.* **12,** 220–225.
11. Dornelas, M. C. and Vieira, M. L. C. (1993) Plant regeneration from protoplast cultures of *Passiflora edulis* var. flavicarpa Deg., *P. amethystina* Mikan. and *P. cincinnata* Mast. *Plant Cell Rep.* **13,** 103–106.
12. Kao, K. N. and Michayluk, M. R. (1975) Nutritional requirements for growth of *Vicia hajastana* cells and protoplasts at a very low population density in liquid media. *Planta* **126,** 105–110.
13. Anthony, P., Lowe, K. C., Davey, M. R., and Power, J. B. (1997) Synergistic effects of haemoglobin and Pluronic® F-68 on mitotic division of cultured plant protoplasts. *Adv. Exp. Med. Biol.* **428,** 477–481.
14. Lowe, K. C., Anthony, P., Davey, M. R., and Power, J. B. (1999) Culture of cells at perfluorocarbon-aqueous interfaces. *Artif. Cells Blood Substit. Immobil. Biotechnol.* **27,** 255–261.
15. Frearson, E. M., Power, J. B., and Cocking, E. C. (1973) The isolation, culture and regeneration of Petunia protoplasts. *Dev. Biol.* **33,** 130–137.
16. Gilmour, D. M., Golds, T. J., and Davey, M. R. (1989) *Medicago* protoplasts: Fusion, culture and plant regeneration, in *Biotechnology in Forestry and Agriculture, vol. 8, Plant Protoplasts and Genetic Engineering I* (Bajaj, Y. P. S., ed.), Springer-Verlag, Heidelberg, Germany, pp. 370–388.
17. Widholm, J. (1972) The use of FDA and phenosafranine for determining viability of cultured plant cells. *Stain Technol.* **47,** 186–194.

18. Murashige, T. and Skoog, F. (1962) A revised medium for rapid growth and bioassays with tobacco tissue cultures. *Physiol. Plant.* **56,** 473–497.
19. Andras, S. C., Hartman, T. P. V., Marshall, J. A., et al. (1999) A drop-spreading technique to produce cytoplasm-free mitotic preparations from plants with small chromosomes. *Chromosome Res.* **7,** 641–647.
20. Raccuia, S. A., Mainolfi, A., Mandolino, G., and Melilli, M. G. (2004) Genetic diversity in *Cynara cardunculus* revealed by AFLP markers: comparison between cultivars and wild-types from Sicily. *Plant Breed.* **123,** 280–284.
21. Nybomb, H. (2004) Comparison of different nuclear DNA markers for estimating intraspecific genetic diversity in plants. *Mol. Ecol.* **13,** 1143–1155.
22. Dziechciarkova, M., Lebeda, A., Dolezalova, I., and Astley, D. (2004) Characterization of *Lactuca* spp. germplasm by protein and molecular markers—a review. *Plant Soil Environ.* **50,** 47–58.
23. Li, W., Sun, G., Lui, J., et al. (2004) Inheritance of plant regeneration from maize (*Zea mays* L.) shoot meristem cultures derived from germinated seeds and the identification of associated RAPD and SSR markers. *Theor. Appl. Genet.* **108,** 681–687.
24. Weiland, J.J. and Yu, M.H. (2003) A cleaved amplified polymorphic sequence (CAPS) marker associated with root-knot nematode resistance in sugarbeet. *Crop Sci.* **43,** 1814–1818.
25. Bardini, M., Lee, D., Donini, P., et al. (2004) Tubulin-based polymorphism (TBP): a new tool, based on functionally relevant sequences, to assess genetic diversity in plant species. *Genome* **47,** 281–291.

18

Isolation, Culture, and Plant Regeneration From *Echinacea purpurea* Protoplasts

Zeng-guang Pan, Chun-zhao Liu, Susan J. Murch, and Praveen K. Saxena

Summary

A plant regeneration system from the isolated protoplasts of *Echinacea purpurea* L. using an alginate solid/liquid culture is described in the chapter. Viable protoplasts were isolated from 100 mg of young leaves of 4-wk-old seedlings in an isolation mixture containing 1.0% cellulase Onozuka R-10, 0.5% pectinase, and 0.3 mol/L mannitol. After isolation and purification, the mesophyll protoplasts were embedded into 0.6% Na-alginate at the density 1×10^{-5} mL and cultured in modified Murashige and Skoog (MS) culture medium supplemented with 0.3 mol/L sucrose, 2.5 µmol/L benzylaminopurine (BA), and 5.0 µmol/L 2,4-dichlorophenoxyacetic acid (2,4-D). The visible colonies were present after 4 wk of culture. The protoplast-derived clones were transferred onto gellan gum-solidified basal medium supplemented with 1.0 µmol/L BA and 2.0 µmol/L indole-3-butyric acid (IBA) and formed compact and green calli. Shoot development was achieved by subculturing the calli onto the same basal medium supplemented with 5.0 µmol/L BA and 2.0 µmol/L IBA. Further subculture onto basal medium resulted in the regeneration of complete plantlets.

Key Words: Callus; *Echinacea purpurea*; medicinal plant; protoplast isolation.

1. Introduction

Echinacea purpurea L. belongs to the family Asteraceae, and has been used extensively in medicinal preparations for the treatments of many diseases including colds, toothaches, snake bites, rabies, and wound infections *(1)*. *Echinacea* preparations are among the best-selling herbs in North America, making up about 10–12% of the natural health product market with an annual value of about $33 million. However, commercial production of natural health products such as *Echinacea* has been limited by a range of issues including contamination of plant materials by microorganisms, pollution from the environment, variability of active components, and lack of pure, standardized plant material for biochemical analysis *(2)*. To address these issues, in vitro technology for this species has been developed recently for the production of high-quality consistent material as well as for genetic improvement *(3–6)*. Genetic improvement of *Echinacea* will be an important strategy to develop genetically superior germplasm for value-added products. In addition to conventional genetic modification, cell manipulation techniques, such as somaclonal variation and somatic hybridization using protoplasts, also provide useful means for genetic improvement *(7)*. A basic requirement for achieving these goals is the successful regeneration of plants from isolated protoplasts. Mesophyll protoplasts from young seedlings grown in vitro can be induced to form cell colonies and plantlets under optimized conditions of isolation and culture *(8)*.

2. Materials

1. *E. purpurea* L. seeds.
2. Plant Preservation Mixture (PPM) (Phytotech Labs): 1%, in sterile water.
3. Ethanol: 70%.
4. Sodium hypochloride solution: 5.4%.
5. Tween-20.
6. Murashige and Skoog (MS) salts.
7. Gamborg B_5 vitamins.
8. Cellulase Onozuka R-10 (Yakult Honsha Co.).
9. Pectinase (Sigma).
10. Gellan gum (Gelrite, Schweitzerhall).
11. Petri dishes.
12. Rotary shaker (Stovall Life Science Inc).
13. Na-alginate (Sigma).
14. Mannitol.
15. Nylon mesh: 50 μm.
16. Benzylaminopurine (BA).
17. Indole-3-butyric acid (IBA).
18. Microscope.
19. Fluorescein diacetate (FDA) (Sigma).
20. Acetone.

3. Methods

3.1. Plant Material Preparation

Seeds of *E. purpurea* were surface sterilized by immersion in 1% PPM for 24 h followed by 70% ethanol for 30 s, 5.4% sodium hypochloride solution containing one drop of Tween-20 per 500 mL for 20 min, and three rinses in sterile distilled water. Surface-sterilized seeds were germinated and maintained on a basal medium (referred to as MSO) containing MS salts *(9)*, B_5 vitamins *(10)*, 30 g/L sucrose, and 2.5 g/L gellan gum (Gelrite, Schweizerhall). PPM (0.3% v/v) was added to the MSO for eliminating fungal contamination prior to autoclaving at 121°C. All cultures were incubated in a growth cabinet in darkness at 26°C. After 2 wk, the geminated seedlings were maintained in a growth room with 16-h photoperiod under cool-white light (50 µmol/m²/s). The plantlets grown for 4 wk were employed for protoplast experiment.

3.2. Protoplast Isolation

One-hundred milligrams of young leaves of 4-wk-old seedlings grown in light were sliced into 1- to 2-mm wide strips and incubated in 5 mL of an isolation mixture containing 1.0% cellulase Onozuka R-10 (Yakult Honsha Co), 0.5 % pectinase (Sigma), 5 mmol/L 2-[*N*-morpholino]ethanesulfonic acid (MES), and 0.3 mol/L mannitol in CPW medium *(11)* in a Petri dish (60 × 15 mm). The pH of the isolation solution was adjusted to 5.6. The incubation was carried out in darkness for 18 h at room temperature on a shaker (Stovall Life Science Inc) at 30 rpm. After digestion, the enzyme mixtures were screened through a 50 µm nylon mesh, and resulting protoplasts were collected by centrifugation at 120*g* for 8 min. The pellet was gently suspended in 6 mL of CPW solution with 0.5 mol/L sucrose, and 2 mL of CPW solution with 0.3 *M* mannitol was gently loaded on top of the sucrose solution. Viable protoplasts formed a band at the interface between sucrose and mannitol solutions following centrifugation at 100*g* for 6 min. The protoplasts were rinsed once with 8 mL of CPW containing 0.3 mol/L mannitol, and centrifuged at 120*g* for 8 min. Viability of protoplasts was assessed by FDA staining method *(12)*. One milliliter of the purified protoplast suspension treated with 25 µL of FDA solution (5 mg of FDA dissolved in 1 mL acetone) for 10 min was used to observe in fluorescence microscopy. The freshly isolated mesophyll protoplasts were green in color with an average diameter of 35 µm (*see* **Fig. 1**).

3.3. Protoplast Culture

For liquid culture, 2.5 mL of the purified protoplasts at a density of 1×10^{-5} mL in a liquid culture medium were pipetted into Petri dish (35–10 mm). For alginate block/liquid culture, the purified protoplasts were suspended in 0.3 mol/L mannitol solution at a density of 2×10^{-5} mL, and mixed with same volume of 0.3 mol/L mannitol solution containing 1.2% (w/v) Na-alginate. The alginate-protoplast suspension (0.8 mL) was poured onto 25-mL solid medium containing 20 mmol/L $CaCl_2$, 0.3 mol/L mannitol, and 1.5% (w/v) agar in a Perti dish, and the resulting solidified protoplast-alginate blocks were transferred into 60 × 15 mm-Petri dishes containing 4 mL of a liquid culture medium. The culture

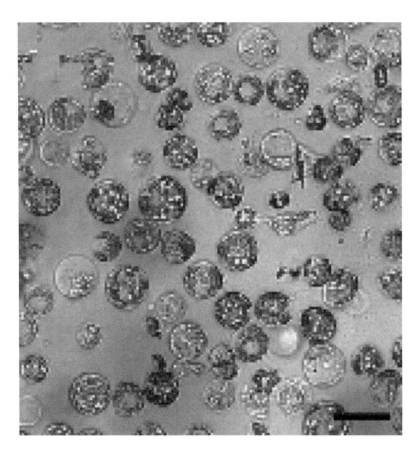

Fig. 1. Freshly isolated mesophyll protoplasts of Echinacea purpurea (bar = 50 mm).

medium contained modified MS (500 mg/L KNO_3 and 500 mg/L NH_4NO_3), 0.3 mol/L sucrose, 2.5 µmol/L BA, and 5.0 µmol/L 2,4-D. The pH was adjusted to 5.6 before autoclaving. The first divisions of protoplasts occurred after 6 d. Sustained divisions and colony formation was observed after 4 wk (*see* **Fig. 2**).

3.4. Plant Regeneration From Cultivated Protoplasts

All cultures were incubated at 26°C in the dark and transferred to dim light (5–10 µmol/m²/s) after 2 wk of culture. Protoplast-derived colonies (1- to 3-mm diameter) were gently transferred onto 0.25% gellan gum-solidified MS medium with 1.0 µmol/L BA, and 2.0 µmol/L IBA, and grown in light to induce callus growth. After 4 wk, the calli were subcultured on MSO supplemented with 5.0 µmol/L 6-benzylaminopurine and 2.0 µmol/L IBA to induce shoot

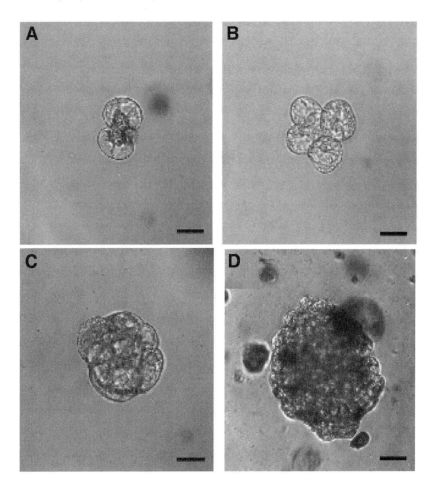

Fig. 2. Subsequent division of *Echinacea purpurea* protoplasts. (**A**) A dividing protoplast-derived cell after 6 d (bar = 25 mm). (**B,C**) Second and third cell division of protoplast (bar = 25 mm). (**D**) Protoplast-derived colony after 6 wk of culture (bar = 100 mm).

formation. Regenerated shoots were separated and transferred to MSO medium for plantlet formation. As shown in **Fig. 3**, calli were visible after 4 wk of transfer of protoplast-derived colonies onto solidified MSO medium with 1.0 µmol/L BA and 2.0 µmol/L IBA. Shoot regeneration was initiated from the calli on MS medium with 5.0 µmol/L BA and 2.0 µmol/L IBA, and regenerated shoots further developed on the same medium after 4 wk. Complete plantlets were obtained from protoplast-derived shoots on MSO medium after 2 mo.

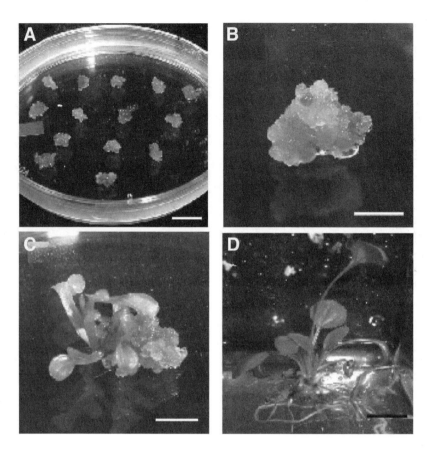

Fig. 3. Callus formation, shoot regeneration and plantlet development from protoplasts of *Echinacea purpurea*. (**A**) Callus formation from protoplast-derived colony (bar = 1 cm). (**B**) Organogenesis from protoplast-derived calli (bar = 1 cm). (**C**) Regenerated shoots from protoplast-derived calli (bar = 1 cm). (**D**) Complete plantlet regenerated from protoplasts of *E. purpurea* (bar = 1 cm).

4. Notes

1. The initial plant material used for protoplast isolation proved to be an important factor for successful isolation and regeneration of *E. purpurea*. A large number of viable protoplasts were obtained from young leaves of vigorous plantlets. The leaves sliced into strips using a sharp knife should be placed into digestion solution quickly. During the purification of isolated protoplasts, the centrifugation speed and time should be well controlled in order to obtain more viable protoplasts.
2. A solid/liquid culture method was found to be suitable for protoplast divisions and colony formation. The Na-alginate block/liquid method clearly improved the colony formation of *E. purpurea* protoplasts. The combination of alginate-protoplast block with liquid medium has been reported to be beneficial for a number

of protoplast culture systems. The suitability of this simple solid/liquid culture method has been attributed to the neutral, and less toxic, nature of alginate that may protect the protoplast-derived cells from chemical and physical damages from the culture microenvironment. In addition, Na-alginate block/liquid cultures provide an easy way to replace medium for accurately decreasing osmoticum during extended culture.

Acknowledgments

The financial support of the Natural Sciences and Engineering Research Council of Canada and the Ontario Ministry of Food Agriculture is gratefully acknowledged.

References

1. Bauer, R. (1999) Chemistry, analysis and immunological investigations of *Echinacea* phytopharmaceuticals, in *Immunomodulatory Agents From Plant*, (Wagner, H., ed.), Birkhauser Verlag, Basel, Boston, Berlin, pp. 41–48.
2. Murch, S. J., Krishnaraj, S., and Saxena, P. K. (2000) Phytopharmaceuticals: problems, limitations and solutions. *Sci. Rev. Alternat. Med.* **4**, 33–38.
3. Choffe, K. L., Victor, J. M. R., Murch, S. J., and Saxena, P. K. (2000) *In vitro* regeneration of *Echinacea purpurea* L.: direct somatic embryogenesis and indirect shoot organogenesis in petiole culture. *In Vitro Cell. Dev. Biol. Plant* **36**, 30–36.
4. Choffe, K. L., Murch, S. J., and Saxena, P. K. (2000) Regeneration of *Echinacea purpurea*: induction of root organogenesis from hypocotyl and cotyledon explants. *Plant Cell Tissue Organ Cult.* **62**, 227–234.
5. Koroch, A., Juliani, H. R., Kapteyn, J., and Simon, J. E. (2002) *In vitro* regeneration of *Echinacea purpurea* from leaf explants. *Plant Cell Tissue Organ Cult.* **69**, 79–83.
6. Zobayed, S. M. A. and Saxena, P. K. (2003) Auxin and a dark incubation increase the frequency of somatic embryogenesis in *Echinacea purpurea* L. *In Vitro Cell. Dev. Biol. Plant* **39**, 605–612.
7. Nagata, T. and Bajaj, Y. P. S. (2001) *Biotechnology in Agriculture and Forestry. Volume 49*: *Somatic Hybridization in Crop Improvement II*. Springer-Verlag, Heidelberg, New York, USA.
8. Pan, Z. G., Liu, C. Z., Zobayed, S. M. A., and Saxena, P. K., (2004) Plant regeneration from mesophyll protoplasts of *Echinacea purpurea*. *Plant Cell Tissue Organ Cult.* **77**, 251–255.
9. Murashige, T. and Skoog, F. (1962) A revised medium for rapid growth and bioassays with tobacco tissue cultures. *Physiol. Plant* **15**, 473–497.
10. Gamborg, O. L., Miller, R. A., and Ojima, K. (1968) Nutrient requirements of suspension cultures of soybean root cells. *Exp. Cell Res.* **50**, 151–158.
11. Patat-Ochatt, E. M., Ochatt, S. J., and J. B. Power, (1988) Plant regeneration from protoplasts of apple rootstocks and scion varieties. *J. Plant Physiol.* **133**, 460–465.
12. Widholm, J. M. (1972) The use of fluorescein diacetate and phenosafranine for determining viability of cultured plant cells. *Stain Technol.* **47**, 189–194.

19

Production of Cybrids in Brassicaceae Species

Maksym Vasylenko, Olga Ovcharenko, Yuri Gleba, and Nikolay Kuchuk

Summary

This chapter describes a method of cytoplasm transfer within the brassicaceae family through Ca-PEG-mediated protoplast fusion. The method includes a protocol of nonmutagenic albinism induction based on spectinomycin-induced plastid ribosome deficiency (PRD). The proposed application of spectinomycin-mediated albinism allows speeding up creation of albino lines, as well as of hybrid production with substituted cytoplasm. According to described method cybrids between *Orychophragmus violaceus* and *Brassica napus*, *O. violaceus* and *Lesquerella fendleri* have been produced. Methods of further molecular analysis are also presented. The time-scale and reliability of described methods are indicated.

Key Words: Albinism; *Brassica* sp.; cybrid; spectinomycin; genome segregation; *Lesquerella fendleri*; *Orychophragmus violaceus*; plastid ribosome deficiency; protoplast fusion; somatic hybridization.

1. Introduction

Discovery of non-Mendelian heredity opened extensive studies in respect to transmission and functioning of cytoplasmic genomes. Plastids and mitochondria that possess their own genomes and crucial metabolic pathways are attractive for scientists as a source for genetic manipulations.

Cytoplasmic male sterility (CMS), temperature tolerance, herbicide and antibiotic resistance, tolerance to some diseases, and other agronomic important traits are transmitted maternally by plastome or mitochondriome *(1–3,12)*. Trait transfer can be carried out by substitution of cytoplasm using multiple sexual backcrosses between the hybrid and the same acceptor species. In particular, cytoplasmic male sterility was generated in cruciferous crops by cytoplasm transfer from several relatives such as *Brassica nigra*, *Raphanus sativus*, *Ogura type*, and so on *(4,5)*. This approach is often time- and effort-consuming, and certain sexual compatibility is also necessary for successful distant crosses. Moreover, only the whole cytoplasm (plastome and mitochondriome together) can be transferred by this way, and at the same time there is no certitude in complete chromosome substitution.

Based on somatic cell fusion, somatic hybridization technology allows for production of new combinations of parent genes and thereby to create unique hybrid products. Somatic hybridization overcomes interspecific incompatibility occurring in distant crossing *(6–8)*. Thus, related wild species can be extensively used as a genetic resource for improvement of cultivated crops.

Independent segregation of nuclear and cytoplasmic genomes as a result of somatic hybridization leads to formation of multiple hybrid forms with different nuclear-cytoplasmic combinations. Highly asymmetric hybrids and cybrids (cytoplasmic hybrids) can be recovered among them. In order to facilitate and to direct the segregation process in a desired way, separate inactivation of cellular genomes is applied. Physical (irradiation with X- or γ-rays) and chemical (iodacetamide, *N*-ethylmaleimide, diethyl pyrocarbonate, and so on) factors are commonly used to produce hybrids with high asymmetry of established genomes *(8–10)*.

Moreover, presence of genetic markers is very useful for subsequent selection of hybrid lines. Genetic and physiological trait complementation (e.g., chlorophyll deficiency, regeneration ability, auxotrophy) is often applied simultaneously with inactivation methods *(11,12)*.

Chlorophyll-deficient mutants are extensively used in cytoplasm transfer experiments *(10–13)*. At the same time application of mutants has some limitations concerned with time-consuming induction, selection, and localization of mutations. On the other hand, there is a way to produce albino plants without mutagen treatment *(14)*. Spectinomycin induces stable plastid ribosome deficiency (PRD) binding with the ribosome subunit and arresting protein synthesis. Plants bleached by this method can maintain chlorophyll deficiency without subsequent antibiotic presence and can be propagated in vitro vegetatively and through sexual crosses *(14)*. Thus, PRD generated with spectinomycin treatment seems to be an alternative and available way for albino line production, and can be efficiently applied in somatic hybridization experiments. The present study provides a cybrids production method for *Brassicaceae* species using *O. violaceus*, *Brassica juncea*, and *Brassica napus* albino lines induced by spectinomycin treatment.

2. Materials

1. Agar-agar.
2. Silver nitrate (silver thiosulphate can also be used as medium additive) *(15)*. To prepare the solution add 2.5 mg/L $AgNO_3$ solution to 14.6 mg/L $Na_2S_2O_3$ solution in equal quantities by drops; 4 mL of filter sterilized solution are usually added per liter of regeneration medium).
3. Spectinomycin.
4. Sucrose, glucose, mannitol.
5. 6-Benzylaminopurine (BA), zeatin.
6. α-Naphthaleneacetic acid (NAA).
7. 2,4-Dichlorophenoxyacetic acid (2,4-D).
8. Murashige and Skoog (MS) medium *(16)*: 1650 mg/L NH_4NO_3, 1900 mg/L KNO_3, 440 mg/L $CaCl_2·2H_2O$, 370 mg/L $MgSO_4·7H_2O$, 170 mg/LKH_2PO_4, 0.83 mg/L KJ, 6.3 mg/L H_3BO_3, 22.3 mg/L $MnSO_4·4H_2O$, 8.6 mg/L$ZnSO_4·7H_2O$, 0.25 mg/L $Na_2MoO_4·2H_2O$, 0.025 mg/L$CuSO_4·5H_2O$, 0.025 mg/L $CoSO_4·6H_2O$, Fe-chelate: (37.3 mg/L Na_2EDTA and 27.8 mg/L $FeSO_4·7H_2O$), 100 mg/L myo-inositol, 10 mg/L thiamine-HCl (B_1), 1 mg/L pyridoxine (B_6), 1 mg/L nicotinic acid (PP), 2 mg/L glycine, and 30 g/L sucrose, pH 5.6–5.8.
9. 50% MS medium (half-strength macro- and micronutrients, complete MS-vitamins and Fe-chelate, 100 mg/L myoinositol, 20 g/L sucrose, pH 5.6–5.8).
10. Callus induction MS medium (CIMS); MS medium supplemented with 4.52 μM 2,4-D, 2.68 μM NAA, 2.22 μM BA, 30 g/L sucrose, pH 5.6–5.8.
11. KM8p medium, filter sterilizing *(17)*: 600 mg/LNH_4NO_3, 1900 mg/L KNO_3, 600 mg/L $CaCl_2·2H_2O$, 300 mg/L$MgSO_4·7H_2O$, 170 mg/L KH_2PO_4, 0.75 mg/L KJ, 3.0 mg/L H_3BO_3, 13.2 mg/L $MnSO_4·4H_2O$, 2.0 mg/L $ZnSO_4·7H_2O$, 0.25 mg/L $Na_2MoO_4·2H_2O$, 0.025 mg/L$CuSO_4·5H_2O$, 0.025 mg/L$CoSO_4·6H_2O$, Fe-chelate: (37.3 mg/L Na_2EDTA and 27.8 mg/L $FeSO_4·7H_2O$), 100 mg/L myo-inositol, 0.5 g/L MES, 20 mg/L sodium piruvate, 40 mg/L citric acid, 40 mg/L malic acid, 40 mg/L fumaric acid, 10 mg/L thiamine (B_1), 2 mg/L ascorbic acid (C), 1 mg/L pyridoxine (B_6), 1 mg/L nicotinic acid (PP), 1 mg/L calcium pantothenate, 1 mg/L choline chloride, 0.4 mg/L folic acid, 0.2 mg/L riboflavin (B_2), 0.02 mg/L p-aminobenzoic acid, 0.02 mg/L cyanocobalamin (B_{12}), 0.01 mg/L biotin (H), 0.01 mg/L retinol [A], 0.01 mg/L cholecalciferol (D3), 4.52 μM 2,4-D, 0.53 μM NAA, 2.22 μM BA, 0.45 M glucose, pH 5.6–5.8.
12. SW1 medium, filter sterilized *(18)*: 134 mg/L NH_4Cl, 950 mg/L KNO_3, 220 mg/L $CaCl_2·2H_2O$, 185 mg/L $MgSO_4·7H_2O$, 85 mg/L KH_2PO_4, 0.75 mg/L KJ, 3.0 mg/L H_3BO_3, 13.2 mg/L $MnSO_4·4H_2O$, 2.0 mg/L$ZnSO_4·7H_2O$, 0.25 mg/L $Na_2MoO_4·2H_2O$, 0.025 mg/L $CuSO_4·5H_2O$, 0.025 mg/L $CoSO_4·6H_2O$, Fe-chelate: (37.3 mg/L Na_2EDTA and 27.8 mg/L $FeSO_4·7H_2O$), 1 g/L MES, 100 mg/L myo-inositol, 100 mg/L glutamine, 10 mg/L thiamine (B_1), 1 mg/L pyridoxine (B_6), 0.9 μM 2,4-D, 10.74 μM NAA, 2.22 μM BA, 0.45 M xylose–glucose, pH 5.6–5.8.

Fig. 1. Regeneration of albino shoots from leaf explants (*left*); plastid ribosome deficient *Orychophragmus violaceus* plant (*right*), axillary buds were used for chlorophyll deficiency induction in *Brassica napus* and *Brassica juncea*. Stem fragments with buds were treated with spectinomycin solution (5 g/L) for 24 h. Then they were planted on the spectinomycin-free MS medium and in the course of cultivation stable chlorophyll deficient plants were selected. Generally, it took 2 mo to pick out stable albino plants which were used in the further hybridization experiments as an acceptor of alien cytoplasm (*see* **Note 1**).

13. Enzyme solution: 5 mM CaCl$_2$, 0.6% Cellulase Onozuka R-10 (Duchefa, the Netherlands), 0.4% Macerozyme R-10 (Duchefa), 0.45 M sucrose.
14. W5 solution (9 g/L NaCl, 0.8 g/L KCl, 18.4 g/L CaCl$_2$·2H$_2$O, 1 g/L glucose).
15. Sucrose solution (0.5 M sucrose, 10 CaCl$_2$).
16. Polyethylene glycol (PEG) solution (30% PEG 4000–6000; 0.3 M glucose and 60 mM CaCl$_2$).
17. High Ca-pH solution for PEG removal. Solution 1: 0.4 M glucose, 60 mM CaCl$_2$; Solution 2: 0.3 M glycine adjusted with NaOH, pH 10.5. Solutions 1 and 2 should be stored at −20°C and mixed (9:1) directly before using.
18. DNA extraction buffer (2 M NaCl, 70 mM ethylenediaminetetraacetic acid [EDTA], 20 mM sodium metabisulphite, 200 mM Tris-HCl, pH 8.0).
19. TE buffer: 70 mM EDTA, 200 mM Tris-HCl, pH 8.0.
20. Endonulease enzymes available for DNA restriction mapping, buffer solutions (for example, DraI, HaeIII, MnlI, and so on).
21. Protein extraction buffer: 0.05 M Tris-HCl pH 8.0; 12% glycerol, 0.2% β-mercaptoethanol.
22. Tris-glycine running buffer (10X solution): 28.8 g/L glycine, 6 g/LTris base.
23. Acrylamide, 30% and 0.8% bis-acrylamide mixture. **Caution:** toxic.
24. 1.5 M Tris-HCl, pH 8.9; 1 M Tris-HCl, pH 6.7.
25. Ammonium persulphate (PSA).
26. TEMED (*N,N,N',N'*-tetramethylethylenediamine).
27. α-Naphthyl acetate.
28. Fast blue RR (4-Benzoylamino-2,5-dimethoxyaniline).
29. Benzidine (*p*-Diaminodiphenyl, cancer suspect reagent).

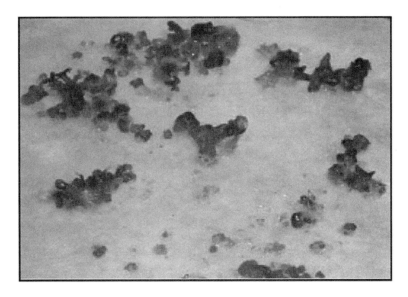

Fig. 2. Development of Orychophragmus-like hybrids with restored green color. The following passages of developing embryos were carried out on the same regeneration medium without mannitol. In approx 6 wk, developing small plants were selected in accordance with their Orychophragmus-like phenotype and restored green color. Selected plants were grown on the hormone-free MS medium.

30. Starch (0.3% water solution).
31. 8-Hydroxyquinoline (2 mM solution).
32. Colchicine (0.01–0.05% water solution, store at –20°C).
33. Acetic alcohol fixative (3 vol ethanol: 1 vol glacial acetic acid); Carnoy's fixative (6 vol ethanol: 3 vol chloroform: 1 vol glacial acetic acid).
34. Ethanol, 70%.
35. Acetic acid, 45%.
36. 1 N-Hydrochloric acid.
37. Acetic orcein (dissolve 1 g of orcein in 45 mL of glacial acetic acid, heat the solution, add 55 mL of distilled water, filtrate the solution and store at 4°C).

3. Methods
3.1. Plant Material

B. napus (cv. Westar, Brutor, Pactol, Kletochny, and Kalinovski), *B. juncea*, *O. violaceus*, and *Lesquerella fendleri* seeds were introduced into in vitro culture. Sterilization was carried out with 70% ethanol for 2 min and then with a mixture of 3% sodium hypochlorite and 0.1% sodium dodecyl sulfate (SDS) solution for 15 min. After three repeated washings with sterile water (for 10 min each step) seeds were transferred on the 50% MS agar-solidified medium

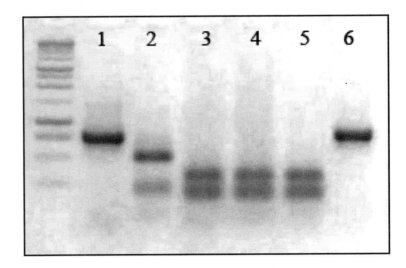

Fig. 3. ITS restriction fragment comparison: *Brassica napus* (*line 2*), *Orychophragmus violaceus* (*line 5*), two hybrid lines (*lines 3* and *4*), and nondigested PCR products (*lines 1* and *6*).

for germination. Seedlings and then plants were cultured at 24–26°C at 4000–6000 lx (16/8 h) and subcultured every 3 wk.

3.2. Albino Line Production

Albinism in *O. violaceus*, *B. juncea*, and *B. napus* was generated by spectinomycin treatment in accordance with Zubko et al. *(14)*. Seeds, buds, and proliferating tissues can be used for exposure with spectinomycin to initiate PRD. Leafstalks of *O. violaceus* were cut into segments 1-cm long and cultured on agar-solidified MS medium supplemented with 1.33 mM BA and 500 mg/L spectinomycin. Regenerated white shoots were cultured then on the spectinomycin free medium for a few weeks to discard green reversions and isolate stable albino lines. Fine structure callus was induced from albino tissues of *O. violaceus* on the CIMS culture medium (*see* **Note 11**).

3.3. Protoplast Isolation and Fusion

Green leaves of *B. napus* and *L. fendleri*, albino tissues of *O. violaceus*, *B. juncea*, and *B. napus* were used for protoplast isolation (*see* **Note 2a**). Leaves of *L. fendleri* were previously irradiated with γ-rays (200 Gy) generated with ^{60}Co-gun. Leaves and callus were cut into narrow strips with a sharp razor blade and incubated in the enzyme solution at 26°C for 14–16 h in the dark. Digested tissues were shacked gently and protoplast suspensions were filtered through

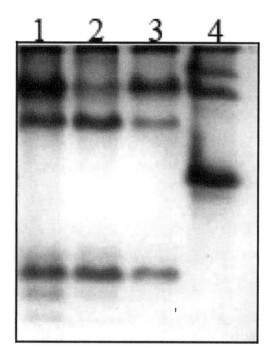

Fig. 4. Comparison of peroxidase isozyme patterns: *Orychophragmus violaceus* (*line 1*), two hybrid lines (*lines 2* and *3*), and *Brassica napus* (*line 4*).

nylon meshes (pore size 50 μm). Additional volume of sucrose solution can be used for more accurate separation of protoplasts and enzyme dilution.

Filtered suspension was transferred into centrifuge tubes and W5 solution (approx 2 mL) was slowly added to cover up protoplast suspension. Then the tubes were centrifuged at 150g for 4–5 min. The viable protoplasts were collected from the interface, diluted with W5 solution and centrifuged at 80g for 4–5 min. The protoplast pellet was resuspended in the fresh W5 solution and centrifuged again. This step was repeated twice. The last centrifugation was realized with the mixture of protoplasts to be fused (approx ratio 1:1). The pellet was slightly diluted with W5 up to dense suspension (approx 10^{-6} cells/mL) and transferred then into plastic Petri dishes by 100 μL drops (approx 3 drops per 60-mm diameter plate), which were left for 15 min to let the protoplasts settle down.

The fusion procedure was carried out in accordance with Menczel et al. *(19)*. After cell sedimentation, an equal volume of PEG solution was gently added to each drop. In 15 min, PEG solution was replaced by the washing solution (*see* **Subheading 2., step 17**), and the plates were left for 20 min. After that KM8p (or SW1) medium was added gradually (to the final volume 5 mL) and replaced by the next volume of the fresh medium after 20 min. The culture was started and maintained at 24–26°C in darkness up to formation of microcolonies.

Somatic hybridization has been performed in the following combinations: *O. violaceus* (albino) and *B. napus*; *O. violaceus* (albino) and *L. fendleri* (γ-irradiated); *B. napus* (albino) and *L. fendleri* (γ-irradiated); *B. juncea* (albino) and *L. fendleri* (γ-irradiated).

3.4. Plant Regeneration and Hybrid Selection

After the first divisions of fusion products the cell culture was diluted with fresh KM8p (or SW1) medium at the same volume and cultured in the scattered light. As the colonies grew, the density of cell population and osmotic pressure were decreased by gradual dilutions with 0.3 M glucose KM8p, (or 0.3 M glucose SW1) medium. The multicellular colonies were transferred onto agar solidified regeneration MS medium supplemented with 0.25 M mannitol, 4.43 μM BA, and 0.54 μM NAA 3 wk after protoplast fusions of *O. violaceus* (albino) and *B. napus* and *O. violaceus* (albino) and *L. fendleri*.

The following passages of developing embryos were carried out on the same regeneration medium without mannitol. Approximately in 6 wk, developing small plants were selected in accordance with their *Orychophragmus*-like phenotype and restored green color. Selected plants were grown on the hormone-free MS medium.

Hybrid calluses obtained after protoplast fusions of *B. napus* (albino) and *L. fendleri* (γ-irradiated) and *B. juncea* (albino) and *L. fendleri* (γ-irradiated) were also selected by chlorophyll synthesis restoration.

Cell colonies formed calluses on the agar-solidified regeneration MS medium supplemented with 30 μM AgNO$_3$, 4.43 μM BA, 2 mg/L zeatin = 9.12 μm zeatin, and 0.53 μM NAA *(20,21)* after being transferred from the liquid medium.

3.5. PCR-RFLP and RAPD Analyses of Nuclear, Plastid, and Mitochondrial DNA

DNA analyses of nucleus and organelles were performed on total DNA extracts from the hybrid lines and parental species. The DNA isolation procedure was carried out according to Cheung et al. *(22)*.

Approximately 30–50 mg of leaf tissue were homogenized using Eppendorf (or similar) grinder machine for 2 min per sample. Five hundred microliters of extraction buffer (*see* **Subheading 2., step 18**) was added to the homogenate; the mixture was incubated at 60°C for 60 min and then centrifuged in a microcentrifuge at top speed for 15 min. Ammonium acetate (45 µL of 10 M solution) and isopropyl alcohol (300 µL were added to the supernatant, and the

Table 1
Primer Sequences and Targeted Regions

Primer	Fragment
5'-GAAGTAGTAGGATTGATTCTC-3' 5'-TACAGTTGTCCATGTACCAG-3'	AtpB-rbcL fragment *(24)*
5'-CAGTGGGTTGGTCTGGTATG-3' 5'-TCATATGGGCTACTGAGGAG-3'	Ndh4 fragment *(25)*
(TTTAGGG) × 3	Telomere repeats *(26,27)*
5'-GGAAGTAAAAGTCGTAACAAGG-3' 5'-TCCYCCGCTTATTGATATGC-3'	Internal transcribing spacer (ITS) of rDNA cluster *(28)*
5'-CAATCGCCGT-3', 5'-TCGGCGATAG-3', 5'-CAGCACCCAC-3'	RAPD analysis; OPA-primers (11–13) from Operon Technologies Inc., USA

samples were incubated at room temperature for 15 min. Then the tubes were centrifuged again at top speed for 15 min. The pellet was washed with 70% ethanol, air dried, and dissolved in 50 μL TE buffer *(23)*.

One microliter of DNA extract was used in a single PCR reaction. Amplification was carried out in a "Terzik" polymerase chain reaction (PCR)-machine (DNA-Technology™, Russia) in 20 μL reaction volume containing: 50 mM KCl; 10 mM Tris-HCl, pH 8.3, 2 mM MgCl$_2$, 0.25 mM of each dNTP; 0.6 U of Taq-polymerase; and 0.2 μM of each primer **(Table 1)**.

Amplification conditions were as follows: 94°C for 2 min, (94°C for 30 s, 54°C for 30 s, 72°C for 2 min) × 35 cycles, 72°C for 3 min.

DNAs of hybrid lines and initial species were compared using restriction fragment length polymorphism (RFLP) analysis. Mapping of PCR products was performed with a range of endonucleases at the conditions recommended by manufacturer. PCR products were separated and analyzed with agarose gel electrophoresis in 1.5% agarose-TBE *(23)*.

3.6. Isozyme Analysis

Selected hybrid lines were also compared with the parent species using isoenzyme analysis. Protein extraction and electrophoresis were performed at 4°C. Plant material was homogenized in extraction buffer (1:1 w/v) (*see* **Subheading 2., step 21**). The homogenate was centrifuged at 15,000*g* for 60 min and the supernatant was used for electrophoresis. The gels were prepared according to the following protocol:

1. Bottom gel (7.5%): 15 mL acrylamide/bis (37.5:1), 15 mL 1.5 M Tris-HCl, pH 8.9, 250 μL PSA, 25 μL TEMED, 30 mL H$_2$O (or 0.3% starch solution).
2. Top gel (3%): 1.5 mL acrylamide/bis (37.5:1), 3.75 mL 1 M Tris-HCl, pH 6.7, 150 μL PSA, 15 μL TEMED, 9.75 mL H$_2$O.

3. Spectrums of isozymes were visualized by termostating of gels with appropriate substrate at 37°C until band staining occurred completely.

3.6.1. Peroxidase Preparation

1. Dilute 50 mg of benzidine in 100 mL H_2O, heat if necessary.
2. Pour the first solution into 400 mL 0.2 M acetate buffer, pH 5.5.
3. Stain the gel at 37°C up to band appearance.

3.6.2. Esterase Production

1. Dilute 25 mg of α-naphthyl acetate in 1 mL acetone and then add 750 mL H_2O.
2. Dilute 40 mg of fast blue RR in 100 mL 0.1 M phosphate buffer, pH 6.0.
3. Add the first solution to the buffer with fast blue RR and filtrate.
4. Stain the gel at 37°C up to band appearance.

3.6.3. Amylase Production

1. Use the starch solution (instead of water) for gel preparation.
2. Wash the gel in 0.2 M acetate buffer, pH 5.5, containing 10 mM $CaCl_2$ at 37°C for 30 min.
3. Stain the gel in a weak iodine solution in 7% acetic acid for 1 h.
4. Wash in 7% acetic acid.

3.7. Chromosome Analysis

Cytogenetic studies were performed according to a basic protocol as stated below. Root tips were isolated from in vitro grown plants and pre-treated with ice water for 6 h and then with 2 mM 8-hydroxyquinoline for 4 h to increase metaphase index. Colchicine pretreatment could also be applied for cell division synchronization. Acetic alcohol or Carnoy's fixative were used then at 2–4°C for 24 h. Roots were washed twice with 70% ethanol for 10 min and were stained immediately. After washing roots can also be stored in 70% ethanol at –20°C for a long period (if necessary).

Before spread preparation, root tips were hydrolyzed with 45% acetic acid (or 1N HCl) at 80°C for 5 min and stained then with 1% acetic orcein at 60°C for 5 min (or overnight without heating).

The stained material was washed with 45% acetic acid (if necessary to reduce the coloration of cytoplasm and to make the chromosomes more contrast), squashed in a drop of 45% acetic acid, and observed using immersion oil.

4. Notes

1. Application of spectinomycin-induced albinism allows to generate stable chlorophyll deficient lines within a short term (2–3 mo). The procedure does not require expensive equipment and skillful work. It was shown that cruciferous plants readily produced albino lines under exposure to spectinomycin, and plastid ribo-

some deficiency state can be maintained for a long period *(14)*. We have produced albino lines of *B. napus*, *B. juncea*, and *O. violaceus* to utilize them as acceptors of cytoplasm in our hybridization experiments. Chlorophyll deficiency was induced through the prolonged action of spectinomycin added to the nutrient medium as well as through the short-term treatment with higher concentrations of the antibiotic (*see* **Subheading 3.2.**). Reversions to the green or yellow-green phenotype were observed after both prolonged and sort-term treatments for all species (to a greater or lesser extent; especially if axillary buds were used), but nonchimeric albino lines were also selected easily. The number of chlorophyll deficient plants correlated with duration and strength of treatment that has been shown earlier *(14)*. Unfortunately, plants of some other species (for example, *N. tabacum*) did not produce well developed albino plants. Thus, this approach can obviously be applied only for cruciferous species now.

2. Since the 1970s, somatic hybridization technology has been further developed for a great number of species. In particular are those methodologies, trends, and issues of cybrid production that were discussed earlier *(29,30)*. Several methods of protoplast fusion and their modifications are usually proposed to achieve somatic cell fusions induced with $NaNO_3$, gelatin, Ca^{2+}-high-pH, electric fields, PEG-Ca^{2+}-high-pH *(8–12,19,20,31,32)*. Electrofusion protocols require special equipment, but they give an opportunity to induce large scale fusions with high efficiency. In the present investigation we have followed Ca^{2+}-PEG-mediated fusion protocol, which is still the most prevalent way of somatic hybrid production.

- In our experiments leaf tissues were mainly used as a source of protoplasts. Quality of initial plants may dramatically influence the digestion and consequently the viable protoplast yield. Therefore, only leaves of well developed and rapidly growing plants, which were subcultured regularly, should be used for protoplast isolation. *O. violaceus* plants with induced plastid ribosome deficiency are usually smaller in size and slower growers. This fact should be taken into account when a fusion experiment is planned. To increase protoplast yield and growth rate of *O. violaceus*, we provided callus culture induced from leafstalks of albino plants (*see* **Subheading 3.2.**). An initial callus was formed within 3 wk. The callus tissue was friable, wet and fine-structured that was a necessary prerequisite for successful digestion.

- Other reliable variations of protoplast fusion method can be applied equally with the protocol presented here. Basically, the procedure of protoplast isolation and fusion includes the following steps: preparation and enzymatic digestion of parental plant material; separation and purification of viable protoplasts; protoplast treatment with a fusion agent; and removal of fusion agent and subsequent cultivation of fusion products.

- Each step should be performed accurately (do not damage the protoplasts by rough pipetting, and avoid dropping and bubbling of protoplasts suspension). It should be noted that duration and strength of treatment may vary to a certain extent in every individual experiment as a result of the type of fusing cells, temperature conditions, and some other factors. Therefore the precise protocol should be defined empirically. Correct choice of nutrient medium for induction of cell divisions and subsequent hybrid development is also important. Sometimes it can provide the first step of hybrid selection, for example, when the physiological complementation is applied. Basic components of KM8p and SW1 media used in our experiments were the same as in the original protocols *(17,18)*. Our modifications concerned composition of plant growth regulators, organic acids and carbohydrates. They are indicated in the present protocol (*see* **Subheading 2.**, **steps 11** and **12**). There are some requirements concerning preparation of protoplast nutrient media such as highly purified water and cell culture tested chemicals. Filter sterilization is the preferable method of preparation of protoplast culture media.

- Selection of hybrid lines is the next necessary stage of any experiment on somatic hybridization. In our work cybrid selection was based on genetic complementation of chlorophyll deficiency of acceptor plant. Inactivation of donor nuclear genome with γ-irradiation was also applied to arrest mitotic activity of donor. We have used this approach to transfer the cytoplasm of *L. fendleri* into *O. violaceus*, *B. napus*, and *B. juncea*. At the same time a high level of asymmetry can be also reached using physiologically different parental protoplasts. High regeneration ability of acceptor species as well as differences in cell cycle rate may result in cybrid formation without inactivation of donor. *O. violaceus* is a wild cruciferous that we have also used as an acceptor of *B. napus* plastids. The high regeneration ability occurring through somatic embryogenesis allowed us to provide successful selection of *O. violaceus* (+*B. napus*) cybrids.

3. As a rule, somatic cell hybridization results in several types of fusion products. Moreover, developing cell populations after cell fusion can contain a range of mutants, polyploid forms, products of chromosome rearrangements, or chimeric colonies that may affect the process of hybrid selection. Thus, biochemical, genetical, and cytological analyses are essential phases of an experiment to confirm the cybrid nature of fusion products and regenerated plants.

References

1. Sjödin, C. and Glimelius, K. (1988) Screening for resistance to blackleg *Phoma lingam* (Tode ex Fr.) Desm. within Brassicacea. *J. Phytopathol.* **123**, 322–332.
2. Thomzik, J. E. and Hain, R. (1988) Transfer and segregation of triazine tolerant chloroplasts in *Brassica napus* L. *Theor. Appl. Genet.* **76**, 165–171.
3. Sacristan, M. D., Gerdemann-Knörk, M., and Schieder, O. (1989) Incorporation of hygromycin resistance in *Brassica nigra* and its transfer to *B. napus* through asymmetric protoplast fusion. *Theor. Appl. Genet.* **78**, 194–200.
4. Pearson, O. (1972) Cytoplasmically inherited male sterility characters and flavor components from the species cross *B. nigra* x *B. oleracea*. *J. Am. Soc. Hort. Sci.* **97**, 397–402.
5. Ogura, H. (1968) Studies of the new male-sterility in Japanese radish with special reference to the utilization of this sterility towards the practical raising of hybrid seeds. *Mem. Fac. Agric. Kagoshima Univ.* **6**, 39–78.

6. Gleba, Y. Y. and Hoffmann, F. (1980) "Arabidobrassica": a novel plant obtained by protoplast fusion. *Planta* **149**, 112–117.
7. Kao, K. N. (1977) Chromosomal behaviour in somatic hybrids of soybean and Nicotiana glauca. *Mol. Gen. Genet.* **150**, 225–230.
8. Skarzhinskaya, M., Landgren, M., and Glimelius, K. (1996) Production of intertribal somatic hybrids between *Brassica napus* L. and *Lesquerella fendleri* (Gray) Wats. *Theor. Appl. Genet.* **93**, 1242–1250.
9. Sidorov, V. A., Menczel, L., Nagy, F., and Maliga, P. (1981) Cloroplast transfer in *Nicotiana* based on metabolic complementation between irradiated and iodoacetate treated protoplasts. *Planta* **152**, 341–345.
10. Thanh, N. D. and Medgyesy, P. (1989) Limited chloroplast gene transfer via recombination overcomes plastome-genome incompatibility between *Nicotiana tabacum* and *Solanum tuberosum*. *Plant Mol. Biol.* **12**, 87–93.
11. Terada, R., Yamashita, Y., Nishibayashi, S., and Shimamoto, K. (1986) Somatic hybrids between *Brassica oleracea* and *B. campestris*: selection by the use of iodoacetamide inactivation and regeneration ability. *Theor. Appl. Genet.* **73**, 379–384.
12. Zubko, M. K., Zubko, E. I., Patskovsky, Y. V., et al. (1996) Novel "homeotic" CMS patterns generated in *Nicotiana* via cybridization with *Hyoscyamus* and *Scopolia*. *J. Exp. Bot.* **47**, 1101–1110.
13. Zubko, M. K., Zubko, E. I., Adler, K., Grimm, B., and Gleba, Y. Y. (2003) New CMS-associated phenotypes in cybrids *Nicotiana tabacum* L. (*Hyoscyamus niger* L.). *Ann. Bot.* **92**, 281–288.
14. Zubko, M. K. and Day, A. (1998) Stable albinism induced without mutagenesis: a model for ribosome-free plastid inheritance. *Plant J.* **15**, 265–271.
15. DeBlock, M. (1988) Genotype-independent leaf-disc transformation of potato (*Solanum tuberosum*) using *Agrobacterium tumefaciens*. *Theor. Appl. Genet.* **76**, 767–774.
16. Murashige, T. and Skoog, F. (1962) A revised medium for rapid growth and bioassays with tobacco tissue culture. *Physiol. Plant.* **15**, 473–497.
17. Kao, K. N. and Michayluk, M. R. (1975) Nutritional requirements for growth of Vicia hajastana and protoplasts at very low population density in liquid media. *Planta*. **126**, 105–110.
18. Sidorov, V. A., Zubko, M. K., and Kuchko, A. A. (1987) Somatic hybridization in potato: use of gamma-irradiated protoplasts of *Solanum pinnatisectum* in genetic reconstruction. *Teor. Appl. Genet.* **74**, 364–368.
19. Menczel, L., Nagy, F., Kiss, Z. R., and Maliga, P. (1981) Streptomycin resistant and sensitive somatic hybrids of *Nicotiana tabacum* + *Nicotiana knightiana*: correlation of resistance to *N. tabacum* plastids. *Teor. Appl. Genet.* **59**, 191–195.
20. Ono, Y., Takahata, Y., and Kaizuma, N. (1994) Effect of genotype on shoot regeneration from cotyledonary explants of rapeseed (*Brassica napus* L.). *Plant Cell Rep.* **14**, 13–17.

21. Hu, Q., Andersen, S. B., and Hansen, L. N. (2000) Plant regeneration capacity of mesophyll protoplasts from *Brassica napus* and related species. *Plant Cell Tissue Organ Cult.* **39,** 189–196.
22. Cheung, W. Y., Hubert, N., and Landry, B. S. (1993) A simple and rapid DNA microextraction method for plant, animal and insect suitable for RAPD and other PCR analyses. *PCR Meths. Applics.* **3,** 69–70.
23. Sambrook, J., Fritsch, E. F., and Maniatis, T. (1989) *Molecular cloning: A Laboratory Manual, 2nd ed.* Cold Spring Harbour, New York.
24. Savolainen, V., Corbar, R., Moncousin, C., Spichiger, R., and Manen, J.-F. (1995) Chloroplast DNA variation and parentage analysis in 55 apples. *Theor. Appl. Genet.* **90,** 1138–1141.
25. Demesure, B., Sodri, N., and Petit, R.J. (1995) A set universal primers for amplification of polymorphic non-coding region of mitochondrial and chloroplast DNA in plant. *Mol. Ecol.* **4,** 129–131.
26. Yong, X., Hong, D., and Chen, S. (1997) Isolation and characterization of six rice telomereassociated sequences. 5th International Congress of Plant Molecular Biology, Singapure, pp. 21–27.
27. Cox, A., Bennett, S., Parokonny, A., Kenton, A., Callimassia, M., and Bennett, M. (1993) Comparison of plant telomere locations using a PCR-generated synthetic probe. *Ann. Bot.* **72,** 239–247.
28. White, T. J., Bruns, T., Lee S., and Taylor, J. (1990) Amplification and direct sequencing of fungal ribosomal RNA genes for phylogenetics, in *PCR Protocols: A Guide to Methods and Applications* (Innis, M., Gelfand, D. H., Sninsky, J. J., White, T. G., eds.), Academic Press, San Diego, pp. 315–322.
29. Pelletier, G. (1990) Cybrids in oilseed *Brassica* crops through protoplast fusion, in *Biotechnology in Agriculture and Forestry, Vol. 10* (Bajaj, Y. P. S., ed.), Springer-Verlag Berlin Heidelberg, pp. 418–452.
30. Medgyesy, P. (1994) Cybrids—transfer of chloroplast traits through protoplast fusion between sexually incompatible Solanaceae species, in *Biotechnology in Agriculture and Forestry, Vol. 27* (Bajaj, Y. P. S., ed.), Springer-Verlag Berlin Heidelberg, pp. 72–85.
31. Nagata, T. (1984) Fusion of somatic cells, in *Encyclopedia of Plant Physiology, Vol. 17* (Linskens, H-F. and Heslop-Harrison, J., ed.), Springer-Verlag Berlin Heidelberg, pp. 491–507.
32. Tempelaar, M. J. and Jones, M. G. K. (1985) Directed electrofusion between protoplasts with different responses in a mass fusion system. *Plant Cell Rep.* **4,** 92–95.

20

Guard Cell Protoplasts

Isolation, Culture, and Regeneration of Plants

Gary Tallman

Summary

Guard cell protoplasts have been used extensively in short-term experiments designed to elucidate the signal transduction mechanisms that regulate stomatal movements. The utility of guard cell protoplasts for other types of longer-term signal transduction experiments is just now being realized. Because highly purified, primary isolates of guard cell protoplasts are synchronous initially, they are uniform in their responses to changes in culture conditions. Such isolates have demonstrated potential to reveal mechanisms that underlie hormonal signalling for plant cell survival, cell cycle re-entry, reprogramming of genes during dedifferentiation to an embryogenic state, and plant cell thermotolerance. Plants have been regenerated from cultured guard cell protoplasts of two species: *Nicotiana glauca* (Graham), tree tobacco, and *Beta vulgaris*, sugar beet. Plants genetically engineered for herbicide tolerance have been regenerated from cultured guard cell protoplasts of *B. vulgaris*. The method for isolating, culturing, and regenerating plants from guard cell protoplasts of *N. glauca* is described here. A recently developed procedure for large-scale isolation of these cells from as many as nine leaves per experiment is described. Using this protocol, yields of $1.5-2 \times 10^7$ per isolate may be obtained. Such yields are sufficient for standard methods of molecular, biochemical, and proteomic analysis.

Key Words: Culture; guard cells; *Nicotiana*; stomata; protoplasts; regeneration,

1. Introduction

The guard cells that flank stomata undergo environmentally induced, turgor-driven cellular movements that regulate stomatal dimensions. In turn, changes in stomatal dimensions regulate rates of transpiration and photosynthetic carbon fixation (*1*). Among the environmental signals that guard cells transduce are light quality, light intensity, intercellular concentrations of leaf carbon dioxide, and apoplastic concentrations of abscisic acid (ABA) (*2*). How guard cells integrate these signals and activate the appropriate signal transduction mechanisms to adjust stomatal dimensions for prevailing environmental conditions is the subject of intense investigation *(2–5)*.

Many fundamental studies of stomata have been performed with detached leaf epidermis, but for many types of experiments, epidermis is not an adequate material. Guard cells are among the smallest and least numerous of cell types in the leaf, and the presence of even a very small number of contaminating cells in epidermal preparations (e.g., epidermal cells that neighbor guard cells and mesophyll cells) can result in significant experimental artifacts *(6)*. This is particularly true of experiments that involve guard cell biochemistry or molecular biology and/or experiments in which guard cell metabolism is measured. Furthermore, the presence of the relatively thick guard cell wall precludes certain types of studies (e.g., electrophysiology).

Good methods for making relatively large (approx 1×10^6 cells), highly purified preparations (<0.01% contamination with other cell types) of guard cell protoplasts (GCP) were first reported more than 20 yr ago *(6–9)*. The availability of these methods *(10)* contributed significantly to a resurgence of interest in the cell biology of guard cells. GCP have been used to demonstrate PSI and PSII activity in guard cells *(6)* and to confirm that guard cells have a functional photosynthetic carbon reduction pathway *(11)*. They have also been used to study mechanisms that underlie blue light-induced activation of the guard cell plasma membrane H^+-translocating ATPase *(12–16)* and to characterize by patch-clamping a host of inward- and outward-rectifying ion channels involved in stomatal opening and closing *(2,17–42)*. GCP have been crucial to our understanding of the integrated regulation of guard cell ion channels by interactions *(43)* among signaling pathways mediated by ABA *(29,37,38,44–51)*, Ca^{2+} *(27,40,45,50,52–55)*, NO *(42)*, reactive oxygen species *(40,41,50,56)*, and sphingosine-1-phosphate *(57,58)* transduced through G-proteins *(51,58–62)*, protein kinases *(14,32,37,48,63,64)*, and protein phosphatases *(23,32,40,63,65,66)*.

Few molecular studies have been performed with guard cell protoplasts *(67)*. Early studies included isolation of two partial-length cDNAs coding for different plasma membrane H^+-ATPase isoforms from GCP of *Vicia faba* L. *(68)*. More recently, methods for preparing GCP of *Nicotiana glauca* have been scaled to provide enough protein for routine western blotting, two-dimensional (2-D) electrophoresis, and proteomic analysis *(69)*, and new methods for preparing GCP of *Arabidopsis thaliana (66,70)* have led to the first gene expression profiling of GCP treated with ABA *(66)*. These reports almost certainly foreshadow the establishment of wild type and mutant cell lines of *A. thaliana* GCP in culture.

Because GCP are used mostly to study signal transduction for stomatal opening or closing, most studies with GCP have been short-term experiments of a few hours duration. Only for a decade have in vitro experimental culture systems existed with which to study the responses of GCP to environmental signals administered over longer periods. However, now that GCP have been established and maintained in culture *(69,71–80)*, the experimental uses and advantages of such a system are becoming obvious. As with any culture system, culture conditions around GCP can be rigorously defined and carefully controlled, but an added advantage of GCP is that the homogeneity and purity of synchronous GCP preparations confers upon them a uniformity of response that is seldom observed in cultures of mixed cell types *(69,76,78,79)*. In the latter, identification of the cell type responding to a change in culture condition may be difficult. If the responding cell type is identified, the basis for any particular response to a change in a culture condition (e.g., an increase or decrease in synthesis of a particular protein or transcription of a particular gene) may still not be identifiable because the condition can evoke from each unique cell type a response that affects the response(s) of each and/or every other cell type.

Monocultures of GCP hold promise for studies of the signal transduction mechanisms underlying the following: (1) hormonal signaling for cell survival in culture; (2) hormone and temperature regulation of cell cycle re-entry in differentiated cells; (3) hormone-dependent reprogramming of genes during dedifferentiation to an embryogenic state; and (4) plant cell thermotolerance *(69)*. All of these processes are activated only after longer periods of exposure to environmental signals, and each can be activated and directed in cultured GCP by manipulating concentrations and/or ratios of plant growth regulators such as auxin, cytokinins, or ABA *(71,74,76)* and/or temperature *(69,76)*. For example, when GCP of *N. glauca* are cultured at 32°C, they dedifferentiate and divide to form a callus from which plants can be regenerated *(74)*. In the process of dedifferentiation, their chloroplasts become chlorotic and revert to proplastids, which then divide *(76)*. When GCP of this species are cultured at 38°C in media containing ±0.1 μM ABA, they acquire thermotolerance within 24 h *(69)*. They survive in high percentages (70–80%) *(69)*, but they no longer require exogenous auxin or cytokinin to survive *(69)*. With or without ABA, at 38°C they do not make the G1-to-S-phase cell cycle transition *(69)*. At 38°C in media containing ABA, GCP remain differentiated, retaining many of the unique physiological characteristics of guard cells *(79)*. Thus, in monocultures of GCP, altering only one or two culture conditions (+ ABA; temperature) is/are sufficient to alter the survival and developmental fates of these cells.

Cultured GCP of *N. glauca* (tree tobacco) *(74)* and *Beta vulgaris* (sugar beet) *(73,75)* have been used to produce friable, embryogenic callus from which plants have been regenerated, demonstrating that GCP are totipotent. Regeneration of plants from cultured GCP of *B. vulgaris* is of commercial importance because callus derived from other cell types, tissues, and organs of this plant are recalcitrant to regeneration *(75)*. Cultured GCP of *B. vulgaris* have been used to produce transgenic plants with enhanced tolerance to herbicides by a relatively rapid protocol *(77)*.

GCP are isolated using a two-step procedure. To remove contaminating mesophyll and epidermal cells, detached epidermis is treated with a mixture of cellulase and pectinase dissolved in a hypotonic solution. Because the cell walls of contaminating epidermal and mesophyll cells are thinner and are of a different chemical composition than those of guard cells, they are digested more quickly than those of guard cells. As protoplasts of contaminating epidermal cells and mesophyll cells are released into the hypotonic medium, they swell and burst. After sufficient time has passed to destroy contaminating cells, the remaining cuticle containing guard cells is collected on a nylon net, rinsed, and transferred to a solution of cellulase and pectinase in a solution that is slightly hypertonic to guard cells. After a few hours of digestion, GCP are released into the medium. The cells are collected by filtration and/or centrifugation, washed, and suspended in culture media (*see* **Fig. 1A**). This procedure can be scaled to digest epidermis from up to nine leaves at a time to produce yields of 1.5–2 × 10^7 GCP *(69)*.

2. Materials

2.1. Plants

1. Seeds of *N. glauca* (Graham), tree tobacco (*see* **Note 1**).
2. Pots (0.16-, 2.0-, 10-L), potting soil, and sand. We have used Supersoil, (Chino, CA) or Pro-Mix "HP" High Porosity Growing Medium (Premier-Western). Sand is 30 grade. Plants grow best in well-drained, porous media.
3. Environmental growth chamber: Conviron Model E7 or equivalent (Conviron, Pembina, ND).
4. Fluorescent light bank: lamps are model FT72T12.CW.1500 (General Electric) or equivalent.
5. Modified Hoagland's nutrient solution (**Table 1**).

2.2. Day Before Culture

1. Four 2-L flasks with cotton stoppers covered with cheesecloth, paper towels, aluminum foil, and autoclave tape.
2. Pyrex® casserole dish: 11 × 7 × 1.5 in.
3. Pyrex casserole dish: 11 × 7 × 1.5 in. Containing: two pairs of fine point forceps wrapped individually in foil, glass plate: 4.5 × 4.5 × 0.13 in., 6-in. cotton swab, single edge razor blade, Petri dish (glass, bottom only), 15 × 60 mm.
4. Plastic beakers: 2-and 1-L.

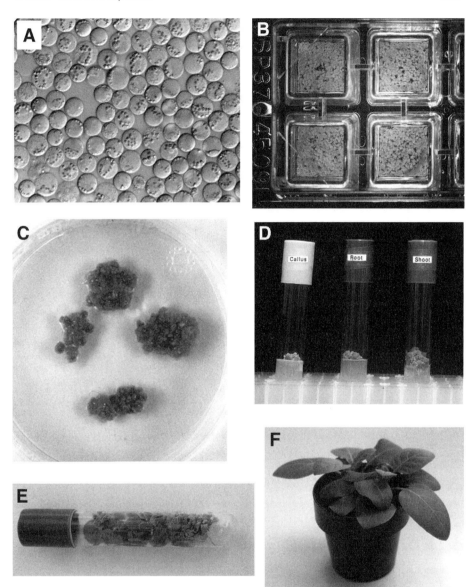

Fig. 1. Regeneration of plants from isolated guard cell protoplasts of *Nicotiana glauca* (Graham), tree tobacco. (**A**) Guard cell protoplasts isolated from leaves of *N. glauca* (Graham), tree tobacco. Differential interference contrast optics; large organelles in protoplasts are chloroplasts. Average diameter approx 15 µm, (**B**) microcolonies in chamber slides, (**C**) primary callus, (**D**) secondary callus, (**E**) shoots derived from secondary callus, and (**F**) regenerated plant.

Table 1
Modified Hoagland's Nutrient Solution (81)

Constituent	Stock solution (g/L)	Final concentration in nutrient solution after dilution of stock (mM)
1. $NH_4H_2PO_4$	21.0	1.0
2. KNO_3	109.0	6.0
3. $Ca(NO_3)_2 \cdot 4H_2O$	170.0	4.0
4. $MgSO_4 \cdot 7H_2O$	43.6	2.0
5. H_3BO_3	0.56	0.4
$MnSO_4 \cdot H_2O$	0.308	10.0
$ZnSO_4 \cdot 7H_2O$	0.042	0.8
$CuSO_4 \cdot 5H_2O$	0.018	0.4
MoO_3	0.011	0.4
NaCl	1.045	100
6. Iron solution:		
First dissolve: Na_2EDTA	6.0	0.09
Then add: $FeSO_4 \cdot 7H_2O$	4.5	0.09
7. KOH	4.0	71.3

To prepare nutrient solution from stock solutions, add 100 mL each of solutions 1–6 and 50 mL of solution 7 to a 20-L carboy. Add water to 18 L; shake to mix.

5. Two-liter plastic beaker containing: three plastic, disposable 125-mL Erlenmeyer flasks with screw caps (loosen caps and cover with foil), plastic funnels with top diameter = 3.5 in., stem diameter = 0.5 in. (one funnel is lined with 220 × 220 μm mesh nylon net; the other is lined with a 30 × 30 μm mesh nylon net; nets are secured to rims of funnels with autoclave tape; wrap funnels in foil). Nylon nets are Nitex® from Sefar America (TETKO), Inc.

2.3. Isolation of Guard Cell Protoplasts

1. Plastic bag for holding leaf in moist paper towels.
2. Gyrotory shaking water bath, pH meter, balance.
3. Solutions—make fresh daily; filter sterilize. All chemicals are reagent-grade. Powdered enzymes are stored refrigerated. Cellulase "Onozuka" RS from Yakult Pharmaceutical Ind. Co., Ltd., 1-1-19, Highashi-Shinbashi, Minato-Ku, Tokyo, 105 Japan. Pectolyase Y-23 from Seishin Pharmaceutical Co., Ltd., 4-13, Koami-cho, Nihonbashi, Chuo-ku, Tokyo 103, Japan. Media containing agar are sterilized by autoclaving at 121°C, 15 psi for 20 min.

   Solution A: 0.5% polyvinylpyrrolidone 40 (PVP40) and 0.05% ascorbic acid, pH 6.5; dissolve 2.5 g of PVP 40, and 0.25 g of ascorbic acid in 450 mL of deionized water; adjust pH to 6.5 with NaOH, bring to final volume of 500 mL with deionized water, and mix.

Solution B: in a small beaker combine 2.533 g sucrose, 0.0054 g $CaCl_2$, 0.4 g Celluase Onozuka RS, 0.003 g Pectolyase Y-23, 0.185 g PVP 40, and 0.075 g bovine serum albumin (BSA). Add 35 mL of deionized water and stir until components are dissolved. The pH of the enzyme solution is adjusted initially to 3.4 with stirring for 7 min, and then raised to 5.5 before filter sterilization (see **Note 2**). Final concentration of sucrose is ca. 0.2 M; final concentration of $CaCl_2$ is approx 0.7 mM.

Solution C: 0.2 M sucrose, 1 mM $CaCl_2$. Dissolve 13.692 g of sucrose and 0.0294 g of $CaCl_2$ in 190 mL of deionized water; adjust pH to 5.5 with HCl and/or NaOH. Bring final volume to 200 mL with deionized water and mix.

Solution D: in a small beaker combine 2.140 g sucrose, 0.0037 g $CaCl_2$, 0.4 g Celluase Onozuka RS, 0.003 g Pectolyase Y-23, 0.125 g PVP 40, and 0.075 g BSA. Add 23 mL of deionized water and stir until components are dissolved. Adjust pH to 5.5 (see **Solution B** and **Note 2**). Final concentration of sucrose is approx 0.25 M; final concentration of $CaCl_2$ is approx 1 mM.

4. Sterile 0.45 µm cellulose nitrate filters in disposable filter units: 115-mL (4); 250-mL (1); 500-mL (1). Enzyme solutions and culture media lacking agar are sterilized by filtration through 0.45 µm cellulose nitrate filters in disposable filter units (115-mL = model 125-0045; 250-mL = 126-0045; 500-mL = 450-0045, Nalgene Co., Rochester, NY). To prevent particulates from plugging filters, a prefilter (Gelman Sciences Type A/E Glass Fiber Filter (50 mm), P/N 6/632, Gelman, Ann Arbor, MI) is used when enzyme solutions are sterilized.
5. Laminar flow cabinet.
6. Sterile latex gloves (see **Note 3**).
7. Disinfectant (see **Note 4**).
8. 5.25% sodium hypochlorite = Clorox bleach (Oakland, CA). Free chlorine concentration = 5.25%.
9. 95% Ethanol.
10. Centrifuge tubes: sterile, disposable, plastic conical with caps; 15-mL (3); 50-mL (3).
11. Syringe: plastic, 60-mL (1), disposable 0.45-µm syringe filter (1); Corning disposable syringe filter: 0.45 µm; 25-mm diameter cellulose acetate membrane in acrylic holder (model no. 21053-25; Corning, Inc., Corning, NY).
12. Glass pipets: sterile, disposable, glass, 10 mL (5), 1 mL (2).
13. Clinical centrifuge.
14. Hemocytometer, microscope, hand tally counter.
15. Incomplete medium I (**Table 2**).

2.4. Primary Cultures and Colony Formation

1. Plastic 8-well microchamber culture slides, Lab-Tek Chamber Slide®, Model 177402, Nunc, Inc., Naperville, IL.
2. Petri dish (1; 2.5 cm deep × 15 cm diameter), Parafilm (American National Can Co., Greenwich, CT).

Table 2
Media Used to Culture and Regenerate Plants From Guard Cell Protoplasts of *Nicotiana glauca* (Graham), Tree Tobacco

Constituent	Medium I, (mg/L)	Medium II, (mg/L)	Tobacco shoot medium, (mg/L)	Tobacco root medium, (mg/L)
Salts				
$Ca(NO_3)_2 \cdot 4H_2O$	180.0			
NH_4NO_3	82.5	825.0	1650.0	1650.0
KNO_3	167.0	950.0	1900.0	1900.0
$CaCl_2 \cdot 2H_2O$	84.7	220.0		
$CaCl_2$			333.3	333.3
$MgSO_4 \cdot 7H_2O$	447.3	1223.0		
$MgSO_4$			181.0	181.0
Na_2SO_4	180.0			
KH_2PO_4	68.0	680.0	170.0	170.0
$NaH_2PO_4 \cdot H_2O$	14.9			
KCl	88.3			
Na_2 EDTA	3.7	37.3		
$FeSO_4 \cdot 7H_2O$	2.8	27.8		
$Fe_2(SO_4)_3$	2.3			
KI	0.76	0.83	0.83	0.83
H_3BO_3	2.0	6.2	6.2	6.2
$MnSO_4 \cdot 4H_2O$	3.1		16.9	16.9
$MnCl_2 \cdot 4H_2O$	2.0	19.8		
$ZnSO_4 \cdot 7H_2O$	2.3	9.2	8.6	8.6
$Na_2MoO_4 \cdot 2H_2O$	0.03	0.25	0.25	0.25
$CuSO_4 \cdot 5H_2O$	0.003	0.025	0.025	0.025
$CoSO_4 \cdot 7H_2O$	0.003	0.03		
$CoCl_2 \cdot 6H_2O$			0.025	0.025
FeNa EDTA			36.7	36.7
Organics				
i-Inositol	39.7	100.00	100.0	100.0
Thiamine·HCl	0.19	1.0	0.4	0.4
Glycine	1.35		2.0	2.0
Niacin	0.45			
Pyridoxine·HCl	0.1		0.5	0.5
Nicotinic acid			0.5	0.5
Casein hydrolysate			1000.0	1000.0
MES	976.0	976.0		
Sucrose	95,840.0	78,729	8557.5	8557.5
IAA				3.0
Kinetin			1.0	
1-NAA	0.3	0.3		
6-BA	0.075	0.075		
Agarose		5000.0		
Agar			8000.0	8000.0

(continued)

3. Hormone stock solution (*see* **Subheading 3.3., step 27**).
4. Incubator (lighting optional).

2.5. Primary Callus

1. Medium II in Petri dishes (1.5 cm deep × 10 cm diameter; approx 20 mL/dish) (**Table 2**).
2. Lighted incubator.

2.6. Secondary Callus

1. Tobacco shoot medium (**Table 2**).
2. Lighted incubator.

2.7. Plant Regeneration

1. Tobacco shoot medium (**Table 2**).
2. Root medium (**Table 2**) in Magenta vessels (Sigma Chemical Co., St. Louis, MO); 75-mL/vessel.
3. Lighted incubator.
4. 0.16-L Small pots; potting soil.

3. Method

3.1. Plants

1. Seeds of *N. glauca* are germinated at high density on the surface of moistened, autoclaved potting soil in small pots (0.16-L) (*see* **Note 1**).
2. Plants are germinated under and maintained on a 16-h light/8-h dark cycle in an environmental chamber. Mean (± SE) temperature during the light cycle is 28 ± 2°C; mean temperature during the dark cycle is 21 ± 2°C. Relative humidity in the chamber is 65–75%. Seedlings are irrigated with tap water daily. The photosynthetic photon flux density (PPFD) at seedling height is 50–70 µmol/m^2/s of photons of photosynthetically active radiation (PAR).
3. After 4–6 wk of growth, transfer plants to 2-L plastic pots containing an autoclaved mixture of 60% soil/40% sand (v:v).

(*From opposite page*) Medium I, Medium II, and root medium are modifications of those described by Shepard and Totten *(82)*. All salts and organics through casein hydrolysate are made as 1000X stocks except CaCl$_2$·2H$_2$O (260X) and KCl (663X). Na$_2$ EDTA, FeSO$_4$·7H$_2$O, and Fe$_2$(SO$_4$)$_3$ are mixed in a single stock solution in that order. Stocks are stored frozen at –20°C; any material precipitated by freezing is resolubilized after thawing. An incomplete Medium I containing all components through 2(*N*-Morpholino) ethanesulfonic acid (MES) is prepared in 1-L batches using stock solutions and solid MES. The incomplete medium is stored frozen in 100-mL aliquots at –20°C until the day GCP are to be isolated. Sucrose and hormones are added to complete wash and culture media on the day of protoplast isolation as described in **Subheading 3.3., step 27**. The final sucrose concentration in Medium I is 0.28 M, in Medium II, 0.23 M, and in shoot differentiation medium, 0.025 M. Concentration of MES in Media I and II is 5 mM; all media are pH 6.1. Tobacco shoot medium is commercially available from Carolina Biological Supply, Burlington, NC.

4. After another 4–6 wk of growth, cull plants are to two plants per pot and allow to grow to a height of 0.2–0.3 m.
5. Transfer plants to 10-L plastic pots containing the same soil/sand mix. Grow on a table under high-intensity fluorescent lights. The PPFD at the top of the canopy is 800–900 µmol/m^2/s of PAR.
6. Water plants three times daily with tap water at 6-h intervals for 4 min and every other day with modified Hoagland's nutrient solution (**Table 1**). Mean temperatures is 27 ± 2°C during the 16-h light cycle and 23 ± 2°C during the 8-h dark cycle. Relative humidity in the room is 45–65%.

3.2. Day Before Culture

1. Autoclave at 121°C, 15 psi for 20 min; exhaust on dry cycle:
   a. Four 2-L Erlenmeyer flasks each containing 1.4 L of deionized water. Stopper flasks with cotton wrapped in cheesecloth. Place a paper towel over each stopper and secure on each side with autoclave tape. Place a double layer of aluminum foil on top of each stopper and secure to the flask with autoclave tape.
   b. One casserole dish, empty: cover the dish with a single layer of foil and secure with autoclave tape. One casserole dish containing: two pairs of fine-tipped forceps wrapped individually in foil, one 6-in. cotton swab, one glass plate, one 15 × 60 mm glass Petri dish bottom, and one single-edge razor blade. Cover the dish with a single layer of foil and secure with autoclave tape.
   c. Two 2-L plastic Nalgene beakers, and one 1-L plastic Nalgene beaker (for waste), empty: cover beakers with a double layer of foil secured with autoclave tape.
   d. One 2-L plastic beaker, empty; one 2-L plastic beaker containing three 125-mL, screw cap disposable plastic Erlenmeyer flasks (caps secured loosely and covered with foil) and two plastic funnels, one lined with 220 × 220 µm nylon netting and the other with 30 × 30 µm nylon netting. Secure netting to rim of each funnel with autoclave tape; wrap funnels in foil. Cover beaker with a double layer of foil secured with autoclave tape.

3.3. Isolation of Guard Cell Protoplasts

1. Turn on shaking water bath, check water level, and bring temperature to 28°C.
2. Thaw 100 mL of incomplete medium I (**Table 2**).
3. For each experiment, harvest one flat leaf from insertion level 4 or 5 from the top of the plant with a blade length of 0.11–0.2 m and with a relatively thick cuticle. Harvest 0.5 to 1.5 h prior to the onset of the light cycle (*see* **Note 5**). Store leaves in moist paper towels in a plastic bag in darkness until detachment of epidermis is initiated.
4. Weigh, mix, and adjust pH of solutions A, B, C, and D; filter sterilize.
5. Turn on laminar flow hood; purge for 10–30 min.
6. Cover hands with sterile, latex gloves (*see* **Note 3**).
7. Wipe the sides and bottom of the laminar flow cabinet with full-strength disinfectant (*see* **Note 4**).
8. Transfer solutions, glassware, and instruments to the laminar flow cabinet. Back

Guard Cell Protoplasts 243

row: solutions; middle row, L to R: 2-L beaker, 2-L beaker, 2-L beaker, 2-L beaker; front row L: casserole dish with implements; front row R: empty casserole dish. Remove foil and contents from beakers. Pour 2 L of sodium hypochlorite into 2-L beaker at left. Fill each of the remaining 2-L beakers with 1.8 L of sterile, deionized water.

9. Remove foil from both casserole dishes.
10. Spray leaf with 95% ethanol, and then lightly buff with tissues to remove some of the wax.
11. Immerse right hand in 5.25% sodium hypochlorite for 3–5 s. Using the right hand, immerse the leaf in the sodium hypochlorite solution for 3–5 s. Immerse leaf in each beaker of water for 3–5 s, starting with the 2-L beaker. Lay the leaf in the empty casserole dish, top (adaxial) side up.
12. Transfer a few milliliters of solution A to the small Petri dish in the other casserole dish; pour the remainder over the leaf.
13. Keeping the leaf beneath the solution and starting at the base of the leaf, break (tear) the leaf near the mid-rib. Bend the torn leaf section toward you at angles of 120–160° to the leaf surface, and "peel" the adaxial epidermis away from the mesophyll. (*See* **Note 5**; it may be necessary to repeat this procedure three to five times to get most of the epidermis from one-half of the leaf.)
14. Transfer each sheet of epidermis, and any attached leaf material to the glass plate in the neighboring casserole dish. Spread the epidermis on the plate with the side that normally faces the mesophyll upward. Brush the epidermis gently with the cotton swab to remove adhering mesophyll. Using the single-edge razor blade, slice the epidermis horizontally into strips, and then vertically to small pieces (approx 5×5 mm). Using fine forceps, transfer pieces to the small Petri dish.
15. Repeat **steps 13** and **14** until most of the epidermis has been peeled from the leaf.
16. Decant the solution in the Petri dish into the casserole dish. Gather epidermal fragments in the Petri dish into a ball with forceps, and drop them down the middle of the neck of a 125-mL Erlenmeyer flask.
17. Add solution B to the flask, rinsing any epidermis adhering to the sides down into the flask with the solution. Cap.
18. Incubate for 15 min at 28°C with shaking at 175 rpm. Depending on leaf age, digestion may take more or less time.
19. Remove used materials from the hood. Leave 2-L beaker of sodium hypochlorite and 2-L beaker of water.
20. After 10 min of incubation, sterilize gloved hands as above.
21. In hood, position funnel with 220×220 μm mesh nylon net over mouth of second 125-mL Erlenmeyer flask.
22. After 15 min, remove the incubating flask from the water bath. In the laminar flow cabinet, remove cap and pour contents of flask over the 220×220 μm mesh nylon net to collect "cleaned" epidermis.
23. Rinse epidermis on net with 100 mL of solution C.
24. Open second set of sterile forceps; collect epidermis in a ball and transfer to third 125-mL Erlenmeyer. Rinse sides of flask during addition of solution D. Cap.

25. Incubate for 3 h 15 min at 28°C with shaking at 25 excursions/min.
26. Remove used material from laminar flow cabinet.
27. During incubation, prepare wash and culture media. Dissolve 9.584 g of sucrose in a total volume of 100 mL using incomplete medium I (**Table 2**). Divide 60 and 40 mL into separate beakers. Adjust pH of 60-mL aliquot to 6.1; adjust 40-mL aliquot to pH 6.8. Dissolve 0.012 g of 1-napthalene acetic acid (NAA) and 0.003 g of 6-benzylaminopurine (BA) in 10 mL of 95% ethanol to make a hormone stock of 6.44 mM NAA and 1.33 mM BA. Add 10 µL of hormone stock to 10 mL of medium at pH 6.1. In laminar flow cabinet, using a 60-mL syringe, filter medium containing hormones through a disposable syringe filter into a 15-mL sterile, plastic conical centrifuge tube. Cap.
28. Filter remaining media at pH 6.1 and 6.8 through separate 0.45 µm disposable filter units; place in laminar flow cabinet along with two 15-mL and 3 50-mL sterile plastic conical centrifuge tubes with caps. Uncap two 50-mL centrifuge tubes. Transfer pH 6.8 medium from filter unit to one 50-mL tube and cap. Transfer pH 6.1 medium to a second 50-mL tube and cap.
29. In hood, unwrap and position funnel with 30× 30 µm mesh nylon net over mouth of an open 50-mL conical centrifuge tube.
30. After 3 h 15 min, remove the incubating suspension of epidermis from the water bath and swirl. In the laminar flow cabinet, remove cap from incubation flask and pour contents over the nylon net.
31. With a 10-mL sterile, disposable pipet, divide filtrate equally between two 15-mL sterile plastic conical centrifuge tubes; cap.
32. Collect GCP by centrifuging the filtrate at 40g for 6–7 min.
33. After centrifugation, remove supernatant with 10-mL pipet and discard in waste container in hood.
34. With a fresh, sterile 10-mL pipet, add 8 mL of medium at pH 6.8 to each tube. Resuspend GCP in wash medium by rolling centrifuge tubes gently between the palms of the hands. Cap tubes, and centrifuge at 40g for 6–7 min.
35. Repeat **steps 33** and **34** two more times, but with medium of pH 6.1 without hormones. After the second centrifugation in this medium, aspirate the supernatant in each tube to 0.5 mL. Use a 1-mL sterile pipet to resuspend GCP in one of the tubes and transfer them into the other (combined total volume of 1 mL).
36. Using a hemocytometer, count GCP to estimate cell density and dilute accordingly with medium (pH 6.1) to give a final value of 1.25×10^5 cells/mL.

3.4. Primary Cultures and Colony Formation

1. Initiate liquid cultures by pipetting 0.3 mL of the cell suspension to wells of eight-well microchamber culture slides. Add to each chamber 0.1 mL of the medium containing NAA and BA.
2. Incubate chamber slides in sterile plastic Petri dishes (2.5 cm deep × 15 cm diameter) containing moist paper towels; seal edges of dishes tightly with Parafilm®.
3. Incubate cultures at 25–32°C (*see* **Note 6**) in darkness or under red light (15–20 µmol/m²/s of photons of PAR) on a 12-h light/12-h dark cycle. For the latter

3.5. Primary Callus

1. After 8–10 wk of culture in microchamber slides, transfer cultured cells to medium II.
2. Seal the dishes with Parafilm, and incubate at 25°C under continuous white fluorescent light (21–27 µmol/m^2/s of photons of PAR).

3.6. Secondary Callus

1. After 8–10 wk of culture, transfer green callus tissue (*see* **Fig. 1C**) to a commercial *N. tabacum* shoot differentiation medium (*see* **Fig. 1D** and **Table 2**).
2. Incubate at 25°C under continuous white fluorescent light (14–23 µmol/m^2/s of photons of PAR).

3.7. Plant Regeneration

1. After 8–10 wk of growth on shoot medium, transfer callus to fresh shoot differentiation medium.
2. When shoots are 0.5–1 cm in height (*see* **Fig. 1E**), transfer them to root medium (**Table 2**) in Magenta vessels, and incubate at 25°C under continuous white fluorescent light (30 µmol/m^2/s of photons of PAR). When roots are sufficiently developed (6–8 wk), transplant plants to small pots, and grow under the conditions described above for seedlings.

3.7.1. Modifications for Large-Scale Isolation

1. Equipment required for isolation from two to nine leaves is listed in **Table 3**. Double the concentrations of ascorbic acid and PVP40. Multiply volumes of solutions B and D by the number of leaves to be used (*see* **Subheading 2.3., step 3**).
2. Use 250-mL disposable Erlenmeyer flasks. Epidermis from as many as 4.5 leaves can be incubated in a single 250-mL flask (*see* **Subheading 3.3., steps 16**, and **24**).
3. For isolates of six to nine leaves, make two preparations of epidermis and stagger the beginning of the first incubation (*see* **Subheading 3.3., step 18**) of the two preparations by 4 min.
4. At the end of the first digestion, collect epidermis from all preparations on net as described (*see* **Subheading 3.3., step 22**), but rinse peels on net with 75 mL of solution C. Transfer peels from net to 125-mL Erlenmeyer containing 75 mL of solution C, swirl vigorously, collect again on the same nylon net, and rinse on net with an additional 50–100 mL of solution C (*see* **Note 7**).
5. Transfer peels into fresh 250-mL flask(s) containing solution D (*see* **Subheading 3.3., step 24**).
6. Incubate for 3.5 h at 55 rpm (*see* **Subheading 3.3., step 25**).

Table 3
Equipment Required for Large-Scale Isolation of Guard Cell Protoplasts From Leaves of *Nicotiana glauca* (Graham), Tree Tobacco

Number of leaves	1	2	4	6	8	9
Equipment						
2-L Beakers	4	4	4	4	4	4
1-L Beaker (for waste)	1	1	1	1	1	1
125-mL Erlenmeyer flasks	4	1	2	3	7	7
250-mL Erlenmeyer flasks	0	3	3	5	5	5
Funnels with 30 × 30 μm mesh nylon net	1	1	1	2	2	2
Funnels with 220 × 220 μm mesh nylon net	1	1	1	2	2	2
2-L Erlenmeyer flasks with 1.4 L sterile, deionized water	4	4	4	4	4	4
Pyrex casserole dishes	2	2	2	3	3	3
Glass plate	1	1	1	1	1	1
Cotton swabs	3	6	10	14	16	18
Razor blades	1	1	2	2	3	3
Forceps	2	2	2	2	2	2
Petri dishes–bottoms only	1	2	2	3	3	4

Preparations with more than four leaves are carried out with two separate, but identical, sets of equipment as described in **Subheading 3.8**.

7. Remove flask(s) from water bath. Swirl gently 10× clockwise, then 10× counterclockwise before filtering through the 30 × 30 μm mesh nylon net. Rinse net with an additional 5–10 mL of incomplete medium I, pH 6.8. (*see* **Subheading 3.3., step 30**).
8. For two to four leaf preparations, after centrifugation, aspirate supernatant to 0.5 mL, resuspend cells by rolling tubes between the palms of the hands, combine contents of all tubes in a single 15-mL conical tube, and collect by centrifugation at 60*g* for 8 min. Aspirate supernatant to 1 mL and count with a hemocytometer. For six to nine leaf preparations, reduce protoplasts from two parallel isolates to a single tube each, count with hemocytometer, and then combine in a single tube if desired (*see* **Subheading 3.3., steps 35 and 36**).

4. Notes

1. Seeds may be obtained on-line from a number of seed companies. Two are: GardenMakers, P. O. Box 65, Rowley, MA 01969-0165, USA (http://www.gardenmakers.com/ seedindx.htm) and Magic Garden Seeds, Moritzstrasse 1, 34127 Kassel, Germany (http:/www.magic-garden-seeds.com). Seeds are sprinkled on the top of the soil and misted daily with an atomizer to prevent

flushing seeds deep into soil. Once seedlings are 5–10 mm in height they are watered directly.
2. Treatment at low pH appears to precipitate some inhibitory, insoluble material that does not return to solution when the pH is raised to 5.5. Each new lot of enzyme should be tested, and the concentration adjusted for any differences in activity from the previous lot. Pectolyase activity is increased dramatically with an increase of only 1–2°C in temperature. The time required for the first enzyme digestion may vary depending on leaf age, with younger leaves taking less time. If leaves are too young, yields will be low, and peanut-shaped remnants of cells will appear in preparations.
3. Gloves are not powdered; air-test for pinholes. Gloves are pulled over lab coat sleeves to cover any of the arm or wrist that might be exposed.
4. We use Lysol Pine Action Cleaner, Household Products Division, Reckitt and Colman, Inc. (Montvale, NJ).
5. Plants are not used after they reached a height of 1 m. Leaf should be fully expanded with a well-developed waxy cuticle. Leaves are removed from darkness so that all stomata are closed. Osmotic potentials of guard cells of leaves with open stomata will not be uniform, and thus solution D may not be hypertonic to all guard cells, reducing yields. Older leaves may give lower yields of GCP. Higher yields of epidermis from older (larger) leaves may saturate enzymes and reduce the number of GCP released over the 3 h 15 min digestion period.
6. Temperature has a dramatic effect on survival; highest survival is at 32°C *(76)*.
7. In large-scale preparations, contaminants (cell wall fragments, cellular debris, and others) get trapped between layers of epidermis on the net. Washing in a separate flask, refiltering, and rewashing on the net reduces this contamination.

Acknowledgments

This work was supported by multiple grants from the U. S. National Science Foundation, most recently, MCB9900525. The author thanks B. Cupples, P. Sahgal, C. Roberts, G. Boorse, and A. Kemper for technical improvement to the protocols described.

References

1. Cowan, I. R. (1982) Regulation of water use in relation to carbon gain in higher plants, in *Encyclopedia of Plant Physiology Vol 12B,* (Lange, O. L., Nobel, P. S., Osmond, C. B., and Ziegler, H., eds.), Springer, Heidelberg, pp 589–613.
2. Outlaw Jr., W. H. (2003) Integration of cellular and physiological functions of guard cells. *CRC Crit. Rev. Plant Sci.* **22,** 503–529.
3. Assmann, S. M. (1993) Signal transduction in guard cells. *Annu. Rev. Cell Biol.* **9,** 345–375.
4. Hetherington, A. M. (2001) Guard cell signaling. *Cell* **107,** 711–714.
5. Schroeder, J. I., Allen, G. J., Hugouvieux, V., Kwak, J. M., and Waner, D. (2001) Guard cell signal transduction. *Annu. Rev. Plant Physiol. Plant Mol. Biol.* **52,** 627–658.

6. Outlaw Jr, W. H., Mayne, B. C., Zenger, V. E., and Manchester, J. (1981) Presence of both photosystems in guard cells of *Vicia faba* L. *Plant Physiol.* **67,** 12–16.
7. Shimazaki, K., Gotow, K., and Kondo, N. (1982) Photosynthetic properties of guard cell protoplasts from *Vicia faba*. *Plant Cell Physiol.* **23,** 871–879.
8. Gotow, K., Kondo, N., and Syono, K. (1982) Effect of CO_2 on volume change of guard cell protoplasts from *Vicia faba* L. *Plant Cell Physiol.* **23,** 1063–1070.
9. Gotow, K., Shimazaki, K., Kondo, N., and Syono, K. (1984) Photosynthesis-dependent volume regulation in guard cell protoplast from *Vicia faba* L. *Plant Cell Physiol.* **25,** 671–675.
10. Weyers, J. D. B., Fitzsimons, P. J., Mansey, G. M., and Martin, E. S. (1983) Guard cell protoplasts-aspects of work with an important new research tool. *Physiol. Plant* **58,** 331–339.
11. Gotow, K., Taylor, S., and Zeiger, E. (1988) Photosynthetic carbon fixation in guard cell protoplasts of Vicia faba L. - evidence from radiolabel experiments. *Plant Physiol.* **86,** 700–705.
12. Assmann, S. M., Simoncini, L., and Schroeder, J. I. (1985) Blue light activates electrogenic ion pumping in guard cell protoplasts of *Vicia faba*. *Nature* **318,** 285–287.
13. Shimazaki, K., Iino, M., and Zeiger, E. (1986) Blue light-dependent proton extrusion by guard-cell protoplasts of *Vicia faba*. *Nature* **319,** 324–326.
14. Kinoshita, T. and Shimazaki, K. (1999) Blue light activates the plasma membrane H^+-ATPase by phosphorylation of the C-terminus in stomatal guard cells. *EMBO J.* **18,** 5548–5558.
15. Kinoshita, T. and Shimazaki, K. (2002) Biochemical evidence for the requirement of 14-3-3 protein binding in activation of the guard-cell plasma membrane H^+-ATPase by blue light. *Plant Cell Physiol.* **43,** 1359–1365.
16. Kinoshita, T., Emi, T., Tominaga, M., et al. (2003) Blue-light- and phosphorylation-dependent binding of a 14-3-3 protein to phototropins in stomatal guard cells of broad bean. *Plant Physiol.* **133,** 1453–1463.
17. Schroeder, J. I., Hedrich, R., and Fernandez, J. M. (1984) Potassium-selective single channels in guard cell protoplasts of *Vicia faba*. *Nature* **312,** 361–362.
18. Schroeder, J. I., Raschke, K., and Neher, E. (1987) Voltage dependance of K+ channels in guard-cell protoplasts. *Proc. Natl. Acad. Sci. (USA)* **84,** 4108–4112.
19. Keller, B. U., Hedrich, R., and Raschke, K. (1989) Voltage-dependant anion channels in the plasma membrane of guard cells. *Nature* **341,** 450–453.
20. Hedrich, R., Busch, H., and Raschke, K. H. (1990) Ca^{2+} and nucleotide dependent regulation of voltage dependent anion channels in the plasma membrane of guard cells. *EMBO J.* **9,** 3889–3892.
21. Lee, Y., and Assmann, S. M. (1991) Diacylglycerols induce both ion pumping in patch-clamped guard-cell protoplasts and opening of intact stomata. *Proc. Natl. Acad. Sci. (USA)* **88,** 2127–2131.
22. Marten, I., Zeilingeer, C., Redhead, C., et al. (1992) Identification and modulation of a voltage-dependent anion channel in the plasma membrane of guard cells by high-affinity ligands. *EMBO J.* **11,** 3569–3575.

23. Luan, S., Li, W., Rusnak, F., Assmann, S. M., and Schreiber, S. (1993) Immunosuppressants implicate protein phosphatase regulation of K^+ channels in guard cells. *Proc. Natl. Acad. Sci. (USA)* **90,** 2202–2206.
24. Hedrich, R., Marten, I., Lohse, G., et al. (1994) Malate-sensitive anion channels enable guard cells to sense changes in the ambient CO_2 concentration. *Plant J.* **6,** 741–748.
25. Li, W., Luan, S., Schreiber, S. L., and Assmann, S. M. (1994) Evidence for protein phosphatases 1 and 2A regulation of K_+ channels in two types of leaf cells. *Plant Physiol.* **106,** 963–970.
26. Schwartz, A., Wu, W.-H., Tucker, E. B., and Assmann, S. M. (1994) Inhibition of inward K^+ channels and stomatal response by abscisic acid: an intracellular locus of phytohormone action. *Proc. Natl. Acad. Sci. (USA)* **91,** 4019–4023.
27. Ward, J. M. and Schroeder, J. I. (1994) Calcium-activated K_+ channels and calcium-induced calcium release by slow vacuolar ion channels in guard cell vacuoles implicated in the control of stomatal closure. *Plant Cell* **6,** 669–683.
28. Müller-Röber, B., Ellenberg, J., Provart, N., et al. (1995) Cloning and electrophysiological analysis of KST1, an inward rectifying K^+ channel expressed in potato guard cells. *EMBO J.* **14,** 2409–2416.
29. Schwartz, A., Ilan, N., Schwarz, M., Scheaffer, J., Assmann, S. M., and Schroeder, J. I. (1995) Anion-channel blockers inhibit S-type anion channels and abscisic acid responses in guard cells. *Plant Physiol.* **109,** 651–658.
30. Schroeder, J. I. (1995) Anion channels as central mechanisms for signal transduction in guard cells and putative functions in roots for plant-soil interactions. *Plant Mol. Biol.* **28,** 353–361.
31. Ward, J. M., Zhen-Ming, P., and Schroeder, J. I. (1995) Roles of ion channels in initiation of signal transduction in higher plants. *Plant Cell* **7,** 833–844.
32. Li, J., Lee, Y-R. J., and Assmann, S. M. (1998) Guard cells possess a calcium-dependent protein kinase that phosphorylates the KAT1 potassium channel. *Plant Physiol.* **116,** 785–795.
33. Baizabal-Aguirre, V. M., Clemens, S., Uozumi, N., and Schroeder, J. I. (1999) Suppression of inward-rectifying K^+ channels KAT1 and AKT2 by dominant negative point mutations in the KAT1 alpha-subunit. *J. Membrane Biol.* **167,** 119–125.
34. Bruggemann, L., Dietrich, P., Dreyer, I., and Hedrich, R. (1999) Pronounced differences between the native K^+ channels and KAT1 and KST1 alpha-subunit homomers of guard cells. *Planta* **207,** 370–376.
35. Hoth, S. and Hedrich, R. (1999) Susceptibility of the guard-cell K^+-uptake channel KST1 to Zn^{2+} requires histidine residues in the S3-S4 linker and in the channel pore. *Planta* **209,** 543–546.
36. Hoth, S. and Hedrich, R. (1999) Distinct molecular bases for pH sensitivity of the guard cell K^+ channels KST1 and KAT1. *J. Biol. Chem.* **274,** 11,599–11,603.
37. Li, J., Wang, X. Q., Watson, M. B., and Assmann, S. M. (2000) Regulation of abscisic acid-induced stomatal closure and anion channels by guard cell AAPK kinase. *Science* **287,** 300–303.

38. Romano, L. A., Jacob, T., Gilroy, S., and Assmann, S. M. (2000) Increases of cytosolic Ca^{2+} are not required for abscisic acid-inhibition of inward K^+ currents in guard cells of *Vicia faba* L. *Planta* **211**, 209–217.
39. Kwak, J. M., Murata, Y., Baizabal-Aguirre, V. M., et al. (2001) Dominant negative guard cell K^+ channel mutants reduce inward-rectifying K^+ currents and light-induced stomatal opening in *Arabidopsis*. *Plant Physiol.* **127**, 473–485.
40. Murata, Y., Pei, Z.-M., Mori, I. C., and Schroeder, J. (2001) Abscisic acid activation of plasma membrane Ca^{2+} channels in guard cells requires cytosolic NAD(P)H and is differentially disrupted upstream and downstream of reactive oxygen species production in *abi1-1* and *abi2-1* protein phosphatase 2C mutants. *Plant Cell* **13**, 2513–2523.
41. Kwak, J. M., Mori, I., Pei, Z.-M., et al. (2003) NADPH oxidase AtrbohD and AtrbohF genes function in ROS-dependent ABA signaling in *Arabidopsis*. *EMBO J.* **22**, 2623–2733.
42. Garcia-Mata, C., Gay, R., Sokolovski, S., Hills, A., Lamattina, L., and Blatt, M. R. (2003) Nitric oxide regulates K^+ and Cl^- channels in guard cells through a subset of abscisic-acid evoked signaling pathways. *Proc. Natl. Acad. Sci. (USA)* **100**, 11,116–11,121.
43. Webb, A. A. R. and Hetherington, A. M. (1997) Convergence of the ABA, CO_2, and extracellular calcium signal tranduction pathways in stomatal guard cells. *Plant Physiol.* **114**, 1157–1160.
44. Lahr, W. and Raschke, K. (1988) Abscisic-acid contents and concentrations in protoplasts from guard cells and mesophyll cells of *Vicia faba*. *Planta* **173**, 528–531.
45. Smith, G. N. and Willmer, C. M. (1988) Effects of calcium and abscisic acid on volume changes of guard cell protoplasts of *Commelina*. *J. Exp. Bot.* **39**, 1529–1539.
46. Schmidt, C., Schelle, I., Liao, Y-U-J., and Schroeder, J. I. (1995) Strong regulation of slow anion channels and abscisic acid signaling in guard cells by phosphorylation and dephosphorylation events. *Proc. Natl. Acad. Sci. (USA)* **92**, 9535–9539.
47. Lee, Y., Choi, Y. B., Suh, S., et al. (1996) Abscisic acid-induced phosphoinositide turnover in guard cell protoplasts of *Vicia faba*. *Plant Physiol.* **110**, 987–996.
48. Mori, I. C. and Muto, S. (1997) Abscisic acid activates a 48-kilodalton protein kinase in guard cell protoplasts. *Plant Physiol.* **113**, 833–839.
49. Pei, Z. M., Kuchitsu, K., Ward, J. M., Schwarz, M., and Schroeder, J. I. (1997) Differential abscisic acid regulation of guard cell slow anion channels in *Arabidopsis* wild-type and *abi1* and *abi2* mutants. *Plant Cell* **9**, 409–423.
50. Pei, Z. M., Murata, Y., Benning, G., et al. (2000) Calcium channels activated by hydrogen peroxide mediate abscisic acid signaling in guard cells. *Nature* **406**, 731–734.
51. Wang, X. Q., Ullah, H., Jones, A. M., and Assmann, S. M. (2001) G protein regulation of ion channels and abscisic acid signaling in *Arabidopsis* guard cells. *Science* **292**, 2070–2072.

52. Schroeder, J. I. and Hagiwara, S. (1989) Cytosolic calcium regulates ion channels in the plasma membrane of *Vicia faba* guard cells. *Nature* **338**, 427–430.
53. Allen, G. J and Sanders, D. (1994) Two voltage-gated, calcium release channels coreside in the vacuolar membrane of broad bean guard cells. *Plant Cell* **6**, 685–694.
54. Lemtiri-Chlieh, F. and MacRobbie, E. A. C. (1994) Role of calcium in the modulation of *Vicia* guard cell potassium channels by abscisic acid: a patch clamp study. *J. Membrane Biol.* **137**, 99–107.
55. Allen, G. J., Chu, S. P., Schumacher, K., et al. (2000) Alteration of stimulus-specific guard cell calcium oscillations and stomatal closing in *Arabidopsis det3* mutant. *Science* **289**, 2338–2342.
56. Kohler, B., Hills, A., and Blatt, M. R. (2003) Control of guard cell ion channels by hydrogen peroxide and abscisic acid indicates their action through alternate signaling pathways. *Plant Physiol.* **131**, 385–388.
57. Ng, C. K., Carr, K., McAinsh, M. R., Powell, B., and Hetherington, A. M. (2001) Drought-induced guard cell signal transduction involves sphingosine-1-phosphate. *Nature* **410**, 596–599.
58. Coursol, S., Fan, L. M., LeStunff, H., Spiegel, S., Gilroy, S., and Assmann, S. M. (2003) Sphingolipid signalling in *Arabidopsis* guard cells involves heterotrimeric G proteins. *Nature* **423**, 651–654.
59. Fairley-Grenot, K. and Assmann, S. M. (1991) Evidence for G-protein regulation of inward K^+ channel current in guard cells of Fava bean. *Plant Cell* **3**, 1037–1044.
60. Wu, W. H. and Assmann, S. (1994) A membrane-delimited pathway of G-protein regulation of the guard-cell inward K^+ channel. *Proc. Natl. Acad. Sci. (USA)* **91**, 6310–6314.
61. Armstrong, F. and Blatt, M. R. (1995) Evidence for K^+ channel control in *Vicia* guard cells coupled by G-proteins to a 7TMS receptor mimetic. *Plant J.* **8**, 187–198.
62. Kelly, W. B., Esser, J. E., and Schroeder, J. I. (1995) Effects of cytosolic calcium and limited, possible dual, effects of G protein modulators on guard cell inward potassium channels. *Plant J.* **8**, 479–489.
63. Li, J. and Assmann, S. M. (1996) An abscisic acid-activated and calcium-independent protein kinase from guard cells of Fava bean. *Plant Cell* **8**, 2359–2368.
64. Pei, Z. M., Ward, J. M., Harper, J. F., and Schroeder, J. I. (1996) A novel chloride channel in *Vicia faba* guard cell vacuoles activated by the serine/threonine kinase, CDPK. *EMBO J.* **15**, 6564–6574.
65. Fricker, M. D. and Willmer, C. M. (1987) Vanadate sensitive ATPase and phosphatase activity in guard cell protoplasts of *Commelina*. *J. Exp. Bot.* **38**, 642–648.
66. Leonhardt, N., Kwak, J. M., Robert, N., Waner, D., Leonhardt, G., and Schroeder, J. I. (2004) Microarray expression analyses of *Arabidopsis* guard cells and isolation of a recessive abscisic acid hypersensitive protein phosphatase 2C mutant. *Plant Cell* **16**, 596–615.
67. Müller-Röber, B., Erhardt, T., and Plesch, G. (1998) Molecular features of stomatal guard cells. *J. Exp. Bot.* (special issue) **49**, 293–304.

68. Hentzen, A. E., Smart, L. B., Wimmers, L. E., Fang, H. H., Schroeder, J. I., and Bennett, A. B. (1996) Two plasma membrane H_+-ATPase genes expressed in guard cells of *Vicia faba* are also expressed throughout the plant. *Plant Cell Physiol.* **37,** 650–659.
69. Gushwa, N., Hayashi, D., Kemper, A., et al. (2003) Thermotolerant guard cell protoplasts of tree tobacco do not require exogenous hormones to survive in culture and are blocked from re-entering the cell cycle at the G1-to-S transition. *Plant Physiol.* **132,** 1925–1940.
70. Pandey, S., Wang, X.-Q., Coursol, S. A., and Assmann, S. M. (2002) Preparation and applications of *Arabidopsis thaliana* guard cell protoplasts. *New Phytol.* **153,** 517–526.
71. Cupples, W., Lee, J., and Tallman, G. (1991) Division of guard cell protoplasts of *Nicotiana glauca* (Graham) in liquid cultures. *Plant Cell Environ.* **14,** 691–697.
72. Herscovich, S., Tallman, G., and Zeiger, E. (1992) Long-term survival of *Vicia* guard cell protoplasts in cell culture. *Plant Sci.* **81,** 237–244.
73. Hall, R. D., Pedersen, C., and Krens, F. A. (1994) Regeneration of plants from protoplasts of sugar beet (*Beta vulgaris* L.) in *Plant Protoplasts and Genetic Engineering*, Vol V. (Bajaj, Y. P. S., ed.) Springer-Verlag, Berlin, Berlin, pp. 16–37.
74. Sahgal, P., Martinez, G. V., Roberts, C., and Tallman, G. (1994) Regeneration of plants from cultured guard cell protoplasts of *Nicotiana glauca* (Graham). *Plant Sci.* **97,** 199–208.
75. Hall, R. D., Verhoeven, H. A., and Krens, F. A. (1995) Computer-assisted identification of protoplasts responsible for rare division events reveals guard-cell totipotency. *Plant Physiol.* **107,** 1379–1386.
76. Roberts, C., Sahgal, P., Merritt, F., Perlman, B., and Tallman, G. (1995) Temperature and abscisic acid can be used to regulate survival, growth, and differentiation of cultured guard cell protoplasts of tree tobacco. *Plant Physiol.* **109,** 1411–1420.
77. Hall, R. D., Riksen-Bruinsma, T., Weyens, G. J., et al. (1996) A high efficiency technique for the generation of transgenic sugar beets from stomatal guard cells. *Nature Biotech.* **14,** 1133–1138.
78. Hall, R. D., Riksen-Bruinsma, T., Weyens, G., et al. (1997) Sugar beet guard cell protoplasts demonstrate a remarkable capacity for cell division enabling applications in stomatal physiology and molecular breeding. *J. Exp. Bot.* **48,** 255–263.
79. Taylor, J. E., Abram, B., Boorse, G., and Tallman, G. (1998) Approaches to evaluating the extent to which guard cell protoplasts of *Nicotiana glauca* (tree tobacco) retain their characteristics when cultured under conditions that affect their survival, growth, and differentiation. *J. Exp. Bot.* **49** (special issue), 377–386.
80. Boorse, G. and Tallman, G. (1999) Guard cell protoplasts: isolation, culture, and regeneration of plants, in *Methods in Molecular Biology: Plant Cell Culture Protocols*, Vol 111 (Hall, R. D. ed.), Humana Press, Totawa, NJ, pp. 243–257.
81. Hoagland, D. and Arnon, D. (1938) The water-culture method for growing plants without soil. Circular of the California Agricultural Experiment Station, No. 347.
82. Shepard, J. F. and Totten, R. E. (1975) Isolation and regeneration of tobacco mesophyll cell protoplasts under low osmotic conditions. *Plant Physiol.* **55,** 689–694.

21

Production of Interspecific Hybrid Plants in *Primula*

Juntaro Kato and Masahiro Mii

Summary

The methods of production of inter-specific hybrids in *Primula* are categorized into four steps: (1) emasculation, (2) pollination, (3) rescue culture of immature embryo, and (4) confirmation of hybridity and ploidy level of the regenerated plants. Although most of the *Primula* species have a heteromorphic self-incompatibility system, an emasculation step is usually needed to avoid self-pollination since self-incompatibility is not always complete. At the rescue culture step, addition of plant hormones (e.g., auxin, cytokinin, and gibberellin) to the culture medium is proved to be effective. The hybridity of the plants is efficiently confirmed at seedling stage by DNA analysis in addition to the comparison of morphological characters. The analysis of relative DNA contents by flow cytometry is easy and rapid technique to confirm hybridity and to estimate ploidy level and genomic combination.

Key Words: Embryo rescue; flow cytometry; interspecific hybrid; plant hormone; *Primula*; RAPD.

1. Introduction

Production of interspecific hybrids is useful not only for producing novel cultivars with intermediate characters between the parental species, but also to induce introgression of the useful genes possessed by one species to another species through alien chromosome addition line. After successful production of interspecific hybrids, several methods have been used for the breeding of novel cultivars. In the case where interspecific F_1 hybrids are fertile, progenies of F_2 and further generations are produced with shuffling of whole genes from both parents through repetition of meiosis and fertilization. For example, *Primula x polyantha* might be originated from fertile interspecific hybrids between *Primura vulgaris* and *Primura veris* (*1*), and numerous seed propagation cultivars have been bred in this hybrid species without any problem. Therefore, the genome difference between these two parental species is considered to be small or none. In contrast, when interspecific F_1 hybrids are completely or incompletely sterile, the genomic composition of the two parental species is considered to be more or less different. In such case, the sterile F_1 progenies with desirable characters will be needed to propagate vegetatively if they are worth to be novel cultivars. Irrespective of the usefulness for the direct usage, sterile interspecific hybrids are also used for production of alien chromosome addition line through chromosome doubling and recurrent backcrossing. It has sometimes been reported that interspecific hybrids with unexpected ploidy level (e.g., triploid) were produced presumably by the fertilization of unreduced gametes *(2,3)*. The triploid hybrids consisting of two different genomes are useful as parental plants for producing alien chromosome addition lines by recurrent back crossing.

In molecular genetics analysis, plant materials originated from interspecific hybridization may be utilized to give various basic informations. If the interspecific hybrid has fertility, differences in various DNA markers between both parents, which may be segregated in the next generation, can be used for making linkage map of the markers at F_2 generation. On the other hand, if interspecific hybrid is sterile, it can be utilized to produce chromosome addition lines, which is useful to identify the located chromosome of objective character *(4)*, through production of its amphidiploid by chromosome doubling treatment followed by recurrent back crossing with either parent.

For the production of interspecific hybrids, however, there exist various reproduction barriers, which are classified into two categories, pre- and post-fertilization barriers, respectively. The pre-fertilization barriers include failure of pollen germination on stigma, and arrest of pollen tube growth on stigma, in style, or in the ovary. The post-fertilization barriers include defectiveness or abnormal growth of endosperm, and abnormal growth or abortion of hybrid embryo by genetic unbalance. Moreover, there exists another type of barriers such as failure of seed germination and lethality of seedlings caused by abnormal development of shoots or roots at early stage of germination. In the cross combinations with post-fertilization barriers, embryo rescue culture techniques including in vitro seeding have been effectively used to obtain interspecific hybrid plants. For example, in interspecific reciprocal crossing between *Primula veris x P. elatior*, unilateral incompatibility is known for the cross using *P. elatior* as a mother plant *(5)*, in which lethality of hybrid embryo is caused by unbalance between rapid embryonic development and slow endosperm formation. When hybrid seeds obtained in interspecific crossing between *Primura sieboldii* and *P. kisoana* were sown on soil, most individuals died after only rooting (Mii unpublished data). These results suggest that

Production of Interspecific Hybrid Plants 255

embryo rescue culture is important for *Primula* interspecific hybridization. In this paper, we described the detailed methods for embryo rescue and confirmation of hybridity in interspecific hybridization in the genus *Primula*.

2. Materials

1. Paraffin paper.
2. Three times diluted woodwork bond with water.
3. Scissors.
5. Wire for flower arrangement.
6. Pollen parent plant.
7. Ovary parent plant.
8. Laminar flow cabinet.
9. 110-mm Filter paper no. 2 (Advantec Co., Japan).
10. 70% EtOH.
11. Forceps (250 mm).
12. Autoclaved sterile distilled water.
13. Sodium hypochlorite solution (1% available chlorine concentration) with 2 drops of Tween-20 (Sigma).
14. 30-mL Glass vial for tissue sterilization.
15. Parafilm.
16. 120-mm Glass dish, containing filter papers, which is sterilized with dry heat at 150°C for 1 h.
17. Chemicals and equipments for preparing Murashige and Skoog (MS) medium *(6)*.
18. Gibberellin A_3, which is stored at 4°C in the dark.
19. 100 mg/L 1-naphtaleneacetic acid (NAA) dissolved by 1 mL ethanol, which is diluted with distilled water and stored at 4°C.
20. 100 mg/L 6-benzylaminopurine (BA) dissolved by 1 mL ethanol, which is diluted with distilled water and stored at 4°C.
21. Sucrose.
22. Gellan gum.
23. Beaker for waste fluid.
24. DNA extraction equipments.
25. Polymerase chain reaction (PCR) equipment.
26. Recombinant *Thermus thermophilus* (rTth) DNA polymerase (Toyobo Co., Japan).
27. Agarose gel electrophoresis equipment.
28. DNA polyacrylamide gel electrophoresis (PAGE) equipment.
29. DNA detection equipment.
30. Type PA (Partec Co.; Germany) flow cytometer (FCM) analysis equipment.

3. Methods

The methods described in **Subheadings 3.1.–3.2.** outline: (1) artificial crossing; (2) embryo and immature seed rescue culture; (3) rescue of abnormal seedling; (4) confirmation of hybridity by RAPD analysis; and (5) confirmation of hybridity and ploidy level by FCM.

3.1. Artificial Crossing

For artificial crossing, two-step procedures (i.e., emasculation and pollination), are required. Because most species of *Primula* have dimorphic heterostylous self-incompatibility system *(2,7)*, legitimate crosses are commonly used for interspecific hybridization.

3.1.1. Emasculation

Emasculation is performed to induce only objective cross combination and to exclude the possibilities of self or natural crossing. For emasculation, flower bud 2–3 d before anthesis are incised and the immature anthers are carefully removed by forceps. Then, each emasculated flower bud is covered with a paraffin paper bag and the base of the bag is tied with a flower arrangement wire. If mother plant shows complete pollen sterility or complete self-incompatibility, this procedure is not necessary (*see* **Note 1**). However the maternal plant must be isolated alone to avoid unexpected natural crossing with pollen of other individuals.

3.1.2. Crossing

After 2–3 d of emasculation, paraffin paper bags are removed and the stigma of the female parent is pollinated with pollen of the male parent by forceps. After pollination, the flower is bagged again, and then tied at the base of the sepals by a flower arrangement wire. After 2 or 3 wk, ovules or immature seeds in enlarged ovary are used for rescue culture (*see* **Notes 2** and **3**).

3.2. Rescue Culture of Embryo or Immature Seed

3.2.1. Preparation of Medium

Appropriate amounts of gibberellin A_3 and the stock solutions of NAA and BA are added to half strength MS medium (1/2 MS) containing 5% (w/v) sucrose to give the final concentrations of 50 mg/L, 0.1 mg/L, and 0.1 mg/L, respectively. After pH of culture medium is adjusted to 5.6–5.8, gellan gum is added to the culture medium at 0.2% (w/v) final concentration. After dissolving gellan gum completely in boiling water bath, medium is dispensed to each culture tube and autoclaved.

3.2.2. Embryo and Immature Seed Rescue Culture

The materials needed for sterile culture are shown in **Fig. 1**. Immature ovaries including ovules with developing hybrid embryos are collected 3–4 wk after pollination in interspecific crosses (*see* **Fig. 2**, *1* and *2*). The timing of

Fig. 1. The materials needed for sterile culture. *(1)* 70% EtOH, *(2)* forceps, *(3)* autoclaved sterile distilled water, *(4)* sodium hypochlorite solution, *(5)* 30-mL glass vial, *(6)* Parafilm, *(7)* 120-mm glass dish containing filter papers, *(8)* medium, and *(9)* beaker for waste fluid.

collection of ovaries is differed according to the cross combinations. Optimum timing should be determined by examining the status of embryo development or by simply changing the timing for the culture. After removing sepals and remnant of petals (*see* **Fig. 2**, *3* and *4*), ovary is put into a vial containing sodium hypochlorite solution (about 15 mL), and the vial is closed with a plastic cap (**Fig. 2**, *5* and *6*). The ovary is sterilized for 15 min with initial handshaking for 2 min (**Fig. 2**, *7* and *8*). After washing with sterile distilled water twice (**Fig. 2**, *9* and *10*), the explant is transferred by a forceps onto wet sterilized filter paper with sterilized distilled water (**Fig. 2**, *11*) and ovary wall is carefully removed by using a forceps and a knife to deliver the placenta with ovules (**Fig. 2**, *12* and *13*). The placenta explant is then plated on culture medium in a tube (**Fig. 2**, *14* and *15*). The cultures are kept at 20°C under weak light condition (approx 300–500 lx) (*see* **Note 4**).

3.2.3. Rescue of Abnormal Seedlings

Seed germination occurs from the ovules on placenta after 1–4 mo of culture. Most of the ovules show normal germination and the seedlings grow normally when transplanted after true leaf development to 1/2 MS medium

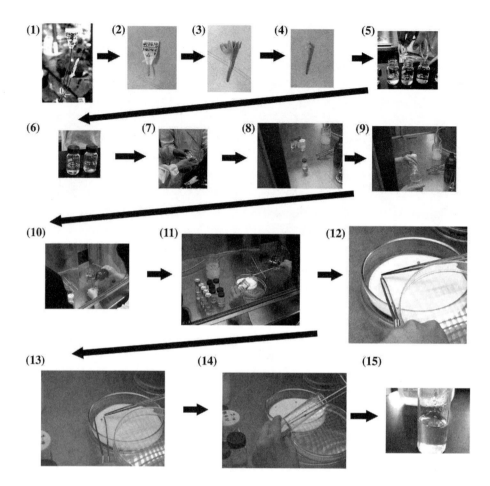

Fig. 2. Procedure of sterile culture.

containing 3% (w/v) sucrose and 0.3% (w/v) gellan gum. However, some of the ovules sometimes show abnormal seed germination and seedling growth such as root growth without shoot formation or growth and direct callus formation from the ovules. It is possible that these abnormal seedlings are converted into normal plantlets by transplanting to MS medium containing auxin and cytokinin to induce redifferentiation of plants after callus induction. Usually 0.1–1 mg/L is considered to be suitable for both NAA as an auxin and any one of the cytokinins such as BA, zeatin (Sigma), and thidiazuron (Sigma).

3.2.4. Confirmation of Hybridity by RAPD Analysis

DNA is extracted using a commercial kit. The rTth DNA polymerase (Toyobo) is employed for the random amplified polymorphic DNA (RAPD) analysis, in which a 25 µL reaction solution containing 1 ng of total plant DNA is subjected to analysis according to the manufacture's protocol. The primers used for RAPD analysis are kits of Operon 10-mer primers (Operon). DNA fragments are amplified by repeating 40 cycles of the following thermal treatments; 92°C for 30 s, 46°C for 1 min, and 72°C for 1 min, in a Takara PCR Personal Thermal Cycler (Takara Co., Japan). Electrophoresis of the amplified DNAs is conducted on a 5% acrylamide gel or 1% agarose gel *(8)*. Hybridity of the plant is confirmed by detecting the specific bands derived from both parents **(Fig. 3)** (*see* **Note 5**).

3.2.5. Confirmation of Hybridity and Ploidy Level by FCM

It is common that different species have different DNA contents each other even in the same genus (e.g., *Lilium [9]*, *Petunia [10]*, and *Primula [3]*). If interspecific hybrids are produced between two species with different DNA contents, they should theoretically show the intermediate value. However, hybrids with unexpected ploidy levels such as triploid and tetraploid are rarely formed in interspecific crosses *(2,3)*. In such cases, it is possible to estimate the genomic combination of the hybrids based on the DNA content of each parental species by using flow cytometric analysis (*see* **Fig. 4**).

For flow cytometric analysis with type PA flow cytometer (Partec, Germany), following procedures are usually adopted. Approximately 0.5-cm^2 segments are excised from a fresh leaf, put into plastic Petri dish with 0.2 mL of Nuclei Extraction Buffer "CyStain UV Precise P" (Partec), and chopped with a razor blade, to which 1 mL of solution B of DAPI Staining "High resolution DNA Kit Type P" (Partec) is added. After 1–2 min of incubation, the solution mixture is filtered with a 45 µm mesh and subjected for measuring DNA content according to the manufacture's protocol. For the accurate comparison of the DNA contents, it is recommended to mix the tissue sample of the hybrid with those of the parents and subject to the analysis at the same time. Estimated ploidy level should also be confirmed by counting the chromosome numbers (*see* **Note 6**).

4. Notes

1. In self-incompatibility plants, it is necessary to confirm the strength of self-incompatibility by self pollination as the strength of the incompatibility varies depending on the individuals. In *Primula*, illegitimate crossing, especially pin/pin, sometimes yields hybrid plants *(2,3)*. In the case of small flower plant, which is difficult to emasculate before anthesis, emasculation should be carefully done immediately after anthesis just before pollination.
2. The degree of difficulties in obtaining interspecific hybrid is mostly related to the distance of relationship between the two species. However, the cross combination between distantly related species sometimes has high cross ability, and dif-

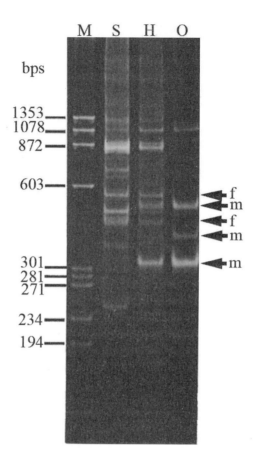

Fig. 3. PCR analysis using the OPR-6 primer on the hybridity of the progeny in interspecific cross between *Primula sieboldii* and *Primula obconica*. M: ϕX174/HaeIII-digestion size marker. H: The progeny obtained from interspecific cross. S: *P. sieboldii* used as maternal parent, O: *P. obconica* used as pollen parent, f: the specific bands of *P. sieboldii*, m: the specific bands of *P. obconica*.

ferences in cross ability between cultivars are commonly found in interspecific crosses *(2,3,11)*. Therefore, it is necessary to make a test cross using many varieties in both parental species.
3. Crossing bags used after pollination are apt to get wet with rain or water, which results in the severe infection with fungi and subsequent difficulty in surface sterilization of the ovary. Therefore, plants used for the crossing must be kept in the glasshouse and watered without causing the bag wet.
4. If immature seeds show good development, they can be cultured as ovule culture after removing from placenta.

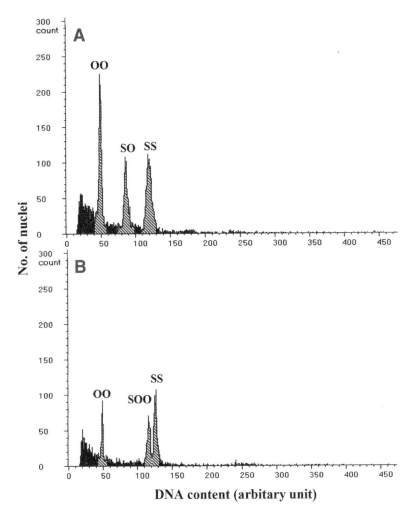

Fig. 4. (**A,B**) Flow cytometric profiles of the plants obtained from interspecific crosses between *Primula sieboldii* used as a maternal parent and *Primula obconica* used as pollen parent. (**A**) Diploid hybrid with an SO genome. (**B**) Triploid hybrid with SOO genome. SS: *P. sieboldii*, OO: *P. obconica*.

5. In RAPD analysis, the annealing temperature of 42°C in our protocol is higher than that in commonly used protocol. If specific marker cannot be obtained from either parent DNA, annealing temperature may be switched to 38°C.
6. In flow cytometry analysis, each sample should always be measured with an internal standard since the peak position may fluctuate according to the various factors such as aging of lamp.

References

1. Richards, J. (1993) *Primula*. Timber Press, Portland, Oregon.
2. Kato, J. and Mii, M. (2000) Differences in ploidy level of inter-specific hybrids obtained by reciprocal crosses between *Primula sieboldii* and *P. kisoana*. *Theor. Appl. Genet.* **101**, 690–696.
3. Kato, J., Ishikawa, R., and Mii, M. (2001) Different genomic combinations in inter-section hybrids obtained from the crosses between *Primula sieboldii* (Section Cortusoides) and *P. obconica* (Section Obconicolisteri) by the embryo rescue technique. *Theor. Appl. Genet.* **102**, 1129–1135.
4. Taketa, S. and Takeda, K. (2001) Production and characterization of a complete set of wheat-wild barley (*Hordeum vulgare* ssp. spontaneum) chromosome addition lines. *Breed. Sci.* **51**, 199–206
5. Woodell, S. R. J. (1960) Studies in British Primulas VIII. Development of seed from reciprocal crosses between *P. vulgaris* Huds. and *P. elatior* (L.) Hill, and between *P. veris* L. and *P. elatior* (L.) Hill. *New Phytol.* **59**, 314–322.
6. Murashige, T. and Skoog, F. (1962) A revised medium for rapid growth and bioassay with tobacco tissue cultures. *Physiol. Plant* **15**, 473–497.
7. Barrett, S. C. H. (ed.) (1992) *Evolution and function of heterostyly*. Springer-Verlag, Berlin Heiderberg.
8. Sambrook, J. and Russell, D. W. (2001) *Molecular Cloning: A Laboratory Manual, 3rd ed.* V. 1, CSHL Press, Cold Spring Harbor, New York.
9. Van Tuyl, J. M. and Boon, E. (1997) Variation in DNA–content in the genus Lilium. *Acta. Hort.* **430**, 829–835.
10. Mishiba, K, Ando, T., Mii, M., et al. (2000) Nuclear DNA content as an index character discriminating taxa in the genus *Petunia* sensu Jussieu (Solanaceae). *Ann. Bot.* **85**, 665–673.
11. Nimura, M., Kato, J., Mii, M., and Morioka, K. (2003) Unilateral compatibility and genotypic difference in crossability in interspecific hybridization between *Dianthus caryophyllus* L. and *Dianthus japonicus* Thunb. *Theor. Appl. Genet.* **106**, 1164–1170.

V

PROTOCOLS FOR GENOMIC MANIPULATION

22

Agrobacterium-Mediated Transformation of Petunia Leaf Discs

Ingrid M. van der Meer

Summary
Many dicotyledonous and also several monocotyledonous plant species are susceptible to *Agrobacterium*-mediated transformation. This current and well-established method has been used successfully with a large number of plant species to mediate gene transfer. This chapter describes an *Agrobacterium*-mediated transformation method of *Petunia hybrida* leaf discs, one of the first species that was routinely transformed using this method.

Key Words: *Agrobacterium*; leaf disks; *Petunia hybrida*; protocol; transformation.

1. Introduction

The transfer of defined DNA segments into plant genomes is an established method not only to study regulation of gene expression but also a major step forward in plant breeding. From the large number of strategies developed to mediate gene transfer, especially the *Agrobacterium*-mediated transfer method is a well-established method that has been used successfully with a large number of plant species. *Agrobacterium*-mediated transformation of plants is applicable to many dicotyledonous and also several monocotyledonous plant species. It can be used to transform many different species based on various factors: the broad host range of *Agrobacterium* (*1*), the regeneration responsiveness of many different explant tissues (*2*) and the utility of a wide range of selectable marker genes (*3*). One of the first species that was routinely transformed using this method besides tobacco was *Petunia hybrida*.

P. hybrida is a very good model plant for the analysis of gene function and promoter activity. It is readily transformed, the culture conditions are easy fulfilled, generation time is 3–4 mo and one can grow up to 100 plants/m^2. Furthermore, its genetic map is well developed and it contains active transposable elements *(4)*.

The protocol presented here is a simplified version of that of Horsch et al. *(5)*. The basic protocol involves the inoculation of surface-sterilized leaf discs with the appropriate disarmed strain of *Agrobacterium tumefaciens* carrying the vector of choice, which in this protocol confers kanamycin resistance. The plant tissue and *Agrobacterium* are then co-cultivated on regeneration medium for a period of 2 or 3 d. During this time, the virulence genes in the bacteria are induced, the bacteria bind to the plant cells around the wounded edge of the explant, and the gene transfer process takes place *(6)*. Using a nurse culture of tobacco or Petunia cells during the co-culture period may increase the transformation frequency. This is probably because of a more efficient induction of the virulence genes. After the co-cultivation period, bacteriostatic antibiotics (cefotaxim or carbenicillin) inhibit the growth of the bacterial population and the leaf tissue is induced to regenerate. The induction and development of shoots on leaf explants occurs in the presence of a selective agent against untransformed plant cells, usually kanamycin. During the next 2–3 wk, the transformed cells grow into callus or differentiate into shoots via organogenesis. After 4–6 wk, the shoots have developed enough to remove them from the explant and induce rooting in preparation for transfer to soil. To speed up the rooting period, the shoots can be rooted without selection on kanamycin. In total, it takes about 2 mo, after inoculation of the leaf disks with *Agrobacterium*, to obtain rooted plantlets that can be transferred to soil.

2. Materials
2.1. Bacteria Media

1. For the growth of *Agrobacterium* use Luria broth (LB) medium.
   a. LB medium: 1% Bacto-peptone (Difco), 0.5% Bacto-yeast extract (Difco), 1% NaCl.
   b. Autoclave, and cool medium to at least 60°C, add appropriate antibiotics to select for presence of plasmids (50 mg/L kanamycin for pBin19 *[7]*).
2. LB-agar: LB medium with 15 g/L agar (Difco). Autoclave, and cool medium to at least 60°C, add appropriate antibiotics to select for presence of plasmids (50 mg/L kanamycin for pBin19). Pour into sterile 20-mm Petri dishes.
3. *Agrobacterium* inoculation dilution medium: Murashige and Skoog (MS) medium (*[7]*, and Appendix A) salts and vitamins (4.4 g/L) (Sigma).

2.2. Stock Solutions

For convenience, most stock solutions are prepared at 1000× the concentration needed for the final media. The antibiotics are added to the media after autoclaving when the temperature of the media has cooled to 60°C.

1. Cefotaxime: 250 mg/mL (Duchefa or Sigma), filter-sterilize, and keep at −20°C.
2. Kanamycin: 250 mg/mL (Duchefa or Sigma), filter-sterilize, and keep at −20°C.
3. 6-Benzylaminopurine (BA) (Sigma): 2 mg/mL. Dissolve 200 mg BA in 4 mL 0.5 N HCl. Add, while stirring, drop by drop H_2O at 80–90°C and make up to 100 mL. Filter-sterilize.
4. I-Naphtalene acetic acid (NAA), (Sigma): 1 mg/mL, dissolved in dimethyl sulfoxide (DMSO), no need to sterilize. Keep at −20°C. DMSO should be handled under a fume hood.

2.3. Plant Culture Media

1. Cocultivation medium: MS (*8*) salts and vitamins (4.4 g/L) (Sigma), 30 g/L sucrose, 2 mg/mL BA, 0.01 mg/mL NAA, adjust pH to 5.8 with 1 M KOH, add 8 g/L agar (Bacto Difco) and autoclave. Pour into sterile plastic dishes that are 20 mm high (Greiner).
2. Regeneration and selection medium: MS salts and vitamins (4.4 g/L) (Sigma), 30 g/L sucrose, 2 mg/L BA, 0.01 mg/L NAA, adjust pH to 5.8 with 1 M KOH, add 8 g/L agar (Bacto Difco). Autoclave and cool media to 60°C, add 250 mg/L cefotaxime to kill off *Agrobacterium* and the appropriate selective agent to select for transformed cells depending on the vector used (100 mg/L kanamycin for pBin19). Pour 25 mL into each sterile 20-mm high Petri dish (Greiner).
3. Rooting medium: MS salts and vitamins (4.4 g/L) (Sigma), 30 g/L sucrose, adjust pH to 5.8 with 1 M KOH, add 7 g/L agar (Bacto Difco). Autoclave and cool media to 60°C, add 250 mg/L cefotaxime to kill off *Agrobacterium* and add the appropriate selective agent to select for transformed shoots depending on the vector used (100 mg/L kanamycin for pBin19). To speed up the rooting process, kanamycin may be omitted from the rooting medium. Pour in Magenta GA7 boxes (Sigma, 80 mL/box).

2.4. Plant Material, Sterilization, and Transformation

1. Petunia hybrida c.v. W115 (Mitchell), grown under standard greenhouse conditions.
2. *Agrobacterium tumefaciens* strain LBA 4404 (as a control) and *A. tumefaciens* LBA 4404 containing pBin19 (in which the gene of interest is inserted).
3. 10% solution of household bleach containing 0.1% Tween or other surfactant.
4. Sterile H_2O.
5. Sterile filter paper (Whatman) and sterile round filters (Whatman, diameter 90 mm).

2.5. General Equipment

1. Sterile transfer facilities.
2. Rotary shaker at 28°C.
3. Cork borer (or a paper punch).
4. Magenta GA7 boxes (Sigma) and 20-mm high Petri dishes (Greiner).

3. Methods
3.1. Plant Material

Young leaves are used as the explant source for transformation. These explants can be obtained from aseptically germinated seedlings or micropropagated shoots, but in this protocol they are obtained from greenhouse-grown material. The genotype of the source material is important in order to obtain high transformation rates. *P. hybrida* cv.W115 gives the best results and is most often used (*see* **Notes 1** and **2**). To obtain plant material suitable for transformation, seedlings should be germinated 4–6 wk prior to transformation.

Sow *P. hybrida* (W115) seeds in soil 4–6 wk prior to transformation, and grow under standard greenhouse conditions in 10 × 10 × 10 cm plastic pots. Use commercially available nutritive solution for houseplants.

3.2. Leaf Disk Inoculation

1. Grow *A. tumefaciens* culture overnight in LB at 28°C on a rotary shaker (130 rpm) with appropriate antibiotics to select for the vector (50 mg/L kanamycin for pBin19) (*see* **Note 3**). The *Agrobacterium* liquid culture should be started by inoculating 2 mL of liquid LB with several bacterial colonies obtained from an *Agrobacterium* streak culture grown on an LB-agar plate at 28°C for 2–3 d. The streaked plate itself can be inoculated from the original –80°C frozen stock of the *Agrobacterium* strain (*see* **Note 4**). This stock is composed of a bacterial solution made from a 1:1 mixture of sterile glycerol (99%) and an overnight LB culture of the *Agrobacterium*.
2. Prepare the culture for inoculation of explants by taking the overnight culture and diluting 1 to 200 with MS salts and vitamins medium (4.4 g/L) (Sigma) to a final volume of 100 mL. The *A. tumefaciens* inoculum should be vortexed well prior to use. Pour the *Agrobacterium* inoculum into four Petri dishes.
3. When the plants are 10- to15-cm high harvest the top leaves to provide explants (leaves 3–8 from the top). Lower leaves and leaves from flowering plants should not be used as they usually have a lower transformation and regeneration response (*see* **Note 1**).
4. Prepare harvested leaves for inoculation by surface sterilization for 15 min in 10% solution of household bleach containing 0.1% Tween or other surfactant. Wash the leaves thoroughly three times with sterile H_2O. Keep them in sterile H_2O until needed (*see* **Note 5**). All procedures following the bleach treatment are conducted in a sterile transfer hood to maintain tissue sterility.
5. Punch out leaf disks with a sterile (1 cm diameter) cork borer (or cut with a scalpel into small squares to produce a wounded edge) in one of the Petri dishes containing the *Agrobacterium* inoculum (20–25 disks per Petri dish) (*see* **Note 6**). Avoid the midrib of the leaf or any necrotic areas. Cut 80–100 disks per construct (*see* **Note 7**).

6. Leave the disks in the inoculum for 20 min. After inoculation, the explants are gently sandwiched between two layers of sterile filter paper (Whatman) to remove excess inoculum.
7. (Optional) Prepare nurse culture plates by adding 3 mL cell suspension culture (e.g., *P. hybrida* cv Coomanche or *Nicotiana tabacum* cv SR1) to Petri dishes containing co-cultivation medium. Swirl the suspension to spread the cells over the surface of the medium and cover with a sterile Whatman filter paper (90-mm diameter) (*see* **Note 8**).
8. Place 20 explants with the adaxial surface downwards on each plate with cocultivation medium (either with or without nurse cells) and incubate for 2–3 d (*see* **Note 9**). Seal the plates with Nescofilm. The culture conditions for the leaf disks are as follows: a temperature of 25°C and a photoperiod of 14-h light (light intensity: 25–40 TE/m^2/s), 10 h dark. The controls to check the transformation protocol are:
   a. Leaf disks inoculated with "empty" *Agrobacterium* (without pBin19 vector) on regeneration medium without selective agent (to check the regeneration).
   b. Leaf disks inoculated with "empty" *Agrobacterium* (without pBin19 vector) on regeneration medium with selective agent (to check the efficiency of antibiotic selection).
9. After cocultivation, transfer the disks to regeneration and selection medium (seal the plates with Nescofilm) and continue incubation until shoots regenerate. Transfer the explants every 2–3 wk to fresh regeneration and selection medium (*see* **Note 10**).

3.3. Recovery of Transformed Shoots

1. After 2–3 wk the first shoots will develop. Cut off the shoots cleanly from the explant/callus when they are 1–1.5 cm long and place them upright in rooting medium in Magenta boxes. The shoots should be excised at the base without taking any callus tissue (*see* **Note 11**). Take only one shoot from each callus on the explant to ensure no siblings are propagated representing the same transformation event (*see* **Note 12**). Shoots from distinctly different calli on the same explant are however likely to be derived from different transformation events and should be transferred separately. Give each shoot a code that allows it to be traced back to specific explants.

   Cefotaxime is kept in the medium to avoid *Agrobacterium* re-growth. The antibiotic used to select the transgenic shoots can be added to the medium to select against escapes, although the rooting process is speeded up when it is omitted.
2. (Optional) Before removing rooted shoots from sterile culture, transfer a leaf to selection medium to test for resistance to kanamycin. If the leaf is obtained from a transformed plantlet it should stay green and form callus on selection medium, if it originates from an untransformed plantlet it should become brown/white and die within a few weeks.

3. After 3–4 wk the shoots will have formed roots. Remove plantlets, wash agar from the base under a running tap, plant the transformants in soil and transfer them to the greenhouse (*see* **Note 13**). To retain humidity, cover the pots with Magenta boxes or place them in a plastic propagation dome. The plants should then be allowed to come to ambient humidity slowly by gradually opening the dome or Magenta box over a period of 7 d (*see* **Note 14**).
4. Fertilize and grow under standard plant growth conditions.

3.4. Analysis of Transformants

Tissue culture can be used to confirm that the putative transgenic shoots produced are expressing the selectable marker gene (*see* **Subheading 3.3., step 2**). However, DNA analysis using Southern blotting or polymerase chain reaction (PCR) will confirm whether regenerants have integrated the antibiotic resistance gene (and also a gene of interest if this was linked to it within the T-DNA). For DNA analysis using Southern blotting *(9)*, leaf DNA can be isolated according to the protocol described by Dellaporte *(10)*. A more rapid method can be used to isolate genomic DNA as described by Wang et al. *(11)* when PCR is going to be used to analyze the presence of foreign DNA in the transformed plants (*see* **Note 15**).

Usually, one to five copies of the foreign DNA are integrated in the plant genome using the *Agrobacterium*-mediated transformation method. However, position effects may silence the expression of the introduced gene (*see* **Note 7**).

More information on the process of T-DNA transfer and integration is described by Gelvin *(14)*, and a review on *Agrobacterium*-mediated information is given by Li et al. *(15)*.

4. Notes

1. This transformation protocol works very well for the often-used *P. hybrida* varieties W115 (Mitchell) and V26. However, some *Petunia* lines show poor regeneration from leaf disk explants and consequently few or no transformants can be obtained from these plants. Uniform, clean and young plants will perform best. It is important not to take leaf material from old, flowering plants.
2. Instead of greenhouse material, aseptically germinated seedlings or micropropagated shoots could also be used as explant source. Then, of course, there is no need to surface sterilize the leaves.
3. The *Agrobacterium* strain that is most often used is LBA4404 (Clontech, *[12]*). Also the more virulent strains C58 or AGLO *(13)* can be used, but these can be more difficult to eliminate after cocultivation. During growth, the *Agrobacterium* culture will aggregate.
4. The streaked plate can be reused for approx 3 wk if kept at 4°C after growth.

5. Be very gentle with the plant material during sterilization since the bleach will easily damage the leaves, especially wounded or weak, etiolated leaves. Damaged tissue should not be used.
6. Disks provide a very uniform explant and are conveniently generated with a cork borer or paper punch. However, square explants or strips can also be used. Avoid excessive wounding during the process. The cork borer should be allowed to cool before use after flaming.
7. This transformation protocol will yield approx 20 transgenic plants from 100 initial explants. The expression level of the construct of interest can be greatly influenced by a position effect owing to its site of integration into the host plant genome. This silencing of expression due to position effects can occur in 20–40% of the transgenic plants, especially if weak promoters are used. Therefore, at least 20 transformants should be generated per construct.
8. The nurse culture is not essential for transformation, but can facilitate the process by increasing frequency and reducing damage to the explant by the bacterium. Any healthy suspension of tobacco or Petunia should work. The suspension cultures can be maintained by weekly transfer of 10 mL into 50 mL fresh suspension culture medium.
9. The cocultivation time may have to be optimized for different *Agrobacterium* strains carrying different vectors.
10. If *Agrobacterium* continues to grow on the regeneration and selection medium (forming slimy gray-white bacterial colonies), 150 mg/L vancomycin can also be added to the medium already containing cefotaxime.
11. Care should be taken that only the stem and none of the associated callus is moved to the rooting medium, otherwise no roots will develop.
12. It is common for multiple shoots to arise from a single transformed cell; therefore it is important to seperate independent transformation events carefully so that sibling shoots are not excised.
13. It is important to wash away the entire agar medium from the roots and to transplant before the roots become too long. When there is still agar left it could enhance fungal growth.
14. Gradual reduction in the humidity is necessary to harden off the plantlets in soil. The roots must grow into the soil and the leaves must develop a protective wax cuticle. If the plantlets start to turn yellow and die from fungal contamination, the lid should be opened faster. If the plantlets begin to wilt, the lid should be opened slower.
15. Confirmation of transformation by PCR may not always be reliable, because of possible carryover of the *Agrobacterium* into the whole plant. DNA analysis using Southern blot hybridization is a better way to confirm whether regenerants are true transformants. Furthermore, the number of inserts can be determined at the same time using this method.

References

1. Richie, S. W. and Hodeges, T. K. (1993) Cell culture and regeneration of transgenic plants, in *Transgenic Plants, vol 1*, (Kung, S. and Wu, R., eds), Academic Press, San Diego, pp. 147–178.
2. Jenes, B., Morre, H., Cao, J., Zhang, W., and Wu, R. (1993). Techniques for gene transfer, in *Transgenic Plants, vol 1*, (Kung, S. and Wu, R., eds), Academic Press, San Diego, pp. 125–146.
3. Bowen, B. A. (1993) Markers for gene transfer, in *Transgenic Plants, vol 1*, (Kung, S. and Wu, R., eds), Academic Press, San Diego, pp. 89–124.
4. Gerats, A. G. M., Huits, H., Vrijlandt, E., Mara a, C., Souer, E., and Beld, M. (1990) Molecular characterization of a nonautonomous transposable element (dTph1) of petunia. *Plant Cell* **2**, 1121–1128.
5. Horsch, R. B., Fry, J. E., Hoffman, N. L., Eichholtz, D., Rogers, S. C., and Fraley, R. T. (1985) A simple and general method for transferring genes into plants. *Science* **227**, 1229–1231.
6. Hooykaas, P. J. J. (1989) Transformation of plant cells via *Agrobacterium*. *Plant Mol. Biol.* **13**, 327–336.
7. Bevan, M. (1984) Binary *Agrobacterium* vectors for plant transformation. *Nucleic Acids Res.* **12**, 8711–8721.
8. Murashige, T. and Skoog, F. (1962) A revised medium for rapid growth and bioassays with tobacco tissue cultures. *Plant Physiol.* **15**, 473–497.
9. Maniatis, T., Fritsch, E.F., and Sambrook, J. (1982) *Molecular Cloning: A Laboratory Manual*. Cold Spring Harbor Laboratory, Cold Spring Harbor, NY.
10. Dellaporte, S. L., Wood, J., and Hicks, J. B. (1983) A plant DNA minipreparation: version II. *Plant Mol. Biol. Rep.* **1**, 19–21.
11. Wang, H., Qi, M., and Cutler, A. J. (1993) A simple method of preparing plant samples for PCR. *Nucleic Acids Res.* **21**, 4153–4154.
12. Hoekema, A., Hirsch, P. R., Hooykaas, P. J. J., and Schilperoort, R. A. (1983) A binary plant vector strategy based on separation of vir and T region of the *Agrobacterium tumefaciens* Ti-plasmid. *Nature* **303**, 179–180.
13. Lazo, G. R., Stein, P. A., and Ludwig, R. A. (1991) A DNA transformation-competent Arabidopsis genomic library in Agrobacterium. *Bio Technology* **9**, 963–967.
14. Gelvin, S. B. (2000) Agrobacterium and plant genes involved in T-DNA transfer and integration. *Annu. Rev. Plant Physiol. Plant Mol. Biol.* **51**, 223–256.
15. Li, W., Guo, G., and Zheng, G. (2000) *Agrobacterium*-mediated transformation: state of the art and future prospect. *Chin. Science Bull.* **45**, 1537–1546.

23

Transformation of Wheat Via Particle Bombardment

Indra K. Vasil and Vimla Vasil

Summary

The protocol described in this chapter was successfully used to produce the first transgenic plants of wheat. It involves high velocity bombardment of explants with DNA-coated microprojectiles (the biolistics procedure). It highlights the importance of selecting the right explants (immature embryos and/or embryogenic callus tissues), pre- and post-osmotic treatment, obtaining high levels of gene expression with the maize ubiquitin promoter, and stringent selection (with *bar*) during culture to obtain stably transformed and normal fertile plants. The protocol is used widely for genetic transformation of wheat.

Key Words: Biolistics; cereals; embryogenic cultures; genetic transformation; *Triticum aestivum*; wheat.

1. Introduction

Wheat (*Triticum aestivum* L.) is the number one food crop in the world based on acreage under cultivation and total production. It is a highly versatile and widely cultivated crop as it has been bred for adaptation to a wide range of ecological conditions and beause of the unique characteristics of its flour that make it possible for processing into a variety of food products. Wheat was first domesticated nearly 10,000 yr ago in the Fertile Crescent of the Tigris-Euphrates basin in Southwestern Asia. As a major commodity in international agriculture, and an important source of nutrition and protein in the human diet, wheat has long played a central role in world food security. From 1965 to 1990, introduction of the Green Revolution high-input and high-yielding varieties led to a nearly threefold increase in world wheat production. However, increases in the productivity of wheat and other major food crops attained through breeding and selection have begun to decline. This is happening at a time when slightly more than one-third (34%) of the wheat crop is lost to pests, pathogens, and weeds *(1)*, in addition to post-harvest losses during storage. Introduction of single genes through genetic transformation into crops such as maize, soybean, potato, canola, and cotton has shown that such losses can not only be greatly reduced or even eliminated, but can also result in greatly reduced use of pesticidies and herbicides, and in soil conservation.

Wheat was last of the major crops to be transformed *(2,3)* because of technical difficulties and the long time it took to establish reliable protocols for high efficiency regeneration and transformation. Owing to the inherent difficulties in the establishment of embryogenic suspension cultures and regeneration of plants from protoplasts, fertile transgenic wheat plants have been obtained primarily by the direct delivery of DNA into regenerable tissue explants by accelerated microprojectiles *(3–6)*. The following protocol describes the genetic transformation of wheat by particle bombardment of immature embryos using the PDS-1000/He (Bio-Rad) system *(3,4,6)*. Although genes for resistance to hygromycin *(7)*, kanamycin *(8)*, and glyphosate *(9)* have been used as selectable markers in wheat transformation, the best results have been obtained with the *bar* gene (phosphinothricin acetyltransferase, PAT, *3–6,8,10,11*), which confers resistance to DL-phosphinothricin (PPT) or glufosinate, the active ingredient in the commercial herbicides Herbiace, Basta and Bialaphos *(12)*. The method described here is based on the use of the plasmid pAHC25 *(4,13)*, containing the *uidA* (β-glucuronidase) and *bar* as reporter and selectable marker genes, respectively *(4,6,14)*. Both genes are under the control of the maize ubiquitin promoter *(4,13)*, which has been shown to provide high levels of gene expression in gramineous species *(15)*. The stability of integration and expression of the transgenes has been shown for several generations *(14,16,17)*.

Although *Agrobacterium*-mediated transformation of wheat has been reported, microprojectile bombardment continues to be the method of choice *(18,19)*. The protocol described below has been successfully used, with little or no adjustment, by numerous groups in academia and industry to produce transgenic wheat plants containing many agronomically useful genes *(18,19)*, such as those for resistance to herbicides (basta, glyphosate), fungal/viral pathogens *(20,21)*, and insects *(22)*, as well as for improved nutritional and breadmaking characteristics *(23)*.

2. Materials
2.1. Culture of Immature Embryos

1. Plant material: wheat (*Triticum aestivum* L., cv Bobwhite) plants are grown either in the field, greenhouse or in growth chambers (first 40 d at 15°C/12°C day/night temperature and 10-h photoperiod at 600 µE/m^2/s, followed by maintenance at 20°C/16°C temperature and 16-h photoperiod) (*see* **Note 1**).

2. Medium for culture of immature embryos: Murashige and Skoog's (MS) *(24)* formulation was used as follows (MS+ medium): MS salts, 4.3 g (Sigma, cat no. M5524); sucrose, 20 g; myo-inositol,100 mg; glutamine, 500 mg; casein hydrolysate, 100 mg; 2,4-dichlorophenoxyacetic acid (2,4-D) from stock: 2.00 mL *(see* **step 4***)*; Gelrite, 2.5 g.
Bring volume to 1 L with distilled water and adjust pH to 5.8 with 4 N NaOH. After autoclaving for sterilization, add 1 mL filter sterilized vitamin stock *(see* **step 5***)*. Pour medium in Petri dishes (100 × 15 mm; Fisher). The sterile media can be stored for 1 mo at room temperature.
3. Medium for osmotic treatment of embryos: add 36.44g of 0.2 M sorbitol, and 36.44g of 0.2 M mannitol to make up 1 L of MS+ medium. Pour in Petri dishes (60 × 15 mm) after autoclaving.
4. 2,4-D stock: dissolve 100 mg 2,4-D in 40 mL 95% ethanol, make up the volume to 100 mL with distilled water. Store in refrigerator.
5. Vitamin stock: dissolve 10 mg thiamin, 50 mg nicotinic acid, 50 mg pyridoxine HCl, and 200 mg glycine, in distilled water and make up volume to 100 mL. Store at –20°C.

2.2. Preparation and Delivery of DNA-Coated Gold Particles

1. Sterile distilled water.
2. Gold particle stock is prepared essentially according to protocol provided by Bio-Rad with PDS/1000 He. Suspend 60 mg gold particles (approx 1 μm diameter) in 1 mL 100% ethanol for 2 min in 1.5-mL Eppendorf tube, with occasional vortexing. Centrifuge for 1 min. Discard supernatant and resuspend in 1 mL sterile distilled water by vortexing. Wash two more times, and finally resuspend in 1 mL sterile water. Aliquot 25 μL or 50 μL in 0.5 mL Eppendorf tubes, with vortexing between aliquots to ensure proper mixing. Store at –20°C.
3. $CaCl_2$ (2.5 M, filter sterilized) in 1 mL aliquots, stored at –20°C.
4. Spermidine free base (0.1 M, filter sterilized) in 1 mL aliquots, stored at –20°C.
5. Ethanol (100%).
6. pAHC25 DNA at 1 μg/mL in TE (10 mM Tris-HCl/1 mM EDTA, pH 8.0).
7. Helium gas cylinder.
8. Sonicator (optional).

2.3. Culture, Selection, and Regeneration

1. Medium for selection: MS+ medium without glutamine and casein hydrolysate, with 3 mg/L bialaphos (4 mg/mL stock) added after autoclaving. Pour in Petri dishes (100 × 15 mm).
2. Medium for regeneration: selection medium without 2,4-D, and supplemented with 1–10 mg/L zeatin.
3. Medium for shoot elongation: MS salts (2.15 gm), sucrose (15 gm), myo-inositol (50 mg), 0.5 mL vitamin stock, and 2.5 gm gelrite. Make volume to 1 L with distilled water, pH 5.8. After autoclaving, add 5 mg/L bialaphos. Pour in Petri dishes (100 ×20 mm).

4. Medium for root elongation: shoot elongation medium (10 mL) with 4–5 mg/L bialaphos in culture tubes (100 × 25 mm).
   All media with bialaphos can be stored for 2 wk at room temperature.
5. Zeatin stock solution: dissolve 40 mg zeatin in few drops of 1N NaOH, and make volume up to 10 mL in distilled water.
6. Bialaphos stock solution: dissolve 40 mg of bialaphos in 5 mL distilled water, and make up volume to 10 mL. Filter, sterilize, and store at –20°C.

2.4. Analysis of Transient and Stable Expression

1. X-gluc (15-bromo-4-chloro-3-indolyl-β-glucuronide) (100 mL): 100 mg X-gluc dissolved in 40 mL distilled water; 20 mL phosphate buffer, pH 7.0 (100 mM); 1 mL potassium ferricyanide (MW 393.3; 0.5 M); 1 mL potassium ferrocyanide (MW 422.4; 0.5 M); 1 mL ethylenediamine tetraacetic acid (EDTA), pH 7.0 (1 M). Bring volume to 100 mL and stir until dissolved. Add 100 µL Triton X-100, filter sterilize (22 µm), and store in small aliquots at –20°C.
2. PAT extraction buffer (EB buffer) (100 mL): 5 mL 1 M Tris-HCl, pH 7.5 (50 mM); 0.4 mL 0.5 M EDTA (20 mM); 15 mg leupeptine (0.15 mg/mL); 15 mg phenylmethylsulphonyl fluoride (PMSF, Sigma) (0.15 mg/mL); 30 mg bovine serum albumin (BSA) (0.3 mg/mL); 30 mg dithiothreitol (DTT) (0.3 mg/mL). Stir to dissolve, filter, sterilize, and store at –20°C.
3. AB buffer (100 mL): 5 mL 1 M Tris-HCl, pH 7.5 (50 mM); 0.4 mL 0.5 M EDTA (2 mM); 100 mg BSA (0.1%). Make up volume to 100 mL with distilled water, and store in refrigerator.
4. Saturated ammonium sulfate: 400 gm $(NH_4)_2SO_4$ in 400 mL distilled water, pH 7.8. Autoclave and store at room temperature. Can be used for at least 3 mo.
5. Thin layer chromatography (TLC) migration solvent (500 mL): 300 mL l-propanol; 178.6 mL ammonium hydroxide. Make up volume to 500 mL with water (saturate the chromatography tank overnight with 3-mm paper). Highly pungent; wear mask.
6. ^{14}C-labeled acetyl-Co-A: wear gloves and work in area designated for radioactive work.
7. Thin layer chromatography plates.
8. Chromatography tank.
9. Kodak (X-Omat) AR X-ray film.
10. Bulk reaction mixture for PAT assay: prepare the necessary volume for two reactions more than the total number of reactions to allow for pipetting error. For one reaction: 0.6 µL PPT (1 mg/mL), 1 µL ^{14}C acetyl Co-A (0.05 µci/µL), 1.4 µL distilled water.

3. Methods

3.1. Culture of Immature Embryos

1. Collect spikes (11–14 d post-anthesis) and wrap in moist paper towels. Spikes can be used fresh or stored for up to 5 d, with cut ends immersed in water, in a refrigerator (*see* **Note 2**).

2. Remove caryopses from the middle half of each spike and surface sterilize with 70% ethanol for 2 min, followed by slow stirring for 15–20 min in 20% chlorox solution (1.05% sodium hypochlorite) containing 0.1% Tween-20. Remove chlorox by washing with four changes of sterile distilled water at 5-min intervals (*see* **Notes 3** and **4**).
3. Aseptically remove the immature embryos (0.8–1.5 mm) under a stereomicroscope and place in Petri dishes containing MS+ medium, with the scutellum exposed and the embryo axis in contact with the medium (*see* **Note 5**). Between 20 and 25 embryos can be cultured in one Petri dish. Seal the culture dish with Parafilm and incubate in the dark at 27°C to induce proliferation of scutellar cells (*see* **Note 6**).
4. Transfer embryos 4–6 h prior to bombardment (4–6 d after culture, when cell proliferation is visible at the edges of the scutellum) to Petri dishes containing the medium for osmotic treatment (*see* **Subheading 2.1.3., step 3**). Arrange 30–40 embryos, with scutellum facing up, in a 2-cm diameter circle in the center of the dish (*see* **Note 7**).

3.2. Preparation and Delivery of DNA-Coated Gold Particles

1. Sterilize the bombardment chamber of PDS 1000/He and the gas acceleration cylinder with 70% ethanol and let dry.
2. Sterilize the rupture and macrocarrier discs by soaking in 100% ethanol for 5 min and air drying in Laminar Flow hood.
3. Sterilize steel screens by autoclaving.
4. Add to two 25 µL aliquots of washed sterile gold particles an equal volume of sterile distilled water, and vortex for 5 s.
5. Add 5 µL plasmid DNA, or 5 µL/construct for cotransformation, to one tube of gold particles (the other tube of gold particles serves as control for bombardments). Vortex immediately for 30 s to ensure good mixing of DNA and gold particles.
6. Add, in quick succession, 50 µL of $CaCl_2$ (*see* **Subheading 2.2., step 3**) and 20 mL of spermidine, (*see* **Subheading 2.2., step 4**) and vortex for 2 min. Let sit on ice for 5–10 min.
7. Centrifuge 10 s, discard supernatant and resuspend in 200 µL 100% ethanol (can sonicate briefly to disperse).
8. Centrifuge 10 s, discard supernatant, and resuspend the pellet in 150–250 µL of 100% ethanol. Leave on ice but use within 2 h.
9. Sonicate and quickly spread 5 µL of gold from control Eppendorf tube (without DNA) or from gold-DNA tube, to the center of a macrocarrier disc placed in the holder (gold concentration 30–50 µg/shot). Air dry. At least two discs should be prepared for control bombardments, and the required amount with gold-DNA based on the number of samples in the experiment. Five macrocarrier discs can be prepared at one time from each suspension (*see* **Note 8**).
10. With the helium tank on, set the delivery pressure to 1300 psi (200 psi above the desired rupture disc value).

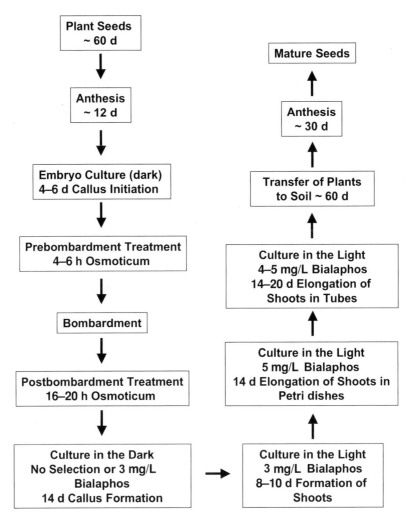

Fig. 1. Time frame for the production of transgenic wheat (cv Bobwhite) plants. Times shown are averages for experiments performed in 1995; those designated by ~ are approximate and vary either with the batch of donor plants or the individual callus line or plant. Transgenic plants were transferred to soil 56–66 d after the initiation of cultures.

11. Turn on the vacuum pump.
12. Place a rupture disc (1100 psi) in the holder and screw tightly in place (*see* **Note 9**).
13. With the steel mesh in place, transfer the coated macrocarrier disc (coated side facing down) to the holder and place the macrocarrier assembly unit in the chamber at level 2 from top.

14. Place a Petri dish containing the cultured embryos on the sample holder at level 4 from top and close door.
15. Follow manufacturer's directions for bombardment, and follow all indicated safety procedures.

3.3. Culture, Selection, and Regeneration

1. Sixteen to twenty hours after bombardment, separate and transfer the embryos to Petri dishes containing either MS+ medium without glutamine and casein hydrolysate (delayed selection) or selection medium for callus proliferation (early selection), for up to 2 wk in the dark. Place 16–20 embryos (from each bombarded dish) in two dishes to avoid overcrowding and cross-feeding (*see* **Note 10**).
2. Transfer the embryogenic calli from callus proliferation medium to regeneration medium for shoot formation under 16 h photoperiod (190 $\mu E/m^2/s$) for 8–10 d. At the end of this period, green areas indicative of shoot formation are visible to the naked eye.
3. Transfer the green shoots, along with the callus, as a unit to shoot elongation medium in light for 1–2 cycles of 2 wk each. Keep each callus-shoot piece well separated from others (*see* **Note 11**).
4. Transfer green shoots (>2 cm in length) to root elongation medium in tubes. The smaller (<1 cm) shoots can be subcultured for one more cycle of shoot elongation in Petri dishes. Calli with no shoots or shoots smaller than 1 cm after two cycles on shoot elongation medium can be discarded (*see* **Note 12**).
5. Plants reaching to the top of the culture tubes and with well developed roots can be tested for PAT activity and transferred to soil in 56–66 d (*see* **Fig. 1**).
   They are grown to maturity in growth chambers under conditions similar to those used for donor plants.

3.4. Histochemical GUS Assay for Transient Expression

1. Remove a sample of embryos (two to four from four dishes each of control and +DNA) 2 d after bombardment, and soak separately in X-gluc solution (50–100 µL/well of a microtiter plate).
2. Seal the plate with Parafilm and incubate overnight at 37°C.
3. Examine embryos under a stereo dissecting microscope to visualize the blue, GUS-expressing units.

3.5. PAT Assay

This assay is based on the detection of ^{14}C labeled acetylated PPT (nonradioactive PPT used as substrate) after separation by TLC.

1. Collect leaf samples on ice in 1.5-mL Eppendorf tubes (prior to transferring putative transformants to soil). Be sure to include ± control samples (optimum sample weight 20–30 mg; freeze in liquid nitrogen and store at –70°C if not to be used the same day).

2. Add about 2 mg Polyvinyl-Pyrrolidone (Sigma PVP40), some silicone powder and 100 µL of ethylene bromide (EB) buffer to each tube. Grind for 30 s with a polytron and keep samples on ice.
3. Centrifuge 10 min at 4°C, collect supernatant in a fresh tube. Repeat centrifugation.
4. Measure protein concentration in the crude leaf extracts following the Bio-Rad protein assay relative to BSA as standard.
5. Adjust protein concentration to 2.5 µg/µL for each sample with EB buffer. Can stop assay here overnight (store at –20°C). The background activity can be eliminated by precipitating the extracts with saturated $(NH_4)_2SO_4$.
6. Use 10 µL of extract for each sample for assay.
7. Aliquot 3 µL of bulk reaction mixture to 10 µL of each sample extract. Mix well and incubate at 37°C for 1 h. Stop reaction on ice.
8. Spot 13 µL of each sample on a TLC plate (can be done by first spotting 6.5 µL, drying with a hair dryer, and spotting the rest to avoid spreading of mixture). Up to 15 samples can be spotted on one plate.
9. Transfer the TLC plate (after all the sample extracts have dried) to a saturated chromatography tank for 1–1.5 h.
10. Remove plate and let dry completely (at least 15 min).
11. Visualize the ^{14}C-labeled acetylated PPT by overnight exposure to X-ray film.
12. Plants testing positive for PAT are regarded as transformed (as the assay tests for expression of the enzyme encoded by the transgene *bar*) and can be transferred to soil.
    Characterization of the transgenes can be accomplished by DNA isolation and Southern hybridization by standard protocols.

4. Notes

1. Because the quality of donor plants directly affects the capacity of the immature embryos to produce embryogenic callus, it is important that the plants be grown under optimal conditions. Pesticide application from pollination to harvest time should be avoided.
2. For optimum response (especially when using field grown material, which is often contaminated) use within 2 d after collection.
3. The size of the caryopsis is variable along the length of the spike. Therefore, a more uniform sample is obtained by collecting caryopses only from the middle half of each spike.
4. The duration of sterilization treatment depends on the quality of the donor plants. Caryopses from growth chamber grown plants tend to be clean and can be sterilized effectively in 10–15 min. Material from field grown plants needs longer (30 min) chlorox treatment.
5. Although embryos ranging in size from 0.5 to 1.5 mm are capable of producing embryogenic callus, the best post-bombardment response is obtained from 1 to 1.2-mm embryos. Embryos at this stage of development are the easiest to dissect, since younger embryos are nearly transparent and difficult to locate whereas the older ones are white owing to stored starch in the scutellum. Holding the cary-

opsis with a forceps and making an incision at the base easily exposes the embryo (the outline of the embryo is visible before dissection).

6. Between 50 and 60 immature embryos can be cultured in the same dish if the donor plants appear healthy and free of infection. Nonetheless, it is better to culture an average of 20 embryos/dish from field material owing to the probability of infection. The cultures should be examined on a daily basis to detect early signs of contamination. The contaminated embryos should be removed along with the medium around them to avoid infecting the adjacent embryos.
7. The highest frequencies of transformation and regeneration are obtained with 4–6 h of pre- and 16 h of post-bombardment osmotic treatment.
8. The concentration of gold particles and DNA in gold/DNA mixture can vary. We recommend 30–50 µg gold/shot for a uniform spread on the macrocarrier disc, to obtain a more even and finer size of blue GUS units, and a consistent efficiency of transformation.
9. Rupture discs in a range of 650–1550 are used, with the 1100 psi disc being most common.
10. Among basta, bialaphos and PPT, bialaphos was found to be most reliable for selection of wheat transformants in the accelerated protocol *(6)*. The time when selection is imposed, the number of explants/dish, and the concentration of the selective agent, are critical variables which depend on the quality of the embryos cultured. Generally, early selection is recommended for high quality embryos, and delayed selection for inferior embryos. Transformation frequencies of 0.1–2.5% were obtained with the protocol shown in **Fig. 1**.
11. Cultures should be examined frequently to assess the concentration of the selective agent required in the next step. For example, with high zeatin in the regeneration medium (10 mg/L), too many shoots are formed. With such cultures, it is safe to use 5 or 6 mg/L bialaphos during shoot elongation. With low (1 mg/L) or no zeatin, when only one or two green areas are seen in each explant, 4 or 5 mg/L bialaphos is sufficient.
12. During the period the explants and the regenerants are in Petri dishes, bialaphos can be used at 5 or 6 mg/L because cross protection is provided by the adjacent explants. However, if very few shoots emerge after the first cycle of shoot elongation, then 4–5 mg/L bialaphos is sufficient because only a single explant is present and there is no cross protection.

References

1. Oerke, E. C., Dehne, H. W., Schönbeck, F., and Weber, A. (1994) *Crop Production and Crop Protection: Estimated Losses in Major Food and Cash Crops.* Elsevier, Amsterdam.
2. Bialy, H. (1992) Transgenic wheat finally produced. *Bio/Technology* **10**, 675.
3. Vasil, V., Castillo, A. M., Fromm, M. E., and Vasil, I. K. (1992) Herbicide resistant fertile transgenic wheat plants obtained by microprojectile bombardment of regenerable embryogenic callus. *Bio/Technology* **10**, 667–674.

4. Vasil, V., Srivastava, V., Castillo, A. M., Fromm, M. E., and Vasil, I. K. (1993) Rapid production of transgenic wheat plants by direct bombardment of cultured immature embryos. *Bio/Technology* **11**, 1553–1558.
5. Weeks, J. T., Anderson O. D., and Blechl, A. E. (1993) Rapid production of multiple independent lines of fertile transgenic wheat (*Triticum aestivum*). *Plant Physiol.* **102**, 1077–1084.
6. Altpeter, F., Vasil, V., Srivastava, V., Stöger, E., and Vasil, I.K. (1996) Accelerated production of transgenic wheat (*Triticum aestivum* L.) plants. *Plant Cell Rep.* **16**, 12–17.
7. Ortiz, J. P. A., Reggiardo, M. I., Ravizzini, R. A., et al. (1996) Hygromycin resistance as an efficient selectable marker for wheat stable transformation. *Plant Cell Rep.* **15**, 877–881.
8. Nehra, N. S., Chibbar, R. N., Leung, N., et al. (1994) Self-fertile transgenic wheat plants regenerated from isolated scutellar tissues following microprojectile bombardment with two distinct gene constructs. *Plant J.* **5**, 285–297.
9. Zhou, H., Arrowsmith, J. W., Fromm, M. E., et al. (1995) Glyphosate-tolerant CP4 and GOX genes as a selectable marker in wheat transformation. *Plant Cell Rep.* **15**, 159–163.
10. Becker, D., Brettschneider, R., and Lörz, H. (1994) Fertile transgenic wheat from microprojectile bombardment of scutellar tissue. *Plant J.* **5**, 299–307.
11. Takumi, S. and Shimada, T. (1996) Production of transgenic wheat through particle bombardment of scutellar tissues: frequency is influenced by culture duration. *J. Plant Physiol.* **149**, 418–423.
12. Vasil, I. K. (1996) Phosphinothricin-resistant crops, in *Herbicide-Resistant Crops* (Duke, S.O., ed.), Lewis Publishers, Boca Raton, FL, pp. 85–91.
13. Christensen, A. H. and Quail, P. H. (1996) Ubiquitin promoter-based vectors for high-level expression of selectable and/or screenable marker genes in monocotyledonous plants. *Transg. Res.* **5**, 213–218.
14. Altpeter, F., Vasil, V., Srivastava, V., and Vasil, I. K. (1996) Integration and expression of the high-molecular-weight glutenin subunit 1Ax1 gene into wheat. *Nature Biotech.* **14**, 1155–1159.
15. Taylor, M. G., Vasil, V., and Vasil, I. K. (1993) Enhanced GUS gene expression in cereal/grass cell suspensions and immature embryos usng the maize ubiquitin-based plasmid pAHC25. *Plant Cell Rep.* **12**, 491–495.
16. Srivastava, V., Vasil, V., and Vasil, I. K. (1996) Molecular characterization of the fate of transgenes in transformed wheat (*Triticum aestivum* L.). *Theor. Appl. Genet.* **92**, 1031–1037.
17. Blechl, A. E. and Anderson, O. D. (1996) Expression of a novel high-molecular-weight glutenin subunit gene in transgenic wheat. *Nature Biotech.* **14**, 875–879.
18. Repellin, A., Baga, M., Jauhar, P. P., and Chibbar, R. N. (2001) Genetic enrichment of cereal crops via alien gene transfer: new challenges. *Plant Cell Tissue Organ Cult.* **64**, 159–183.

19. Sahrawat, A. K., Becker, D., Lutticke, B., and Lorz, H. (2003) Genetic improvement of wheat via alien gene transger: an assessment. *Plant Sci.* **165,** 1147–1168.
20. Bliffeld, M., Mundy, J., Potrykus, I., and Futterer, J. (1999) Genetic engineering of wheat for increased reisistance to powdery mildew disease. *Theor. Appl. Genet.* **98,** 1079–1086.
21. Sivamani, E., Brey, C. W., Dyer, W. E., Talbert, L. E., and Qu, R. (2000) Resistance to wheat streak mosaic virus in transgenic wheat expressing the viral replicase (Nib) gene. *Mol. Breed.* **6,** 469–477.
22. Altpeter, F., Diaz, I., McAuslane, H., Gaddour, K., Carbonero, P., and Vasil, I. K. (1999) Increased insect resistance in transgenic wheat stably expressing trypsin inhibitor CMe. *Mol. Breed.* **5,** 53–63.
23. Vasil, I. K., Bean, S., Zhao, J., et al. (2001) Evaluation of baking properties and gluten protein composition of field grown transgenic wheat lines expressing high molecular weight glutenin gene 1Ax1. *J. Plant Physiol.* **158,** 521–528.
24. Murashige, T. and Skoog, F. (1962) A revised medium for rapid growth and bioassays with tobacco tissue cultures. *Physiol. Plant* **15,** 473–497.

24

Chloroplast Transformation

Xiao-Mei Lu, Wei-Bo Yin, and Zan-Min Hu

Summary

In this chapter we briefly review the developmental history and current research status of chloroplast transformation and introduce the merits of chloroplast transformation as compared with the nuclear genome transformation. Furthermore, according to the chloroplast transformation achieved in oilseed rape (*Brassica napus*), we introduce the preparation of explants, transformation methods, system selection, identification methods of the transplastomic plants, and experimental results. The technical points, the bottleneck, and the further research directions of the chloroplast transformation are discussed in the notes.

Key Words: *Bacillus thuringiensis* (Bt); bombardment; chloroplast transformation; oilseed rape; spectionomycin.

1. Introduction

Plant genetic transformation is a core research tool in plant biology and crop improvement. The introduction of novel genes into nuclear genomes has become a common method in genetic transformation. However, the introduction of novel genes into the nuclear genome has led to a growing public concern of the possibility of causing genetic pollution to other crop and/or wild plants. Chloroplast transformation might be a potential solution to these problems because maternal inheritance of the chloroplast genome prevents the introduced genes from escaping through pollen grains in most plants *(1)*. Therefore, chloroplast transformation is an environmentally friendly approach to plant genetic engineering that minimizes out-crossing of transgenes to related weeds or crops *(2)*. In addition, chloroplast genetic engineering offers several other advantages over nuclear transformation *(2–4)*. First, the plastids of higher plants have their own double-stranded genomes of 120–160 kb in size. A remarkable feature of the chloroplast genome is its extremely high ploidy level: a single tobacco leaf cell contains as many as 100 chloroplasts, each harboring approx 100 identical copies of the plastid genome, resulting in an extraordinarily high ploidy number of up to 10,000 plastid genomes per cell *(5)*. Therefore, the chloroplast genome of higher plants is an attractive target for crop engineering, owing to the feasibility to express foreign genes at high levels. Operon-derived *Cry 2Aa2* protein accumulates in transgenic chloroplasts as cuboidal crystals, to a level of 45.3% of the total soluble protein and remains stable even in senescing leaves (46.1%) *(6)*. Second, foreign gene expression is more stable in plastome chloroplast than in nuclear genome. Nuclear transformation in plants occurs by the random integration of transgenes into unpredictable locations in the genome by nonhomologous recombination and can result varying level of expression and, often led to gene silencing. Transgenes are integrated into chloroplast genomes by homologous recombination and are not affected by gene silencing. Third, chloroplast transformation eliminates "the position effect" often observed in nuclear transformation. Fourth, chloroplasts have the capacity to express multiple transgenes from a single operon (transgene staking) as a result of efficient translation of polycistronic messenger RNAs (mRNAs) in plastid. "It's like a bacterial fermenter in a plant cell; you solve a lot of problems in one shot," says research director Peter Heifetz of the Torrey Mesa Research Institute in San Diego *(7)*.

The first successful chloroplast transformation was in *Chlamydomonas reinhardtii (8,9)*. Whereas the stable transformation of plastids in higher plants was firstly reported in tobacco *(4)*. Chloroplast transformation in higher plants has been so far reported in tobacco *(1–18)*, *Arabidopsis (19)*, potato *(20,21)*, rice *(22)*, tomato *(3,23)*, and oilseed rape *(24,25)*. Chloroplast transformation has numerous potent applications in developing plants resistant to biotic and abiotic stresses *(26–29)*, and for production of therapeutic proteins *(3,12,30–32)*, vaccines, and lipids *(33)*. The insertion sites of foreign genes used in these studies included the *rbcL-ORF512 (13,28)*, the 16S*trnV-rps12rps7 (12,19)*, the *trnI-trnA (6)*, the *rps7* and *ndhB (1)* intergenic region and other regions. The choice of marker gene and selective agent are critical for successful transformation. Four dominant selectable genes have been used for plastid transformation in higher plants: the *aadA* gene (aminoglycoside adenyltransferase, AAD) inactivating spectinomycin and streptomycin *(13)*, the *nptII* and *aphA-6* genes (aminoglycoside phosphotransferases) detoxifying kanamycin *(10,34–36)* and the *badh* gene (betaine aldehyde dehydrogenase) in combination with betaine aldehyde as a selection agent *(37)*.

To date, biolistic delivery and polyethylene glycol (PEG)-induced DNA uptake can yield stable plastid transformation *(39–41)*. Microinjection is an approach for the plastid transformation, which yields transient gene expression

(42). Transformating DNA into plastids of suspension cells *(15)* or isolated chloroplast has not yet been widely used.

Protocols developed in tobacco chloroplast transformation should be applicable to the tomato without modification *(40)*, and they could also be used in other plants that utilize leaves as the explants except that the culture mediums, selecting and regenerated conditions are different. Seeing that the tobacco chloroplast transformation protocol was already discussed in the first edition and other reports *(14–18)*, here we only discuss the oilseed rape chloroplast transformation.

Among the insecticidal genes, *Bacillus thuringiensis* (*Bt*) crystal protein genes have been proven effective in controlling insect larvae in many crop plants. The use of commercial, nuclear transgenic crops expressing Bt toxins has escalated in recent years because of their advantages over traditional chemical insecticides *(16,20,25,28,43–45)*. Expressing the Bt insecticidal protein genes in the nucleus were unstable, yielding many truncated polyadenylated mRNAs and only a small fraction of full-size transcript *(40)*. It is confirmed that the expression of *Bt* in the tobacco plastid transformants is 20- to 30-fold higher than that of current commercial nuclear transgene plants, with multiple toxins and tissue-specificity *(28)*. However, in crops with several target pests with varying degrees of susceptibility to Bt, there is concern regarding the suboptimal production of toxin, resulting in reduced efficacy and increased risk of *Bt* resistance. Here, we introduce a protocol to transform *Bt cryAa10* into oilseed rape (*Brassica napus*) chloroplast to express Bt protein against *Plutella xylostera*.

2. Materials
2.1. Bombardment (46)

1. Sterile disposable pipets.
2. Autoclaved Whatman filter paper disks.
3. Triangular flasks.
4. Tweezers (autoclaved; flamed before each use).
5. Petri dishes.
6. Autoclaved tips.
7. Eppendorf tubes.
8. Sterile rupture disks.
9. Stopping screens.
10. Ethanol: 100%.
11. Microcarry holder.
12. Rupture disk retaining caps.
13. Microcarrier launch assembly with 70% ethanol.
14. Sterile gold microcarriers.
15. Ethanol (100% and 70%).

16. 2.5 M Calcium chloride (weigh 1.84 g of $CaCl_2$ and dissolve in 5 mL of water, filter sterilize. Make fresh just before use. May be stored at 4°C for short periods; do not freeze).
17. 1 M Spermidine-freebase. Highly hygroscopic—therefore, take a 1 g unopened bottle of spermidine stored at 4°C and add 6.8 mL of sterile water and filter sterilized, store as 25 µL aliquots at −20°C.
18. 100 mg/mL Spectinomycin. Take a 1 g unopened bottle of spectinomycin stored at 4°C and add 10 mL sterile water and filter sterilized; store as 1-mL aliquots at −20°C.

2.2. Tissue Culture

2.2.1. Plant Materials

1. Oilseed rape (*Brassica napus*).

2.2.2. Medium

1. Methyl sulfoxide (MSO) medium (1 L): 4.3 g Murashige and Skoog (MS) salts, 100 mg I-inositol, 5 mg pyridoxine HCl, 5 mg nicotinic acid, 30 g sucrose (molecular biology grade), 1 mg thiamine HCl (100 µL from a stock solution of 10 mg/mL), Phytagar 6 g. Adjust pH to 5.8 with NaOH (initial pH will be around 4.4) and autoclave for 20 min.
2. Oilseed rape selection medium (1 L). (M6): MSO medium with 6 mg/L 6- benzylaminopurine (6-BA) and 10 mg/L spectinomycin
3. (M1): MSO medium with 1 mg/L 6-BA and 10 mg/L spectinomycin.
4. Oilseed rape rooting medium (1 L): MSO medium with 0.1 mg/L α-naphthyl acetic acid and 0.8 mg/L spectinomycin for rooting and finally transferred to soil.
5. Hypertonic medium (1 L): MSO complement with 0.5 mg/L mannitol.

2.3. Preparation for Bombardment

2.3.1. Plant Tissue

Oilseed rape seeds are washed with 70% ethanol for 30 s. Wash off ethanol with sterile distilled water, three times. Sterilize with sodium hypochlorite for 15 min and wash three times with sterile distilled water. Oilseed rape is grown aseptically by germinating treated seeds in MSO solid medium for 3–4 d. The green cotyledon petioles (1–2 mm in length, attached on cotyledon) are harvested for biolistic bombardment. Before bombardment, the harvested green cotyledon petioles are placed on oilseed rape hypertonic medium for 4 h in dark *(44)*.

2.3.2. Preparation of Plasmid DNA

Large-scale preparations of plasmid DNA was performed as described in **ref. 47**.

3. Methods

3.1. Preparation of Microcarriers (48)

The following procedures prepare gold (1.0 µm) for 120 bombardments using 500 µg of the microcarrier per bombardment.

1. Weigh out 30 mg of microparticles into a 1.5 mL microfuge tube. Add 1 mL of 70% ethanol (v/v).
2. Vortex vigorously for 3–5 min.
3. Allow the particles to soak in 70% ethanol for 15 min.
4. Pellet the microparticles in a microfuge tube by centrifugation for 5 s.
5. Remove and discard the supernatant.
6. Repeat the following wash steps for three times:
   a. Add 1 mL of sterile water.
   b. Vortex vigorously for 1 min.
   c. Allow the particles to settle for 1 min.
   d. Pellet the microparticles by briefly spinning in a microfuge.
   e. Discard the liquid.
7. After the third wash, add 500 µL sterile 50% glycerol to bring the microparticle concentration to 60 mg/L (assume no loss during preparation).
8. The microparticles can be stored at room temperature for up to 2 wk. Tungsten aliquots should be stored at –20°C to prevent oxidation. Gold aliquots can be stored at 4°C or room temperature.

3.2. Bombardment of Tissues

3.2.1. Coating Washed Microcarriers With Vector DNA

The following procedure is sufficient for six bombardments; if fewer bombardments are needed, adjust the quantities accordingly.

1. Vortex the microcarriers prepared in 50% glycerol (30 mg/mL) for 5 min on a platform vortex mixer to resuspend and disrupt agglomerated particles. When removing aliquots of microcarriers, it is important to continuously vortex the tube containing the microcarriers to maximize uniform sampling. When pipetting aliquots, hold the microcentrifuge tube firmly at the top while continually vortexing the base of the tube.
2. Remove 50 µL (3 mg) of microcarriers to a 1.5-mL microcentrifuge tube.
3. Continuous agitation of the microcarriers is needed for uniform DNA precipitation onto microcarriers. For added convenience and/or multiple samples, use a platform attachment on your vortex mixer for holding microcentrifuge tubes.
4. While vortexing vigorously, add in order:
   a. 10 µL DNA (1 µg/µL).
   b. 50 µL 2.5 M $CaCl_2$.
   c. 20 µL 0.1 M spermidine (free base, tissue culture grade).
5. Continue vortexing for 2–3 min.

6. Allow the microcarriers to settle for 1 min.
7. Pellet microcarriers by spinning for 2 s in a microfuge.
8. Remove the liquid and discard. Add 150 µL of 70% ethanol (high performance liquid chromatography [HPLC] or spectrophotometric grade). Remove the liquid and discard.
9. Add 150 µL of 100% ethanol. Remove the liquid and discard.
10. Add 60 µL of 100% ethanol. Gently resuspend the pellet by tapping the side of the tube several times, and then by vortexing at low speed for 2–3 s. Use 10 µL for each bombardment.

3.2.2. Performing Bombardment

1. Select bombardment parameters (useful is 1100 psi, target shelf levels are level 2 = 6 cm or level 3 = 9 cm) for gap distance between rupture disk retaining cap and microcarrier launch assembly. Place the stopping screen support in proper position inside fixed nest of microcarrier launch assembly.
2. Clean/sterilize: clean the chamber and equipment (rupture disk retaining cap, microcarrier launch assembly) with 70% ethanol. Consumables (macrocarrier, macrocarrier holders, stopping screen, and rupture disks) with 100% ethanol. Allow time for drying. Insertion of the macrocarrier into macrocarrier holder with plastic insertion tool.
3. Load DNA-coated microcarriers onto a macrocarrier/macrocarrier holder and allow time for drying.
4. Unscrew the macrocarrier cover lid from the assembly. Place a sterile stopping screen support. Install the macrocarrier/macrocarrier holder on the top rim of the fixed nest. The dried microcarriers should be facing down, toward the stoping screen. Replace the macrocarrier cover lid on the assembly and turn clockwise until snug. Do not over-tighten.
5. Place the microcarrier launch assembly in the top slot inside the bombardment chamber. Place the sample (usually contained within a Petri dish) on the Target Shelf. Close and latch the sample chamber door.
6. Bombard the sample (leaf or green cotyledon petioles). Leaves are placed abaxial side up on medium in Petri dish for bombardment. The cut end of green cotyledon petioles should be bombarded to increase the efficiency of chloroplast transformation in oilseed rape.

3.3. Construction of Oilseed Rape Chloroplast Expression Vectors

The chloroplast transformation vector pNRAB, a plasmid pBluescript SK (Stratagene, Germany) derivative, was constructed by our lab as described previously *(49)* and shown in **Fig. 1**.

The insertion site of foreign genes used in most studies is the rbcL-ORF512 *(14,28)*, the *16StrnV-rps12rps7* *(18,19)*, and the *trnI-trnA* intergenic regions *(6)*. In this study, we insert the *Bt* between *rps7* and *ndhB*.

In the construction, the 1.0 kb *rps7* gene (KpnI-ApaI fragment, GenBank accession AF124376) *(50)* and the 2.4 kb *ndhB* gene (SacII-NotI fragment, GenBank accession AF126026) of oilseed rape chloroplast were used as targeting sequences and firstly constructed into pBluescript SK, yielding the plasmid pSNR4. A 1.3 kb spectinomycin resistance gene cassette, consisting of the aminoglycoside 3'-adenylyltransferase (*aadA*) gene under the control of the tobacco plastid *16S rRNA* operon promoter *(Prrn)* and the 3'-untranslated region (UTRs) of the *psbA* gene (*psbA3'*), was from a plasmid pZS197 (a gift from Dr. P. Maliga, *13*). *Prrn* promoter is a strong plastid rRNA operon, which ensures high levels of mRNA. *Prrn* can be fused with translation control sequences of plastid and phage origin to facilitate the translation of recombinant *(40,51)*.

The *aadA* cassette was excised as XhoI-NotI fragment from vector pSZB8 (subcloned from pZS197 by laboratory of Dr. G.F. Shen) and constructed into the XhoI-NotI site of pSNR4 between *rps7* and *ndhB*, yielding the plasmid pNRA8 (**Fig. 1A**). The 3.5 kb insect resistance gene *cry1Aa10* (GenBank accession AF154676) *(27)* was under the control of the tobacco plastid *16S rRNA* operon promoter (*Prrn*) and the 3'-untranslated region (a gift from Dr. H. Shimada) of the rice plastid *psbA* gene (*psbA3'*). A 4.0-kb XhoI-NotI fragment containing the *prrn-cry1Aa10* cassette was isolated from a plasmid pTPBt9 *(44)* (**Fig. 1B**) and the XhoI site was filled in with Klenow DNA polymerase. The fragment was then inserted at the SpeI-NotI site of the plasmid pNRA8 (SpeI site was filled in) between *aadA* cassette and the *ndhB* fragment, yielding the chloroplast transformation vector pNRAB (**Fig. 1C**).

3.4. Culture of the Bombarded Oilseed Rape Tissue

The bombarded explants (green cotyledon petioles) are incubated for 2–3 d on the M6 medium without selective antibiotic under dark. Explants were then transferred to shoot induction medium M6 (with spectinomycin 10 mg/L). The explants should be transferred to fresh medium every 2 wk. Green shoots that formed are carefully removed and placed into shoot outgrowth medium M1. Shoots that remained green on M1 medium are considered to be transformants and transferred to rooting medium. Rooted shoots are transferred to soil under glasshouse conditions *(44)*.

Shoot start to develop from cut ends of cotyledon petioles 5–7 wk after transfer of bombarded explants to selection medium. Thirty-six putative transformants were obtained from 1000 cotyledon. Culture temperature is 25°C.

3.5. The Homoplasmy of Chloroplast Transformants

The challenge of plastome engineering is to uniformly alter all genome copies because genetically stable transgenic plants are obtained only if all the chloroplast genome copies are identical. Each transgenic clone was subjected to a

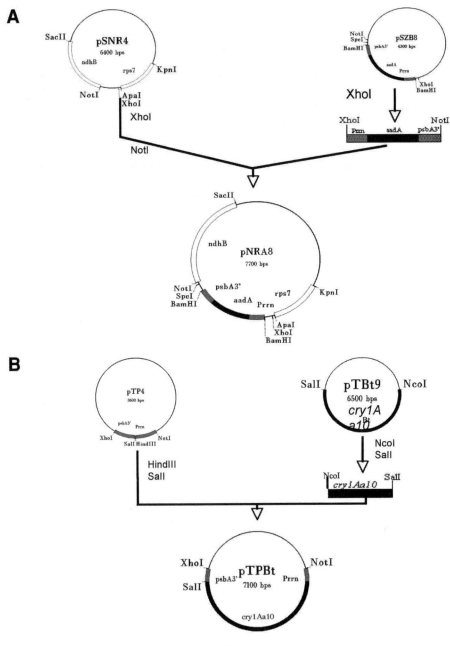

Fig. 1.

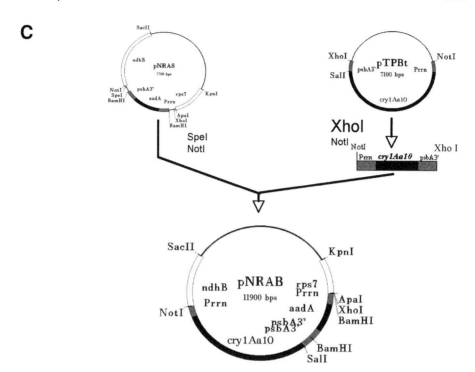

Fig. 1. *(From opposite page)* Oilseed rape chloroplast transformation vectors. **(A)** The plasmid vector pNRA8 contains a 1.3-kb spectinomycin resistance gene cassette from plasmid pZS197 and cloned oilseed rape chloroplast targeting sequences from plasmid pSNR4. **(B)** The plasmid vector pTPBt contains 3.5-kb insect resistance gene *cry1Aa10* under the control of the tobacco plastid *16S rRNA* operon promoter (*Prrn*) and the 3'-untranslated region of *psbA* gene. **(C)** The *prrn-cry1Aa10* cassette inserted at the SpeI-NotI site of the plasmid pNRA8 between *aadA* cassette and the *ndhB* fragment, yielding the chloroplast transformation vector pNRAB.

second round of spectinomycin selection and extending the selection phase *(17,23)*. The leave were cut into small pieces (2 × 2 mm) and placed onto M1 medium with selective antibiotic. Many laboratories reported that homoplasmy of plastid transformants typically is achieved by subjecting the primary transformant to one to two additional rounds of regeneration on antibiotic-containing medium *(52)*.

3.6. Evaluation of Results

Confirmation of transformation was done using leaf disc assay against *Plutella xylostera* and DNA gel blot analysis.

3.6.1. Polymerase Chain Reaction Southern Analysis of the Transformants

Total DNA of putative transgenic plants is isolated with CTAB DNA extraction protocol (*see* **Note 7**). Polymerase chain reaction (PCR) is performed by the Taq DNA polymerase (TaKaRa, Japan). Primer P1 (5'-gagat ccacc ctaca atatg-3') is designed to land on the upstream of the native *rps7* of the chloroplast genome and primer P2 (5'-TCCTG AAAAG CTAAT CATAG G-3') landed on the *ndhB* gene. Because the upstream sequences of *rps7* in oilseed rape have not been determined so far, here we use the oligomers of corresponding region in tobacco as the sequences of primer P1. Samples were carried through 30 cycles of PCR by using the following temperature regimens: 95°C for 1 min, 47°C for 1 min, and 72°C for 8 min. The PCR products are electrophoresed on 0.8% agarose gel. Southern hybridizations of PCR products were carried out with either *rps7* targeting sequence or the 0.6 kb cryAa10 fragment as the probe. All probes are DIG-labeled by using the probe synthesis kit.

Putative chloroplast transformants are initially identified by PCR. Because the primer P1 lands on the upstream of the chloroplast genomic *rps7* gene and the primer P2 lands on the *ndhB* gene (**Fig. 2A**), a 1.1-kb PCR product would be observed in wild-type chloroplast genome, and a 6.4-kb PCR product only in the transformed chloroplast genome. After PCR analysis, we found that four of the 36 putative transformants yield both the 1.1 and 6.4 kb products while the other resistant plants and untransformed plants yielded only the 1.1 kb PCR product (**Fig. 2B**). The presence of the 6.4-kb product in the 4 transgenic plants confirmed the site-specific integration of the foreign gene cassettes between *rps7* and *ndhB*. However, homoplasmy was not accomplished.

The four positive plants were subjected to further analysis by Southern blot hybridization. When probed with the *rps7* targeting sequence, both the 1.1- and 6.4-kb bands of PCR products gave positive hybridization signals (**Fig. 3, *I***). When probed with the *cry1Aa10* gene, only the 6.4-kb band gave a hybridization signal (**Fig. 3, *II***). These results demonstrated that the four positive plants had the *aadA* and *cry1Aa10* cassettes inserted between *rps7* and *ndhB*, which increased the size of the PCR product to 6.4 kb.

3.6.2. Insect Resistance Bioassay of Chloroplast Transformants

Leaf bioassays are conducted against *Plutella xylostera*, a major pest of oilseed rape, on about 6 cm^2 leaf material in 90 × 12 mm plastic Petri dishes. Ten second-instar lavae are assayed per sample (with three replications) and evaluated daily for mortality of the insect and consumption of leaf area, using wild type leaf tissue as control. After 5 d, mortality rates of larvae on different samples are determined and surviving larvae are weighted (**Table 1**).

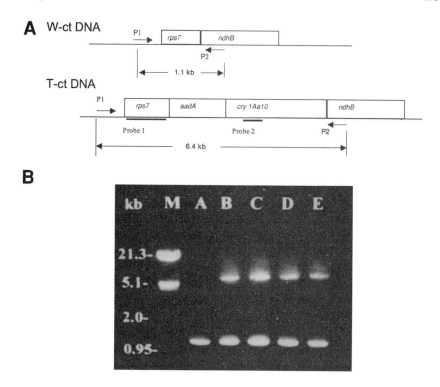

Fig. 2. PCR analysis of transgenic plants. (**A**) Sites of PCR primers in the oilseed rape chloroplast genome. W-ct DNA: wild-type chloroplast DNA; T-ct DNA: transformed chloroplast DNA. P1: primer 1, designed to land upstream of the native rps7 targeting sequence. P2: primer 2, designed to land in the *ndhB* gene. Probe 1 and probe 2 denote the targeting sequence probe and the cry1Aa10 probe, respectively, used in Southern blot analysis. (**B**) The results of PCR amplification by using primers P1 and P2. *M:* molecular weight marker; *A:* wild type oilseed rape; *B–E:* transgenic plants CBT4, CBT9, CBT11, and CBT28, respectively.

The insect-resistant activity of these transgenic plants was tested with second-instar *P. xylostera* larvae. The results showed that leaf material from four transplastomic lines were toxic to test insects, causing 33–47% mortality in 5 d but with slight leaf damage (**Fig. 4**). By contrast, the negative control leaf samples were severely devoured within the first 48 h of the assay, with no mortality in 5 d (**Fig. 4**). The surviving larvae feeding on the transgenic samples had a significantly lower weight than those feeding on the wild type samples (**Table 1**). These results suggested that the *cry1Aa10* gene had been expressed in the transplastomic plants and conferred appreciable insect-resistance *(1)*.

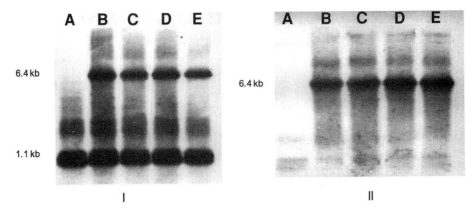

Fig. 3. Southern hybridization analyses. *(I)* Hybridization of PCR products with the *rps7* targeting sequence probe. *(II)* Hybridization of PCR products with the *cry1Aa10* probe. M: molecular weight marker; A: wild type oilseed rape; B–E: transgenic plants CBT4, CBT9, CBT11 and CBT28, respectively.

Fig. 4. Leaf bioassay of control *(left)* and *cry1Aa10* chloroplast transgenic oilseed rape leaves *(right)* assayed against *Plutella xylostera*. Photographs were taken on d 5 of the assay.

4. Notes

1. Here, we bombard the plant tissue with DNA-coating gold microcarriers. Tungsten microcarriers (0.6 µm) are widely used in chloroplast transformation *(46,53)*. The procedure preparing gold microcarriers is the same as tungsten.
2. Plastid transformation vectors most commonly contain a spectinomycin or streptomycin resistance (*aadA*) gene, flanked by plastid DNA sequences targeting

Table 1
The Differences of Insect Resistance Between Transformants and Wild Plants *(44)*

No.	The number of larvae	The number of death larvae	Mortality (%)	Average weight (mg)
CK	30	0	0	8.7
CBT4	30	10	33	5.2
CBT9	30	12	40	6.3
CBT11	30	14	47	5.1
CBT28	30	10	33	7.1

insertion of the marker gene by homologous recombination into the plastid genome. These antibiotics are commonly used to control bacterial infection in humans and animals. There is concern that their over-use might lead to the development of resistant bacteria. In addition, the marker gene protein present may make up as much as 10% of total soluble protein, a significant metabolic burden on the plant *(54)*.

3. Therefore, several studies have explored strategies for engineering chloroplasts that are free of antibiotic-resistance *(35,54,55)*. It is described that green fluorescent protein (gfp) can be used as a reporter gene in chloroplast transformation *(21,22,53)*. The spinach betaine aldehyde dehydrogenase (*badh*) gene has been developed as a plant-derive selectable marker to transform chloroplast genomes *(37)*. The transformation study showed that *badh* selection was 25-fold more efficient than spectinomycin, exhibiting rapid regeneration of transgenic shoots within 2 wk. Antibiotic gene can be excised by direct repeats *(55)*. It is demonstrated that the CRE-lox site-specific system can excises the antibiotic marker gene (flanked by two directly oriented 34 bp lox-site) *(56,57)*. Earlier experiment with *Chlamydomonsa reinhardtii* showed that it is possible to exploit endogenous chloroplast recombinases to eliminate introduced selectable marker genes *(35)*. A more efficient approach is to utilize plastid mutants as source tissue for chloroplast transformation. These mutants are photosynthesis-deficient plastid mutants with an albino phenotype. Using the deleted plastid sequences and foreign gene co-transformed homoplastomic mutant. These transformation/selection systems re-inducing the deleted gene restore the wild-type phenotype. Furthermore, the approaches have the additional advantage that reconstitution restores photosynthesis, which rapidly drives the transformants towards homoplasmy *(58)*.
4. Selection and adjustment of system bombardment parameters: Factors affecting bombardment efficiency are numerous, and interact in complex ways. The rupture pressure determines the power of the shock wave entering the bombardment chamber. Each of the nine different rupture disks is available to rupture at a specific pressure, ranging from 450 to 2200 psi. Increasing helium pressure will increase particle acceleration and subsequent target tissue penetration by the

DNA-coated microcarriers. Since the shock wave or resulting acoustic wave may cause damage to the target cells or tissue, use the lowest helium pressure that still gives high transformation efficiency.
5. One of the most important parameters to optimize is target shelf placement within the bombardment chamber-distance between stopping screen and target shelf (microcarrier flight distance). This placement directly affects the distance that the microcarriers travel to the target cells for microcarrier penetration and transformation. Four target shelf levels are available in the bombardment chamber: level 1 for 3 cm, level 2 for 6 cm, level 3 for 9 cm, and level 4 for 12 cm below the stopping screen. In this research, we chose the level 2 (6 cm).
6. Compare with wild type without antibiotic challenge, the regeneration frequency on selection medium was not reduced, while green shoot regeneration frequency was reduced to about 5% of wild type: 36 putative green transformations were obtained from 1000 cotyledon petioles. The frequency is similar to the frequency obtained in *Lesquerella fendleri*, a wild oilseed species *(24)*.
7. Additional translational-enhancing elements have been engineered from nonplant sequences. Trans-activation expression systems using completely heterologous RNA polymerases-T7 RNA polymerase have been developed for plastids. Phage T7 RNA polymerase has been used extensively in *Escherichia coli* for high-level expression of selected genes laced under the control of the phage T7 gene 10 promoter. The advantage of this approach is that it allows the imposition of developmental, tissue-specific, or chemically inducible regulation upon the expression of plastid transgenes through control of the polymerase by nuclear promoters of the desired specificity *(59–61)*.
8. Until now, there have some bottlenecks in plant chloroplast transformation. (1) It is unclear for the chloroplast sequences in most plants, which include some important crops. So we cannot ascertain the suitable inserted point for the foreign gene. (2) In a great many plants, especially important standing grain and cereal crops, regenerative tissues, including callus and immature embryo and others, have no mature chloroplasts, but have small former plastids. These former plastids are considered smaller than microparticles and difficult to shot. And that it affects the expression of the marker gene for the fewness of the DNA molecule number, which not benefit for the homogeneous filtration. (3) Only bombardment is proved to be efficient in chloroplast transformation of different plant species. PEG method and microinjection were already successful, but still need to be improved. Areas of improvement include, its use in chloroplast genome sequencing in different plants, the possibility of the transformation of white former plastids using bombardment, and other techniques and methods in chloroplast transformation that can be applied more widely in genome research and genetic optimization.

Some often used protocols and solutions in the experiment:
a. Total DNA isolation:
- Warm 2X CTAB buffer to 60°C.
- Put 0.1 g leaf tissue in a mortar, grind gently to a fine powder.
- Add 1.5 mL of warm 2X CTAB buffer and homogenize gently. If the slurry gets very sticky start over with less tissue.

- Transfer homogenate to 2-mL Eppendorf tube, incubate at 65°C for 30 min.
- Add equal volume of phenol, vortex briefly and invert centrifuge for 10 min (20,000g). Collect top layer to new tube.
- Add equal volume of CIA (chloroform-isoamyl alcohol, 24:1), vortex briefly, and invert.
- Centrifuge for 10 min (20,000g). Collect top layer to new tube.
- Add 2.5 mL RNAse (10 mg/mL), shake briefly, and incubate at 37°C for 30 min.
- Add equal volume of CIA. Vortex briefly and invert for 5 min.
- Centrifuge for 10 min (20,000g). Collect top layer to new tube.
- Add 2/3 volume of ice-cold ethanol. Invert several times; incubate for 15 min at 4°C.
- Centrifuge for 10 min (20,000g), with tube hinges facing away from center. Carefully remove supernatant.
- Wash DNA pellet twice with 250 mL of 70 % ethanol. Air-dry DNA pellet.
- Add 50–100 mL TE buffer, shake briefly, and leave on shaker for 30 min or overnight at 4°C to resuspend.
- Run 3 mL DNA together with 1–2 mL 6X loading buffer on 0.8% agarose gel, to estimate DNA concentration.

b. 2X CTAB Buffer:
- 2% CTAB (2 g CTAB).
- 1.4 M NaCl (8.19 g NaCl).
  100 mM Tris-HCl, pH 8.0; 10 mL of 1 M Tris-HCl, pH 8.0.
- 20 mM EDTA (4 mL of 0.5 M EDTA).
- 0.2% 2-Mercaptoethanol (200 mL 2-mercaptoethanol).
- Make to 100 mL with dH$_2$O.

c. TE Buffer: 10 mM Tris, 0.1 mM EDTA, adjust pH 8.0 with HCl.

d. RNAse Stock solution (stored at –20°C):
- Stock-I: 10 mg/mL in dH$_2$O.
- Stock-II: 1% of Stock I (i.e., 10 mL of Stock I, make to 1 mL).

e. Gel preparation and electrophoresis:
- 0.8% Agarose (e.g., 0.64 g agarose in 80 mL TAE).
- Gel buffer, 0.5X TBE.
- Tank buffer, 0.5X TBE.

Acknowledgments

We thank Prof. Gao Xi-Wu for kindly providing the test insect *P. xylostera*, Dr. Pal Maliga for kindly providing *aadA* cassette, and Dr. Richard Wang for critically reviewing the manuscript. This project was supported by the MOST of China, NSFC, and CAS.

References

1. Hou, B. K., Zhou, Y. H., Wan, L. H., et al. (2003) Chloroplast transformation in oilseed rape. *Trans. Res.* **12,** 111–114.

2. Daniell, H., Khan, M. S., and Allison, L. (2002) Milestones in chloroplast genetic engineering: an environmentally friendly era in biotechnology. *Trends Plant Sci.* **7,** 84–91.
3. Ruf, S., Hermann, M., Berger, I. J., Carrer, H., and Bock R. (2001) Stable genetic transformation of tomato plastids and expression of a foreign protein in fruit. *Nat. Biotechnol.* **19,** 870–875.
4. Su, N., Wu, Y. M., Sun, B. Y., and Shen, G. F. (2001) A new way of plant genetic engineering: chloroplast transformation. *Biotechnol. Infor.* **4,** 9–13.
5. Maliga, P. (1993) Towards plastid transformation in flowering plants. *Trends Biotechnol.* **11,** 101–107.
6. Cosa, B. D., Moar, W., Lee, S. B., Miller, M., and Daniell., H. (2001) Overexpression of the Bt cry2Aa2 operon in chloroplasts leads to formation of insecticidal crystals. *Nat. Biotechnol.* **19,** 71–74.
7. Gewolb, J. (2002) Plant scientists see big potential in tiny plastids. *Science* **295,** 258–259.
8. Boynton, J. E., Gillham, N. W., Harris, E. H., et al. (1988) Chloroplast transformation in *Chlamydomonas* with high velocity microprojectiles. *Science* **240,** 1534–1538.
9. Kindle, K. L., Richards, K. L., and Stern, D. B. (1991) Engineering the chloroplast genome: Techniques and capabilities for chloroplast transformation in *Chlamydomonas reinhardtii. Proc. Natl. Acad. (USA)* **88,** 1721–1725.
10. Carrer, H., Hockenberry, T.N., and Maliga, P. (1993) Kanamycin resistance as a selectable marker for plastid transformation in tobacco. *Mol. Gen. Genet.* **241,** 49–56.
11. Staub, J. M. and Maliga, P. (1992) Long regions of homologous DNA are incorporated into the tobacco plastid genome by transformation. *The Plant Cell* **4,** 39–45.
12. Staub, J. M., Garcia, B., Graves, J., et al. (2000) High-yield production of a human therapeutic protein in tobacco chloroplasts. *Nat. Biotechnol.* **18,** 333–338.
13. Svab, Z. and Maliga, P. (1993) High-frequency plastid transformation in tobacco by selection for a chimeric aadA gene. *Proc. Natl. Acad. (USA)* **90,** 913–917.
14. Svab, Z., Hajdukiewicz, P., and Maliga, P. (1990) Stable transformation of plastids in higher plants. *Proc. Natl. Acad. (USA)* **87,** 8526–8530.
15. Ye, G. N., Daniell, H., and Sanford, C. (1990) Optimization of delivery of foreign DNA into higher-plant chloroplasts. *Plant Mol. Biol.* **15,** 809–819.
16. Zhang, Z. L., Chen, X., Qian, K. X., and Shen, G. F. (1999) Studies on insect resistance of Bt transplastomic plants and the phenotype of their progenies. *Acta Bota. Sin.* **41,** 947–951.
17. Zhang, Z. L., Qian, K. X., and Shen, G. F. (2000) Transplastomic plants homoplasmic for foreign transgenes. *Acta Biochim. Biohys. Sin.* **32,** 620–626.
18. Zoubenko, O. V., Allison, L. A., Svab, Z., and Maliga, P. (1994) Efficient targeting of foreign genes into the tobacco plastid genome. *Nucl. Acids Res.* **22,** 3819–3824.
19. Sikdar, S. R., Serino, G., Chaudhuri, S., and Maliga, P. (1998) Plastid transformation in *Arabidopsis thaliana. Plant Cell Rep.* **18,** 20–24.
20. Perlak, F. J., Stone, T. B., Muskopf, Y. M., et al. (1993) Genetically improved potatoes: protection from damage by Colorado potato beetle. *Plant Mol Biol.* **22,** 313–321.

21. Sidorov, V. A., Kasten, D., Pang, S. Z., Hajdukiewicz, P. T., Staub, J. M., and Nehra, N. S., (1999) Stable chloroplast transformation in potato: use of green fluorescent protein as a plastid marker. *Plant J.* **19**, 209–216.
22. Khan, M. S. and Maliga, P. (1999) Fluorescent antibiotic resistance marker for tracking plastid transformation in higher plants. *Nat. Biotechnol.* **17**, 910–915.
23. Maliga, P. (2001) Plastid engineering bears fruit. *Nat. Biotechnol.* **19**, 826–827.
24. Skarjinskaia, M., Svab, Z., and Maliga, P. (2003) Plastid transformation in *Lesquerella fendleri*, an oilseed *Brassicacea*. *Transg. Res.* **12**, 115–122.
25. Xiang, Y., Wong, W. K. R., Ma, M. C., and Wong, R. S. C. (2000) *Agrobacterium*-mediated transformation of *Brassica campestris* ssp. parachinensis with synthetic *Bacillus thuringiensis* cry1Ab and cry1Ac genes. *Plant Cell Rep.* **19**, 251–256.
26. Daniell, H., Datta, R., Varma, S., Gray, S., and Lee, S. B. (1998) Containment of herbicide resistance through genetic engineering of the chloroplast genome. *Nat. Biotechnol.* **16**, 345–348.
27. Hou, B. K., Dang, B. Y., Zhang, Y. M., and Chen, Z. H. (2000) Cloning and sequence analysis of *cry1Aa10* gene from *Bacillus thuringiensis* subsp. kurstaki HD-1-02 and its expression in *E. coli*. *J. Agri. Biotechnol.* **8**, 289–293.
28. Kota, M., Daniell, H., Varma, S., Garczynski, S. F., Gould, F., and Moar, W. J. (1999) Overexpression of the *Bacillus thuringiensis* (Bt) *Cry2Aa2* protein in chloroplasts confers resistance to plants against susceptible and Bt-resistant insects. *Proc. Natl. Acad. USA* **96**, 1840–1845.
29. Lutz, K. A., Knapp, J. E., and Maliga, P. (2001) Expression of bar in the plastid genome confers herbicide resistance. *Plant Physiol.* **125**, 1585–1590.
30. Daniell, H., Lee, S. B., Panchal, T., and Wiebe, O. P. (2001) Expression of the native cholera toxin B subunit gene and assembly of functional oligomers in transgenic tobacco chloroplast. *J. Mol. Biol.* **311**, 1001–1009.
31. Kang, T. J., Loc, N. H., Jang, M. O., et al. (2003) Expression of the B subunit of *E. coli* heat-labile enterotoxin in the chloroplasts of plants and its characterization. *Transg. Res.* **12**, 683–691.
32. Tregoning, J., Nixon, P., Kuroda, H., et al. (2003) Expression of tetanus toxin fragment C in tobacco chloroplasts. *Nucl. Acids Res.* **31**, 1174–1179.
33. Madoka, Y., Tomizawa, K. I., Mizoi, J., Nishida, I., Nagano, Y., and Sasaki, Y. (2002) Chloroplast transformation with modified accD operon increases acetyl-CoA carboxylase and causes extension of leaf longevity and increase in seed yield in tobacco. *Plant Cell Physiol.* **43**, 1518–1525.
34. Bateman, J. M. and Purton, S. (2000) Tools for chloroplast transformation in *Chlamydomonas*: expression vectors and a new dominant selectable marker. *Mol. Gen. Genet.* **263**, 404–410.
35. Hare, P. D. and Chua, N. H. (2002) Excision of selectable marker genes from transgenic plants. *Nat. Biotechnol.* **20**, 575–580.
36. Huang, F. C., Klaus. S. M. J., Herz, S., Zou, Z., Koop, H. U., and Golds, T. J. (2002) Efficient plastid transformation in tobacco using the aphA-6 gene and kanamycin selection. *Mol. Gen. Genet.* **268**, 19–27.

37. Daniell, H., Muthukumar, B., and Lee, S. B., (2001) Marker free transgenic plants: engineering the chloroplast genome without the use of antibiotic selection. *Curr. Genet.* **39**, 109–116.
38. Golds, T., Maliga, P., and Koop, H. U. (1993) Stable plastid transformation in PEG-treated protoplasts of *Nicotiana tabacum*. *Biotechnology* **11**, 95–97.
39. Koop, H. U., Steinmüller, K., Wagner, H., R^ssler, C., Eibl, C., and Sacher, L. (1996) Integration of foreign sequences into the tobacco plastome via PEG-mediated protoplast transformation. *Planta* **199**, 193–201.
40. Maliga, P. (2003) Progress towards commercialization of plastid transformation technology. *Trends Biotechnol.* **21**, 20–28.
41. O'Neill, C., Horváth, G. V., Horváth, É, Dix, P. J., and Medgyesy, P. (1993) Chloroplast transformation in plants: polyethylene glycol (PEG) treatment of protoplast is an alternative to ballistic delivery systems. *Plant J.* **3**, 729–736.
42. Haring, M. A. and De Block, M. (1990) New roads towards chloroplast transformation in higher plants. *Physiol. Plant.* **79**, 218–220.
43. Cheng, X. Y., Sardana, R., Kaplan, H., and Altosaar, I. (1998) *Agrobacterium*-transformed rice plants expressing synthetic cry1Aband cry1Ac genes are highly toxic to striped stem borer and yellow stem borer. *Proc. Natl. Acad. Sci. USA* **95**, 2767–2772.
44. Hou, B. K. (2000) Cloing of insecticidal protein gene from *Bacillus thuringiensis* and construction of chloroplast expression Vector and genetic transformation in *Brassica napus* L. Thesis of PhD. 122–123.
45. Perlak, F. J., Fuchs, R. L., Dean, D. A., McPherson, S. L., and Fischoff, D. A. (1991) Modification of the coding sequence enhances plant expression of insect control proteins. *Proc. Natl. Acad. Sci. USA* **88**, 3324–3328.
46. Daniell, H. (1997) Transformation and foreign gene expression in plants mediated by microprojectile bombardment. *Meth. Mol. Biol.* **62**, 463–489.
47. Sambrook, J., Fritsch, E. F., and Maniatis, T. (1989) *Molecular Cloning: A Laboratory Manual, second edition*, Cold Spring Harbor Laboratory Press, New York.
48. Bio-Rad Introduction of Biolistic PDS-1000/He Particle Delivery System.
49. Hou, B. K., Zhang, Z. L., Zhou, Y. H., Zhang, Y. M., Shen, G. F., and Chen, Z. H. (2000) Construction of vector for oilseed rape chloroplast transformation and its insecticidal toxicity. *High Technol. Lett.* **10**, 5–11.
50. Hou, B. K., Dang, B. Y., Zhou, Y. H., and Chen, Z. H. (2000) Cloning and sequence analysis of the gene rps7 coding for chloroplast ribosomal protein CS7 from *Brassica napus*. *Acta Phyto Physiol. Sin.* **26**, 350–358.
51. Sriraman, P., Silhavy, D., and Maliga, P. (1998) Transcription from heterologous rRNA operon promoters in chloroplasts reveals requirement for specific activating factors. *Plant Physiol.* **117**, 1495–1499.
52. Maliga, P. and Nixon, P. (1998) Judging the homoplastomic state of plastid transformants. *Trends Plant Sci.* **3**, 4–6.
53. Hibberd, J. M., Linley, P. J., Khan, M. S., and Gray, J. C., (1998) Transient expression of green fluorescent protein in various plastid types following microprojectile bombardment. *Plant J.* **16**, 627–632.

54. Maliga, P. (2002) Engineering the plastid genome of higher plants. *Curr. Opin. Plant Biol.* **5,** 164–172.
55. Iamtham, S. and Day, A. (2000) Removal of antibiotic resistance genes from transgenic tobacco plastids. *Nat. Biotechnol.* **18,** 1172–1176.
56. Cormeille, S., Lutz, K., Svab, Z., and Maliga, P. (2001) Efficient elimination of selectable marker genes from the plastid genome by the CRE-lox site-specific recombination system. *Plant J.* **27,** 171–178.
57. Hajdukiewicz, P. T. J., Gilbertson, L., and Staub, J. M. (2001) Multiple pathways for Cre/lox-mediated recombination in plastids. *Plant J.* **27,** 161–170.
58. Klaus, S. M. J., Huang, F. C., Eibl, C., Koop, H. U., and Golds, T. J. (2003) Rapid and proven production of transplastomic tobacco plants by restoration of pigmentation and photosynthesis. *Plant J.* **35,** 811–821.
59. Heifetz, P. B. (2000) Genetic engineering of the chloroplast. *Biochimie.* **82,** 655–666.
60. McBride, K. E., Schaaf, D. J., Daley, M., and Stalker, D. M. (1994) Controlled expression of plastid transgenes in plants based on a nuclear DNA-encoded and plastid-targeted T7 RNA polymerase. *Proc. Natl. Acad. Sci. USA* **91,** 7301–7305.
61. Peter, B. H. and Ann, M. T. (2001) Protein expression in plastids. *Curr. Opin. Plant Biol.* **4,** 157–161.

25

The Biochemical Basis for the Resistance to Aluminum and Their Potential as Selection Markers

César De los Santos-Briones and Teresa Hernández-Sotomayor

Summary

Aluminum (Al) toxicity is one of the most widespread agronomic problems in world agriculture. The cellular mechanisms that some plant species use to tolerate Al are still not understood today. This knowledge is essential in order to develop crop species that can be cultivated on acid soils. Plant cell culture has been used for the investigation of both: Al-toxicity and -tolerance. The method used to obtain Al-tolerant cell line will be shown; as well as a protocol to measure Al-free concentration in the culture medium by using Morin (a fluorescent histochemical indicator for Al). Because Al affects different processes involved in the signal transduction pathway such as phospholipase C (PLC) activity, which could serve as selection marker, the PLC activity determination will be introduced.

Key Words: Aluminum tolerance; aluminum toxicity; cell suspension; Morin; phospholipase C.

1. Introduction

The important economic effects produced by aluminum (Al) toxicity in soils, have encouraged plant breeders to develop Al-tolerant plants. On the last years, the research has been focused on molecular and biochemical studies about damages caused by Al accumulation inside plant cells; a range of alternative toxicity mechanisms have been proposed. Research on plants has indicated that one of the Al toxicity mechanisms could involve interactions between Al and components of the phosphoinositide signal transduction pathway *(1)*. Phospholipase C (PLC) plays an important role in this signal transduction pathway.

Although Al tolerance has been observed in several species, the pathways leading to this tolerance are not well understood. It has been proposed that tolerant plants can detoxify Al within the plant cell, or exclude Al from the root *(2)*. The internal mechanism of tolerance requires that Al be immobilized in the cytoplasm by chelating compounds, polypeptides or by compartmentation in the vacuole. The exclusion mechanism proposes that Al be immobilized in the cell wall by the formation of a rhizosphere pH barrier or by root exudation of Al-chelating compounds such as organic acids or polypeptides *(2)*. The latter mechanism has been demonstrated in tolerant plants where the release of citrate and/or suggests a signaling process mediated by Al malate *(3–5)*. Plant tissues have been examined for their capability to exclude Al by using the Al-indicator dye morin *(6,7)*. Morin is a fluorescent dye with a high sensitivity for Al and has been used as a means to detect the presence of Al in the culture media *(8)*.

Most studies to screen Al tolerance in plants have been carried on whole plants. In vitro cell and tissue cultures may provide adequate systems in which to perform studies on Al tolerance. Only a few studies have focused on screening Al tolerance from suspension cultures of plant cells. However, the major problem in obtaining cellular lines displaying Al tolerance in culture is the composition of the medium. The method described here uses a modification of the culture medium in which Al is soluble and toxic to the cells and was used to obtain the Al-tolerant cell line.

Previously, it has been shown that Al inhibits the specific activity of PLC *(9)*. When tolerant and control cells were incubated in the presence of different concentrations of Al the pattern of inhibition of PLC activity was the same for both lines. However, the basal specific activity of a Al-tolerant cell line was higher than the control line and, even when the cells were incubated with Al, there was still twofold higher specific activity in the Al-tolerant cell line than the control line *(10)*. Therefore, the determination of PLC activity could be used as a potential selection marker in order to obtain Al-tolerant lines.

A review covering various aspects of interaction in plant cells of Al and different effectors of signaling pathways such as phosphoinositide, protein phosphorylation, polyamines, and anion channels have appeared recently *(11)*.

2. Materials
2.1. Initiation of an Aluminum-Tolerant Line
2.1.1. Suspension Cell Culture

1. Murashige and Skoog half ionic strength (MSHIS) basal medium *(12)*. This medium is used as follows:
   Bring volume to 1 L with distilled water, and adjust pH to 4.3 with KOH. Pour 50 mL of medium in Erlenmeyer flask and autoclave for 20 min at 15 psi (*see* **Note 2**).

2.165 g	MS salts (Sigma, cat. no. M-5524; *see* **Note 1**)
30.0 g	Sucros
100 mg	Myo-inositol
10 mg	Thiamine
25 mg	Cysteine
3.0 mL	2,4-dichlorophenoxyacetic acid (2,4-D) from stock
1 mL	6-BAP from stock

2. 2,4-D stock: 1 mg/mL. Dissolve 50 mg in 20 mL 95% ethanol and make up the volume to 50 mL with distilled water. Store at 4°C.
3. 6-benzylamine purine (6-BA) stock: 1 mg/mL. Dissolve 50 mg in distilled water. Add, while stirring, drop-by-drop 1 M KOH. Make up the volume to 50 mL. Store at 4°C.
4. Aluminum chloride ($AlCl_3$; Aldrich, cat. no. 20,691-1) stock: 100 mM. Dissolve 0.2 g in distilled water and make up the volume to 15 mL. Store at 4°C.

2.2. Determination of PLC Activity

2.2.1. Protein Samples

1. Extraction buffer: 50 mM NaCl, 1 mM EGTA, 50 mM Tris-HCl, pH 7.4, 250 mM sucrose, 10% glycerol, 1 mM phenylmethylsulfonyl fluoride, 10 mM sodium pyrophosphate, 0.2 mM orthovanadate, 1 mM β-mercaptoethanol.
2. 74 kBq [^{3}H]-Phosphatidylinositol-4,5-bisphosphate (NEN Life Science Products. NET-895 [*see* **Note 3**]).
3. L-α-Phosphatidyl-D-myo-inositol-4,5-bisphosphate (1 mg) from bovine brain, ammonium salt (Roche Molecular Biochemicals, cat. no. 1 016 598).

2.2.2. Stock Solutions

1. 50 mM NaH_2PO_4, 100 mM KCl, pH 6.8.
2. 2% Deoxicolic acid (w/v).
3. 50 mM NaH_2PO_4, 100 mM KCl, 1 mM EGTA, pH 6.8.
4. 10 mM $CaCl_2$.
5. 1% (w/v) bovine serum albumin (BSA).
6. 10% (w/v) trichloroacetic acid (TCA).
7. Scintillation liquid.
8. General equipment: probe sonicator, water bath, scintillation counter.

2.3. Determination of Free Aluminum in Culture Medium

1. 2', 3', 4', 5', 7'-pentahydroxyflavone (morin). 100 mM stock: dissolve 0.3022 g in 10 mL of dimethylsulfoxide (DMSO) (Baker Analized cat. no. 9224). 1 mM stock: mix 100 µL from 100 mM stock in 9.9 mL of DMSO. Store at –20°C.

2. Aluminum chloride (AlCl$_3$; Aldrich, cat. no. 20,691-1) stock: 100 mM; dissolve 0.2 g in distilled water and make up the volume to 15 mL. Store at 4°C.
3. General equipment: Fluorometer (Bio-Rad VersaFluor™ Fluorometer System).

3. Methods

3.1. Initiation of an Aluminum-Tolerant Line

1. Determine the culture cycle of cell suspension line by measuring both the fresh and dry weight every 2 or 3 d over a period of 30 d. For one flask, transfer 5 mL of a parent culture (approx 0.5 g fresh cells) to 50 mL of MSHIS medium. Seal flask covering the rubber stopper with Nescofilm and incubate in the dark on a rotatory shaker at 28°C. Measure fresh weight by taking the total cellular weight of five flasks per day through filtration of cell suspension with Whatman no. 1 paper and vaccum. Determine the dry weight by incubating the fresh cells at 55°C during 2 d and after measuring its weight. A typical growth curve must show a lag, linear, and stationary phase. The pattern must be the same with both fresh and dry weight (**Fig. 1**).
2. Determine the lethal dose 50 (LD) (*see* **Note 4**) by measuring the reduction in growth within a range of Al concentrations. Prepare MSHIS medium including different concentrations of AlCl$_3$, ranging from 25 to 1000 µM (5 flasks per treatment). Culture the cell suspensions (as in **step 1**) for a period corresponding to a specific day within the linear phase and measure the fresh weight. The growth rate of the cells must be diminished in a dose-responsive fashion in the presence of Al.
3. Initiate tolerance in the cell line by transferring 5 mL of the parent culture (approx 0.5 g fresh cells growing in MSHIS medium) to 50 mL of MSHIS medium containing the LD50 of AlCl$_3$. Culture the cell suspensions for the specific day within the linear phase.
4. When the cells are in the specific day within the linear phase, subculture the cell suspension in MSHIS medium without AlCl$_3$ and grow them during a culture cycle. Incubate in the dark on a rotatory shaker at 28°C.
5. When the cells are in the specific day within the linear phase, subculture the cell suspension in MSHIS medium in the presence of AlCl$_3$ (as in **step 3**) and grow them during a culture cycle. Incubate in the dark on a rotatory shaker at 28°C.
6. The process must be repeated 25 times at least, altering the presence of Al in every other subculture. After several subculture steps in the presence or absence of Al, a tolerant cell line should be obtained. This cell line should be able to maintain its tolerance despite continuous subculture in media without Al.

3.2. Determination of PLC Activity

3.2.1. Obtention of Protein Sample

1. Freeze cells (0.5–1.0 g) quickly with liquid nitrogen and pulverize them in a mortar with a pestle. The cells can be kept frozen during pulverization by the addition of liquid nitrogen.
2. Transfer the powdered cells to a glass vial containing 1.25–2.5 mL (1:2.5; 1 g of cells in 2.5 mL) of extraction buffer.

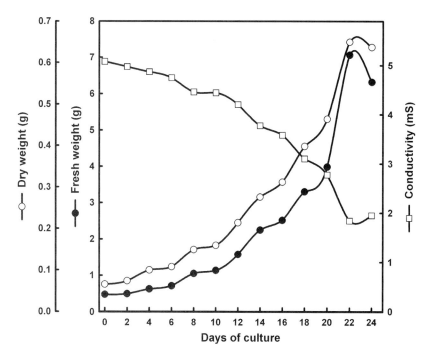

Fig.1. Typical growth curve of a cell suspension line of *Coffea arabica*. The experiment was initiated with 14-d-old cells of a suspension cycle. Cells were maintained in culture for 24 d.

3. Homogenize the cells for 1 min at 4°C with a polytron homogenizer.
4. Filter the extract by using a piece of gauze or cheesecloth.
5. Transfer the homogenate to a 1.5-mL Eppendorf tube and centrifuge at 20,000g in an Eppendorf microcentrifuge for 30 min at 4°C.
6. Recover the supernatant and use it as total protein extract (the concentration of protein is usually 3.5–5.0 mg/mL). Store the protein sample at –70°C.
7. Measure protein concentration following the BCA protein assay (Pierce, cat. no. 23227) relative to BSA as standard.

3.2.2. Preparation of Substrate [^{3}H]PIP$_2$/PIP$_2$ Stock: 2.65 mM

1. Transfer 1 mg of PIP$_2$ ammonium salt to cryogenic vial. Add 74 kBq of [^{3}H]PIP$_2$ (approx 200 µL) and dissolve powder by vortexing. Dry solvent by using a direct nitrogen stream over sample for 1 min or until it is approx 10 µL.
2. Add 180 µL of 50 mM NaH$_2$PO$_4$; 100 mM KCl, pH 6.8. Mix and sonicate 2 min.
3. Add 180 µL of 2% deoxicolic acid (w/v). Mix and sonicate 2 min.
4. Transfer 4 µL to 5 mL of scintillation liquid in a counting vial. Mix and quantitate by liquid scintillation counting. At this point, total counts must be approx 18,000–20,000 cpm (*see* **Note 5**).
5. Store in refrigerator (4°C) until its use.

3.2.3. Preparation of Reaction Mix

1. Add the following components to a 1.5-mL microcentrifuge tube sitting on ice: 22 µL 50 mM NaH$_2$PO$_4$; 100 mM KCl; 1 mM EGTA, pH 6.8; 4 µL 2.65 mM [^{3}H]PIP$_2$, PIP$_2$; 4 µL 10 mM CaCl$_2$. Total volume of mix reaction per sample should be 30 µL.
2. If desired, a Master Mix can be prepared for multiple reactions. Preparation of this Master Mix eliminates the need to repeatedly pipet volumes, resulting in increased consistency between samples. Include in the preparation of Master Mix two more samples to obtain a background of the reaction mix and to calculate the total counts (*see* **Note 6**). Mix contents of tube and sonicate 1 min.
3. If a Master Mix has been prepared, subsequently make aliquots of 30 µL into 1.5 mL microcentrifuge tubes just before to initiate reaction. Each one of these aliquots corresponds to one reaction mix for one sample.

3.2.4. Measurement of PLC Activity

1. Initiate reaction by adding 20 µL of sample protein (*see* **Note 7**). If desired, small volumes of sample protein can be used but is necessary to add distilled H$_2$0 to make up 20 µL of total volume of sample. Total volume of the reaction must be 50 µL (30 µL of reaction mix + 20 µL of sample protein). Briefly vortex the mixture (*see* **Note 8**).
2. Incubate tubes in a water bath at 30°C for 10 min (*see* **Note 9**).
3. Stop reaction by adding 100 µL of 1% BSA and 250 µL of 10% TCA. Mix the contents by vortexing.
4. Centrifuge tubes in an Eppendorf microcentrifuge at 20,000g for 10 min,
5. Take all-supernatant avoiding remains of pellet and transfer it to 5 mL of scintillation liquid in a counting vial. Mix and quantitate by liquid scintillation counting for 5 min.
6. To obtain the total counts of 30 µL of mix reaction transfer it to 5 mL of scintillation liquid in a counting vial. Mix and quantitate by liquid scintillation counting for 5 min.
7. Use the following formulas to calculate the specific activity in pmol/min/mg^{-1} of protein:

$$\text{pmol per count} = \frac{10{,}000 \text{ pmol}}{\text{Total cpm in 30 µL of reaction mix}}$$

$$\text{pmol/min sample} = \frac{(\text{cpm sample} - \text{cpm background}) \times \text{pmol per count}}{10}$$

$$\text{pmol/min mg/protein} = \frac{\left(\text{pmol/min}^{-1}\right) \text{ of sample}}{\text{mg of protein in 20 µL of sample}}$$

3.3. Determination of Aluminum Concentration in Culture Medium

Obtain a standard curve by using increased $AlCl_3$ concentrations ranging from 2 to 100 µM. Add the volume of 1 mM morin stock indicated to 4 mL of media in which the cell suspension is cultured (*see* **Note 10**):

Final Morin concentration (µM)	µL of 1 mM Morin stock
2	8
5	20
10	40
20	80
40	160
60	240
80	320
100	400

Measure the flourescence by using a fluorometer. Fit the Bio-Rad VersaFluor® Fluorometer System with a 390–410 nm excitation filter and a 515–525 nm emission filter so that the reflective face of the filters faced towards the incident beam. Use the fluorometer according to the manufacturer's instructions. Take aliquots of 5 mL of media with different concentrations of $AlCl_3$, whose free aluminum concentration is required to be known. Add Morin from 100 mM stock at 0.1 mM final concentration (5 µL in 5 mL of media).

Incubate for 1 h at room temperature and measure fluorescence.

Use the standard curve to determine the concentration of free aluminum in the culture media of each unknown sample.

4. Notes

1. This medium can be purchased in powdered form to which sucrose, vitamins and growth regulators are added.
2. Use 250-mL Erlenmeyer flask with a rubber stopper (which has a hole in center and covered with a cotton ball).
3. The [^{3}H]PIP$_2$ is shipped in a sealed glass ampoule dissolved in CH_2Cl_2: $EtOH:H_2O$ (20:10:1) and under inert gas. Open the ampoule (snap off the neck of the ampoule) and make aliquots of 200 µL in cryogenic vials. Once the ampoule has been opened, the radiochemical is no longer being stored under optimum conditions and the rate of decomposition may increase. If possible, the air in the cryogenic vials should be replaced by nitrogen gas before resealing. Store the cryogenic vials at –70°C to prevent loss of the solvent by evaporation.

4. The LD can be determined by calculating the Al concentration at which there is a 50% of reduction in the cell growth compared with control.
5. If there are a less number of counts, sonicate one more minute. Alternatively, add 30–50 µL of [^{3}H]PIP$_2$, mix, sonicate 2 min and verify the number of counts by liquid scintillation counting.
6. An example will be used to illustrate the preparation of a Master Mix for eight protein samples including one sample for background and one sample to calculate the total counts in the reaction mix; subsequently take 30 µL for each sample:

Components	Volume per 1 sample	Volume per 10 samples
50 mM NaH$_2$PO$_4$, 100 mM KCl, 1 mM EGTA, pH 6.8	22 µL	220 µL
2.65 mM [^{3}H]PIP$_2$, PIP$_2$	4 µL	40 µL
10 mM CaCl$_2$	4 µL	40 µL
Total volume of Master Mix	30 µL	300 µL

7. The concentration of protein sample can vary. In order to carry out an enzyme assay, it is first necessary to demonstrate that the response of the assay is linearly proportional to protein concentration. Assay a concentration range (1, 5, 10, 15, and 20 µL of protein sample) to identify the range in which this proportionality holds true and choose a concentration inside this range.
8. For multiple reactions use the following table to initiate and stop the reactions. Use a stopwatch to start time measurement:

Sample no.	Initiate reaction at	Stop reaction at
Sample 1	0 min 30 s	10 min 30s
Sample 2	1 min 0 s	11 min 0 s
Sample 3	1 min 30 s	11 min 30 s
Sample 4	2 min 0 s	12 min 0 s
Sample 5	2 min 30 s	12 min 30 s
Sample 6	3 min 0 s	13 min 0 s
Sample 7	3 min 30 s	13 min 30 s
Sample 8	4 min 0 s	14 min 0 s

Continue initiating and stopping reactions every 30 s.

9. Another key consideration in an enzyme assay is to identify reaction conditions that allow the full working linear range to be used during the standard assay. In this case is necessary to know, as protein concentration, the range of temperature and time at which the reaction holds a linear range. Perform the assay at different time intervals (2, 4, 6, 8, and 10 min) and temperature (26, 30, and 37°C).
10. The Morin compound fluoresces only when it forms a complex with free aluminum *(13)*. The media must turn yellow when Morin forms a complex with free

aluminum. Take account of pH is a key factor with respect to the availability of free aluminum.

Acknowledgments

We thank Dr. Manuel Martínez-Estévez, José Armando Muñoz-Sánchez, and Angela Ku-González for their experimental contributions that allow us to obtain the present protocol. Work in Hernández-Sotomayor's laboratory was funded by a grant 33646-N from Consejo Nacional de Ciencia y Tecnología (CONACYT).

References

1. Jones, D. L. and Kochian, L. V. (1997) Aluminum interaction with plasma membrane lipids and enzyme metal binding sites and its potential role in Al cytotoxicity. *FEBS Lett.* **400,** 51–57.
2. Kochian, L. V. (1995) Cellular mechanisms of aluminum toxicity and resistance in plants. *Ann. Rev. Plant Physiol. Plant Mol. Biol.* **46,** 237–260.
3. Jorge, R. A. and Arruda, P. (1997) Aluminum-induced organic acids exudation by roots of an aluminum-tolerant tropical maize. *Phytochemistry* **45,** 675–681.
4. Pellet, D. M., Grunes, D. L., and Kochian, L. V. (1995) Organic acid exudation as an aluminum-tolerance mechanism in maize (*Zea mays* L.). *Planta* **196,** 788–795.
5. Delhaize, E., Ryan, P. R., and Randall, P. J. (1993) Aluminum tolerance in wheat (*Triticum aestivum* L.). II. Aluminum-stimulated excretion of malic acid from root apices. *Plant Physiol.* **103,** 695–702.
6. Larsen, P. B., Tai, C. Y., Kochian, L. V., and Howell S. H. (1996) Arabidopsis mutants with increased sensitivity to aluminum. *Plant Physiol.* **110,** 743–751.
7. Larsen, P. B., Degenhardt, J., Tai, C. Y., Stenzler, L. M., Howell, S. H., and Kochian, L. V. (1998) Aluminum-resistant Arabidopsis mutants that exhibit altered patterns of aluminum accumulation and organic acid release from roots. *Plant Physiol.* **117,** 9–18.
8. Martínez-Estévez, M., Muñoz-Sánchez, J. A., Loyola-Vargas, V. M., and Hernández-Sotomayor, S. M. T. (2001) Modification of the culture medium to produce aluminum toxicity in cell suspensions of coffe (*Coffea arabica* L.). *Plant Cell Rep.* **20,** 469–474.
9. Martínez-Estévez, M., Racagni-Di Palma, G., Muñoz-Sánchez, J. A., Brito-Argáez, L., Loyola-Vargas, V. M., and Hernández-Sotomayor, S. M. T. (2003) Aluminium differentially modifies lipid metabolism from the phosphoinositide pathway in *Coffea arabica* cells. *J. Plant Physiol.* **160,** 1297–1303.
10. Martínez-Estévez, M., Ku-González, A., Muñoz-Sánchez, J. A., et al. (2003) Changes in some characteristics between the wild and Al-tolerant coffee (*Coffea arabica* L.) cell line. *J. Inorg. Biochem.* **97,** 69–78.
11. Martínez-Estévez, M., Echevarría-Machado, I., and Hernández-Sotomayor, S. M. T. (2003) Aluminum toxicity and plant signal transduction. *Recent. Res. Devel. Plant Biol.* **3,** 15–29.

12. Murashige, T. and Skoog, F. (1962) A revised medium for rapid growth and bioassays with tobacco tissue cultures. *Physiol. Plant.* **15,** 473–497.
13. Browne, B. A., McColl, J. G., and Driscoll, C. T. (1990) Aluminum speciation using Morin: Morin and its complexes with aluminum. *J. Environ. Qual.* **19,** 65–72.

26

Transformation of Maize Via *Agrobacterium tumefaciens* Using a Binary Co-Integrate Vector System

Zuo-yu Zhao and Jerry Ranch

Summary

This chapter describes a stepwise protocol to achieve success in genetic transformation of maize using *Agrobacterium tumefaciens* as a DNA delivery system. Researchers will be able to effectively transform immature embryos of Hi-II and related genotypes with this protocol. The outcome of the transformation process will be transgenic embryogenic callus tissue, transgenic plants, and transgenic progeny seeds. Recommendations for molecular confirmation and evaluation of transgenic tissue/plants are also provided.

Key Words: *Agrobacterium tumefaciens*; binary vector system; genetic transformation; maize; monocot; super-binary vector; transgenic corn; transgenic maize; transgenic plants.

1. Introduction

The rapid deployment of plant biotechnology has resulted in the development of genetic transformation systems for economically important crops. Genetic transformation of crops provides a powerful avenue for germoplasm improvement and production of superior commercial products as well as affording a tool for basic and applied research in plant sciences.

Agrobacterium, a natural plant pathogen, has been genetically modified and widely adopted as a DNA delivery system for genetic transformation in dicotyledonous and monocotyledonous plants. Maize, the third most planted crop in the world, can be transformed with *Agrobacterium* (1–4). Highly efficient transformation of maize was achieved with *Agrobacterium* strains carrying a "Super-binary" vector (1–3). Recently an *Agrobacterium* standard binary vector system was also successfully used to transform maize (4). In general, *Agrobacterium*-mediated plant transformation offers the advantage that T-DNA integration into the genome often occurs in single or low copy number with rearrangements being relatively rare *(1,5–7)*.

Various factors have been shown to affect plant transformation efficiency when *Agrobacterium* is used for plant transformation. Those factors include explant type and age, *Agrobacterium* strains and types of binary vectors, culture media and conditions in each of transformation steps, and so on *(8–16)*. In this chapter, we describe the most efficient method of *Agrobacterium*-mediated maize transformation we have seen so far and we also provide some alternatives in major areas or critical steps of the protocol.

High-II (High Type II) maize is derived from A188 and B73 and it produces friable, rapidly growing, and embryogenic Type II callus (embryogenic tissue) *(17)*. High-II has been widely used in biolistic and *Agrobacterium* mediated transformation of maize in both academic *(4,18,19)* and industrial labs *(3,20,21)*.

2. Materials

2.1. Plant Material

2.1.1. Size of Immature Embryos

Immature embryos of Hi-II are the target explants for transformation. Source ears can be derived from maize plants grown under greenhouse, growth chamber, or field conditions. Immature ears are harvested 9–12 d post-pollination depending on growing conditions. The size of the immature embryos used in the transformation ranges from 1.0 to 1.5 mm in length.

2.1.2. Surface Disinfections of Immature Ears

Immature ears are immersed in 30–50% commercial bleach mixed with 1% Tween-20 for 10–30 min under house vacuum. Subsequently, ears are rinsed with sterile water three times and immature embryos are isolated for *Agrobacterium* infection.

2.2. Medium

2.2.1. Media for Agrobacterium Preparation

1. Minimal AB: 50 mL/L stock A, 50 mL/L stock B, 5 g/L glucose, 9 g/L Phytagar, antibiotic at appropriate concentration
    a. Stock A: 60 g/L K_2HPO_4, 20 g/L NaH_2PO_4, pH 7.0.
    b. Stock B: 20 g/L NH_4Cl, 6 g/L $MgSO_4·7H_2O$, 3 g/L KCl, 0.2 g/L $CaCl_2$, 0.5 g/L $FeSO_4·H_2O$.
2. YP medium: 5 g/L yeast extract, 10 g/L peptone, 5 g/L NaCl, 15 g/L Bacto-agar, antibiotic at appropriate concentration *(1)*.

Transformation of Maize 317

2.2.2. Media for Plant Transformation

2.2.2.1. STOCK SOLUTIONS

1. Acetosyringone stock: to make 0.1 M Acetosyringone stock, dissolve 490 mg 3',5'-dimethoxy-4'-hydroxyacetophenone (Aldrich, cat. no. D13,440-6) in 25 mL dimethyl sulfoxide (DMSO), filter sterilize, and freeze in 1 mL aliquots.
2. Bialaphos stock: Herbiace herbicide obtained from Meiji Seika K. K., Japan; containing 20% active ingredient, bialaphos.
   a. Mix 20 mL Herbiace with 80 mL DI water.
   b. Prepare BAKERBOND spe column (VWR JT7020-13): add 1.5–2 mL absolute MeOH to each of 12 columns held in a column processor and collect samples with Falcon 15 mL tubes (VWR21008-935), verify that columns are empty of methanol (no drips); then flush each column with 2.5 mL DI water, remove Falcon tubes, and replace with fresh tubes.
   c. Add 2 mL of Herbiace dilution to each of the conditioned columns; do not apply vacuum.
   d. When the green front of the herbicide reaches the fritted disk turn stopcock off. Combine eluates from columns; the bialaphos fraction in the tubes should be straw colored.
   e. Determine bialaphos concentration: sample 5 µL of eluate and dilute with 1995 µL DI water (1:400 dilution), measure optical density (OD) at 205 and 280 nm and compute bialaphos concentration with formula $E = 27 + 120(OD280/OD205)$ in mg/mL; multiply by 400 for the original concentration.
   f. Store bialaphos frozen.
   g. Dilute bialaphos to 1 mg/mL for use in media and store in refrigerator no longer than 2 mo.

2.2.2.2. MEDIA (see **NOTE 1**)

1. PHI-A: 4 g/L CHU basal salts, 1.0 mL/L 1000X Eriksson's vitamin mix, 0.5 mg/L thiamine HCl, 1.5 mg/L 2,4-D, 0.69 g/L L-proline, 68.5 g/L sucrose, 36 g/L glucose, pH 5.2. Add 100 µM acetosyringone, filter-sterilized before using.
2. PHI-B: PHI-A without glucose, increased 2,4-D to 2 mg/L, reduced sucrose to 30 g/L and supplemented with 0.85 mg/L silver nitrate (filter-sterilized), 3.0 g/L gelrite, 100 µM acetosyringone (filter-sterilized), pH 5.8.
3. PHI-C: PHI-B without gelrite and acetosyringone, reduced 2,4-D to 1.5 mg/L and supplemented with 8.0 g/L agar, 0.5 g/L Ms-morpholino ethane sulfonic acid (MES) buffer, 100 mg/L carbenicillin (filter-sterilized).
4. PHI-D: PHI-C supplemented with 3 mg/L bialaphos (filter-sterilized).
5. PHI-E: 4.3 g/L of Murashige and Skoog (MS) salts (Gibco, BRL 11117-074), 0.5 mg/L nicotinic acid, 0.1 mg/L thiamine HCl, 0.5 mg/L pyridoxine HCl, 2.0 mg/L glycine, 0.1 g/L myo-inositol, 0.5 mg/L zeatin (Sigma, cat no. Z-0164), 1 mg/L indole acetic acid (IAA), 26.4 µg/L abscisic acid (ABA), 60 g/L

sucrose, 3 mg/L bialaphos (filter-sterilized), 100 mg/L carbenicillin (filter-sterilized), 8 g/L agar, pH 5.6.
6. PHI-F: PHI-E without zeatin, IAA, ABA; sucrose reduced to 40 g/L; replacing agar with 1.5 g/L gelrite; pH 5.6.

2.3. Agrobacterium *Strain and Vector*

The engineered *Agrobacterium tumefaciens*, in strain LBA4404 or EH101/EH105 yields high frequency transformation for Hi-II maize. A binary vector should contain a selectable marker gene, such as bar *(22)* or Pat *(23)* and your favorite gene(s) within the two borders of the T-DNA. Transformation frequency of about 40% can be routinely obtained with such an engineered vector. Other marker genes, such as intron-green fluorescent protein (GFP) *(24)* or intron-GUS *(25–27)*, may be required for optimal event recovery. Other types of vectors have also been used in maize transformation and evaluation of T-DNA delivery efficiency (*see* **Notes 2** and **3**).

3. Methods
3.1. Preparation of Agrobacterium
3.1.1. Preparation of Agrobacterium *Master Plate*

A. tumefaciens strains are stored as glycerol stocks at –70°C. Using standard microbiological technique, streak a loop-full of bacteria to produce single colonies on minimal AB medium in a 100× 15 Petri dish and incubate the plate, inverted, at 28°C in the dark for 2–3 d. Bacteria on a master plate are usable for up to 4 wk if the plates are sealed with Parafilm and stored in the cold (4°C).

3.1.2. Preparation of a Working Plate

1. From a master plate, pick one to three colonies to streak a fresh plate of YP medium. Incubate bacterial plate inverted, at 28°C, in darkness, for 1–3 d.

*3.1.3. Preparation of Bacteria for Embryo Infection (see **Note 4**)*

1. PHI-A infection medium is warmed to room temperature.
2. Using sterile bacteria loop scrape bacteria off working plate and place in PHI-A medium with 100 μ*M* acetosyringone.
3. Vigorously shake and/or vortex until clumps are broken up to form a uniform suspension as determined by visual inspection.
4. Use 1 mL of agro-suspension to determine optical density at 550 nm. If OD is over 1.0, dilute with PHI-A until OD is between 0.35 and 1.0. When the OD is between 0.35 and 1.0, the *Agrobacterium* suspension is diluted to OD 0.35 with PHI-A.

3.2. Immature Embryo Preparation

1. Aseptically dissect embryos from caryopses and place in a 2 mL microtube containing 2 mL PHI-A following standard methods for maize.

3.3. Agrobacterium *Infection and Co-Cultivation of Embryos*

3.3.1. Infection Step

1. Remove PHI-A with 1 mL micropipettor and add 1 mL *Agrobacterium* suspension. Gently invert tube to mix.
2. Incubate 5 min at room temperature.

3.3.2. Co-Culture Step (see **Note 5**)

1. Remove *Agrobacterium* suspension from infection step with 1 mL micropipettor.
2. Scrape the embryos from the tube using a sterile spatula.
3. Transfer immature embryos to plate of PHI-B medium in a 100 × 15 mm Petri dish. Orient the embryos with embryonic axis down on the surface of the medium.
4. Culture plates with embryos at 20°C, in darkness, for 3 d.
5. Transfer embryos to PHI-C with the same orientation and incubate at 28°C for 3 d.

3.4. Selection of Putative Transgenic Events

1. Transfer 10 embryos to each plate of PHI-D selection medium in a 100 × 15-mm Petri dish, maintaining orientation. Seal dishes with Parafilm.
2. Incubate plates in darkness at 28°C. Actively growing putative events, as pale yellow embryogenic tissue, should be visible in 6–8 wk. Embryos that produce no events may be brown and necrotic, and little friable tissue growth is evident.
3. Subculture putative transgenic embryogenic tissue to fresh plates of PHI-D at 2–3 wk intervals, depending on growth rate. Record events.

3.5. Regeneration of T_0 Plants

1. From embryogenic tissue propagated on PHI-D, subculture tissue to somatic embryo maturation medium, PHI-E, in 100 × 25 mm Petri dishes.
2. Incubate plates at 28°C, in darkness, until somatic embryos mature, for about 10–18 d.
3. Individual, matured somatic embryos with well-defined scutellum and coleoptile are transferred to PHI-F embryo germination medium and incubated at 28°C in the light (about 80 µE from cool white or equivalent fluorescent lamps).
4. After shoots and roots emerge, about 7–10 d, individual plants are transferred to PHI-F medium in 150 × 25 mm glass tubes covered with closures, and incubated at 28°C in the light (about 80 µE from cool white or equivalent fluorescent lamps).
5. In 7–10 d, regenerated plants, about 10 cm tall, are potted in horticultural mix and hardened-off using standard horticultural methods.

3.6. Confirmation of Transformation

Putative transgenic events should be subjected to analyses to confirm their transgenic nature. The choice of specific analytical test performed on any transgenic is dependent on the transgene. In general, almost all events are tested for the presence of the gene of interest by polymerase chain reaction (PCR). Transgene product can also be assayed. For example, in those events produced with the GUS gene, tissues are stained with GUS histochemical assay reagent. Additionally, T_0 plants can also be painted with bialaphos herbicide (1% v/v Liberty). The subsequent lack of herbicide-injury lesion indicates the presence and action of the BAR/PAT transgene product, which conditions for herbicide resistance. Usually, Southern blotting is used to determine copy number, insertion pattern, rearrangement, and integration vector backbone DNA into maize genome *(7)*.

4. Notes

1. MS medium can be used in infection, co-cultivation and selection steps of *Agrobacterium*-mediated maize transformation *(1)*. In a paired comparison of MS and N6 medium for Hi-II transformation, transgenic maize plants were recovered with both media. However, our data suggested that N6 medium was superior to MS medium for Hi-II when carbenicillin was used as the counter-selective agent *(3)*.
2. Hi-II immature embryos can be transformed with the standard *Agrobacterium* binary vector *(4)*. Compared to the super-binary vector (about 40% of the immature embryos produce events) *(3)*, transformation efficiency of the standard vector was much lower (5.5%) *(4)*.
3. A co-transformation vector *(29)* can be used in *Agrobacterium*-mediated maize transformation *(30)* for independent segregation of selectable marker gene and agronomic trait gene (your favorite gene). In progeny of co-transformed events, transgenic plants containing only (your favorite gene) agronomic trait gene(s) can be recovered.
4. *Agrobacterium* suspension for infection can also be prepared in liquid shake culture. One day prior to transformation, about 30 mL of minimal AB medium in a 30 mL baffle flask containing 50 µg/mL spectinomycin is inoculated with a 1/8 loop-full of *Agrobacterium* from a 1 to 2 d old working plate. The *Agrobacterium* is grown at 28°C at 200 rpm in darkness overnight (about 14 h). In mid-log phase, the *Agrobacterium* cells are harvested and re-suspended at 3 to 5×10^8 CFU/mL in PHI-A medium + 100 µM acetosyringone using standard microbial techniques and standard curves.
5. L-Cysteine can be used in the co-cultivation phase *(4,28)*. With the standard binary vector, co-cultivation medium supplied with 100–400 mg/L L-cys is critical for recovering stable transgenic events *(4)*.

Acknowledgments

The authors thank Dr. Deping Xu and Mr. Gary Sandahl for critical review of this manuscript.

References

1. Ishida, Y., Saito, H., Ohta, S., Hiei, Y., Komari, T., and Kumashiro, T. (1996) High efficiency transformation of maize (*Zea mays* L.) mediated by *Agrobacterium tumefaciens*. *Nat. Biotechnol.* **14,** 745–750.
2. Negrotto, D., Jolley, M., Beer, S., and Wenck, A.R. (2000) The use of phosphomannose-isomerase as a selectable marker to recover transgenic maize plants (*Zea mays* L.) via *Agrobacterium transformation*. *Plant Cell Rep.* **19,** 798–803.
3. Zhao, Z. Y., Gu, W., Cai, T., et al. (2001) High throughput genetic transformation mediated by *Agrobacterium tumefaciens* in maize. *Molec. Breeding* **8,** 323–333.
4. Frame, B. R., Shou, H., Chikwamba, R. K., et al. (2002) *Agrobacterium tumefaciens*-mediated transformation of maize embryos using a standard binary vector system. *Plant Physiol.* **129,** 13–22.
5. Hiei, Y., Ohta, S., Komari, T., and Kumasho, T. (1994) Efficient transformation of rice (*Oryza sativa* L.) mediated by *Agrobacterium* and sequence analysis of the boundaries of the T-DNA. *Plant J.* **6,** 271–282.
6. Cheng, M., Fry, J. E., Pang, S., et al. (1997) Genetic transformation of wheat mediated by *Agrobacterium tumefaciens*. *Plant Physiol.* **115,** 971–980.
7. Zhao, Z. Y., Gu, W., Cai, T., et al. (1998) Molecular analysis of T0 plants transformed by Agrobacterium and comparison of *Agrobacterium*-mediated transformation with bombardment transformation in maize. *Maize Genet. Coop. Newsl.* **72,** 34–37.
8. Boase, M. R., Bradley, J. M., and Borst, N. K. (1998) An improved method for transformation of regal pelargonium (Pelargonium Xdomesticum Dubonnet) by Agrobacterium. *Plant Sci.* **139,** 59–69.
9. Cao, X., Liu, Q., Rowland, L. J., and Hammerschlag, F. A. (1998) GUS expression in blueberry (*Vaccinium* ssp.): factors influencing *Agrobacterium*-mediated gene transfer efficiency. *Plant Cell Rep.* **18,** 266–270.
10. Cervera, M., Pina, J. A., Juarez, J., Navarro, L., and Pena, L. (1998) *Agrobacterium*-mediated transformation of citrange: factors affecting transformation and regeneration. *Plant Cell Rep.* **18,** 271–278.
11. Guo, G., Maiwald, F., Lorenzen, P., and Steinbiss, H. (1998) Factors influencing T-DNA transfer into wheat and barley cells by *Agrobacterium tumefaciens*. *Cereal Res. Commun.* **26,** 15–22.
12. Sangwan, R. S., Bourgeois, Y., and Sangwan-Norreel, B. S. (1991) Genetic transformation of Arabidopsis thaliana zygotic embryos and identification of critical parameters influencing transformation efficiency. *Mol. Gen. Genet.* **230,** 475–485.

13. Vergauwe, A., Geldre, E., Van Inze, D., Montagu, M., and Van den Eeckhout, E. (1998) Factors influencing *Agrobacterium tumefaciens* -mediated transformation of *Artemisia annua* L. *Plant Cell Rep.* **18,** 105–110.
14. Nadolska-Orczyk, A. and Orczyk, W. (2000) Study of the factors influencing Agrobacterium-mediated transformation of pea (*Pisum sativum* L.). *Molec. Breeding* **6,** 185–194.
15. Uze, M., Potrykus, I., and Sautter, C. (2000) Factors influencing T-DNA transfer from Agrobacterium to precultured immature wheat embryos (*Triticum aestivum* L.) *Cereal Res. Commun.* **28,** 17–23.
16. Cheng, M., Lowe, B. A., Spencer, M., Ye, X., and Armstrong, C. L. (2004) Invited review: factors influencing *Agrobacterium*-mediated transformation of monocotyledonous species. *In Vitro Cell. Dev. Biol. Plant* **40,** 31–45.
17. Armstrong, C. L., Green, C. E., and Phillips, R. L. (1991) Development and availability of germplasm with high type II culture formation response. *Maize Genetics Cooperation Newsletter* **65,** 92–93.
18. O'Kennedy, M. M., Burger, J. T., and Watson, T. G. (1998) Stable transformation of Hi-II maize using the particle inflow gun. *S. Afr. J. Sci.* **94,** 188–192.
19. Southgate, E. M., Davey, M. R., Power, J. B., and Westcott, R. J. (1998) A comparison of methods for direct gene transfer into maize (*Zea mays* L.). *In Vitro Cell Dev. Biol. Plant* **34,** 218–224.
20. Songstad, D. D., Armstrong, C. L., Peterson, W. L., Hairston, B., and Hinchee, M. A. W. (1996) Production of transgenic maize plants and progeny by bombardment of Hi-II immature embryos. *In Vitro Cell Dev. Biol. Plant* **32,** 179–183.
21. Zhong, G. Y., Peterson, D., Delaney, D. E., et al. (1999) Commercial production of aprotinin in transgenic maize seeds. *Molec. Breeding* **5,** 345–356.
22. Thompson, C., Movva, N. R., Tizard, R., et al. (1987) Characterization of the herbicide-resistance gene bar from streptomyces hygroscopicus. *EMBO J.* **6,** 2519–2523.
23. Wohlleben, W., Arnold W., Broer, I., Hillemann, D., Strauch, E., and Punier, A. (1988) Nucleotide sequence of the phosphinothricin N-acetyltransferase gene from Streptomyces viridochromogenes Tu494 and its expression in *Nicotiana tabacum. Gene* **70,** 25–37.
24. Chalfie, M., Tu, Y., Euskirchen, G., Ward, W. W., and Prasher, D. C. (1994) Green fluorescent protein as a marker for gene expression. *Science* **263,** 802–805.
25. Jefferson, R. A., Burgess, S. M., and Hirsh, D. (1986) β-Glucuronidase from *Escherichia coli* as a gene-fusion marker. *Proc. Natl. Acad. Sci. USA* **83,** 8447–8451.
26. Vancanneyt, G., Schmidt, R., O'Connor-Sanchez, A., Willmitzer, L., and Rocha-Sosa, M. (1990) Construction of an intron-containing marker gene: splicing of the intron in transgenic plants and its use in monitoring early events in *Agrobacterium*-mediated plant transformation. *Mol. Gen. Genet.* **220,** 245–250.
27. Ohta, S., Mita, S., Hattori, T., and Nakamura, K. (1990) Construction and expression in tobacco of a β-glucuronidase (GUS) reporter gene containing an intron within the coding sequence. *Plant Cell Physiol.* **31,** 805–813.

28. Olhoft, P. M. and Somers, D. A. (2001) L-Cysteine increases Agrobacterium-mediated T-DNA delivery into soybean cotyledonary-node cells. *Plant Cell Rep.* **20,** 706–711.
29. Komari, T., Hiei, Y., Saito, Y., Murai N., and Kumashiro T. (1996) Vectors carrying two separate T-DNAs for co-transformation of higher plants mediated by *Agrobacterium tumefaciens* and segregation of transformants free from selection markers. *Plant J.* **10,** 165–174.
30. Miller, M., Tagliani, L., Wang, N., Berka, B., Bidney, D., and Zhao, Z. Y. (2002) High efficiency transgene segregation in co-transformed maize plants using an *Agrobacterium tumefaciens* 2 T-DNA binary system. *Transgenic Res.* **11,** 381–396.

VI

ACCUMULATION OF METABOLITES IN PLANT CELLS

27

Capsaicin Accumulation in *Capsicum* spp. Suspension Cultures

Neftalí Ochoa-Alejo

Summary

Fruits of chili peppers (*Capsicum* spp.) specifically synthesize and accumulate a group of analogs known as capsaicinoids in the placenta tissues. These secondary metabolites are responsible for the hot taste of chili pepper fruits. Capsaicinoids are of economic importance because of their use in the food, cosmetic, military, and pharmaceutical industry. Several efforts have been focused to investigate the biosynthetic capacity of in vitro chili pepper cells and tissue cultures in order to determine the production feasibility of these compounds at the industrial level under controlled conditions. A description of techniques for the establishment of in vitro cultures of chili pepper, the addition of precursors and intermediates to the culture medium, and the selection of cell lines as a means to increase the production of capsaicinoids as well as the extraction, separation, and quantification of capsaicinoids from chili pepper cell cultures is reported in this chapter.

Key Words: Callus cultures; *Capsicum* spp.; capsaicinoids; cell cultures; chili pepper.

1. Introduction

Chili pepper fruits produce a group of compounds known as capsaicinoids that are responsible for their characteristic hot taste. Capsaicin is the major analog followed by dihydrocapsaicin, nordihydrocapsaicin, homodihydrocapsaicin, and homocapsaicin (**Fig. 1**); among them, the two former compounds account for approx 90% of total capsaicinoids content in the fruits (*1*). Capsaicinoids are important items in the food industry (preparation of hot sauces) as well as in the pharmacological (elaboration of pads for relieving muscular pain, rubefascient creams, products for alleviating the pain caused by arthritis, post-herpetic neuralgia, diabetic neuropathy, and nonallergic rhinitis), cosmetic (shampoos to prevent hair loss), and military industry (sprays repellents against dogs and thieves).

From: *Methods in Molecular Biology, vol. 318: Plant Cell Culture Protocols, Second Edition*
Edited by: V. M. Loyola-Vargas and F. Vázquez-Flota © Humana Press Inc., Totowa, NJ

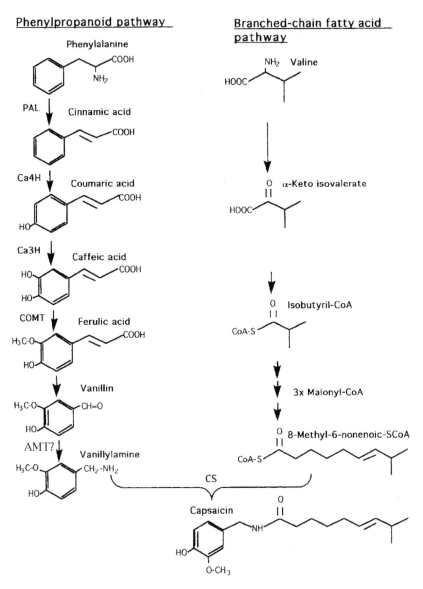

Fig. 1. Capsaicin biosynthetic pathway showing the phenylpropanoid and the branched chain fatty acids vias. The involved enzymes are: phenylalanine ammonia lyase (PAL), cinnamic acid 4-hydroxylase (Ca4H), coumaric acid 3-hydroxylase (Ca3H), caffeic acid O-methyltransferase (COMT), and capsaicinoid synthase (CS). The possible participation of an amino transferase (AMT) is also indicated.

Because of the industrial value of capsaicinoids, different efforts have been carried out to try to produce them using chili pepper cell or tissue cultures (*1*). In this chapter a description is made on the in vitro cultures establishment, capsaicinoids extraction, separation, and quantification as well as some approaches to increase the production of these compounds by cell cultures.

2. Materials

1. Acetic acid.
2. Acetonitrile (ACN).
3. Agar/Gelrite.
4. Benzyladenine (BA).
5. Capsaicin.
6. Chloroform.
7. Chromium trioxide
8. Commercial bleach.
9. 2,4-dichlorophenoxyacetic acid (2,4-D).
10. Ethanol.
11. Ethyl acetate.
12. Inorganic salts and organic compounds of Murashige and Skoog (MS) medium.
13. Methanol.
14. Trifluoroacetic acid (TFA).
15. Tween-20.
16. Flasks, glass bottles, and Petri dishes.
17. Millipore FH membranes (0.5 µm).
18. Millipore membranes (0.22 µm).
19. Whatman no. 41 filter paper.
20. Gas chromatography (GC) equipment.
21. High performance liquid chromatography (HPLC) equipment.
22. Orbital shaker.
23. Oven.
24. pH-meter.
25. Prodigy ODS2 5 column.
26. Tekmar Tissuemizer.
27. µBondapak C_{18} column.

3. Methods

The methods outlined in the next section describe the establishment of callus cultures, the preparation of cell suspensions, the extraction, separation, and quantification of capsaicinoids by HPLC, the addition of precursors and intermediates to cell cultures, and the selection of cell clones.

3.1. Callus Cultures

Establishment of callus cultures has been previously described *(2)*. Wash seeds of chili pepper (*Capsicum* spp.) with abundant tap water to eliminate dust and then surface sterilize by immersion in 96% ethanol for 5 min followed by 20 min in 20% (v/v) bleaching solution containing 0.1% Tween-20. Wash the seeds thoroughly in sterile deionized water, and sew them on the MS basal medium *(3)* without growth regulators and incubate at 25 ± 2°C under photoperiod (16-h light/8-h dark; daylight fluorescent tubes; 50 µmol/m^2/s) for 4 wk. Excise hypocotyl explants (1 cm in length) from the seedlings and transfer the segments onto glass bottles (100 mL) containing 20 mL of MS basal medium supplemented with 12.5 µ*M* 2,4-D (*see* **Note 1**). The culture medium contains sucrose (3%) as the carbon source. The pH is adjusted at 5.7 before autoclaving (121°C; 20 min) and the medium is gelled with 0.8% agar or 0.3% Gelrite (*see* **Note 2**). Incubate the cultures for 8 wk under the same conditions described previously for seed germination and seedlings production. At this time, friable callus tissue can be observed at the cut ends of the explant. Calli are usually subcultured every month by transferring few portions of friable callus tissue onto fresh medium (**Fig. 2**). Callus can be also induced from cotyledon and leaf explants.

3.2. Cell Suspensions

Chili pepper cell suspensions are established from callus tissue cultures as previously described *(4)*. Inoculate callus (2.5 g fresh weight) into 125-mL Erlenmeyer flasks containing 25 mL of MS liquid culture medium supplemented with 12.5 µ*M* 2,4-D. Incubate the cultures on a rotary shaker (125 rpm) at 25 ± 2°C under photoperiod (16-h light/8-h dark; Cool white fluorescent tubes; 50 µE/m^2/s). Subculture the cell suspensions at 15-d intervals (**Fig. 2**). Maintain the cell cultures in MS liquid culture medium supplemented with 6.25 µ*M* 2,4-D and 0.66 µ*M* BA, pH 5.7, and at a starting cell inoculum size equivalent to 5 mg dry weight (dw) (100°C/5 h) per mL of fresh medium.

3.3. Extraction, Separation, and Quantification of Capsaicinoids

Capsaicinoids are extracted from either cells or culture medium from cell suspension samples collected during the culture cycle *(5)*. Homogenize lyophilized cell samples (500 mg dw.) with a Tekmar Tissuemizer in 10 mL absolute methanol and filter the homogenate through Millipore FH membranes (0.5 µm). Extract the filtrate with chloroform (3 × 10 mL), evaporate the organic phase to dryness over anhydrous Na_2SO_4 at 50°C under vacuum and suspend the residue in 3 mL methanol (HPLC grade). In the case of culture

Fig. 2. Chili pepper callus tissue *(left)* growing on MS medium supplemented with 12.5 μ*M* 2,4-D, and a cell suspension culture established in Murashige and Skoog medium with 6.25 μ*M* 2,4-D + 0.66 μ*M* benzylaminopurine *(right)*.

medium, extract the capsaicinoids from 10 mL samples starting at the chloroform extraction step as described for cell samples. Separate the capsaicinoids by HPLC using a Hewlett-Packard chromatograph Model 1084B equipped with a μBondapak C_{18} column (10 μm particle size, 300 × 3.9 mm) operating at 40°C *(4)*. Inject samples (20 μL) and separate the capsaicinoides with a solvent phase consisting of 6.6% acetic acid in water (phase A) and a methanol-acetic acid (50:1) mixture (phase B) pumped isocratically (73% phase A and 27% phase B) at a flow rate of 0.5 mL/min. Use 280 nm to detect the capsaicinoids. Under these conditions capsaicinoids appeared as a peak with a mean retention time of approx 7.25 min which corresponds to a standard of capsaicin (98%; Sigma Chemical Co.). Calculate total capsaicinoid contents from the sum of capsaicinoids in the culture medium (the amount [μg] released into the medium per gram of cells [fresh weight; fw]) plus capsaicinoids in cell extracts (the amount [μg] per gram of cells [fw]) during the culture cycle (18 d).

Capsaicinoids can be also extracted, separated, and quantified as previously described *(6)*. Separate the cells and the culture medium from a complete flask culture (50 mL/flask) using vacuum filtration through Whatman no. 41 filter paper. Wash the biomass three times with cold water, ground the cells in a mortar under liquid nitrogen, and extract with methanol for 1 h. Filter the methanolic extract through Whatman no. 41 filter paper and concentrate the filtrate to 300 µL for the analysis of capsaicinoids profile by HPLC. On the other hand, extract the filtered culture medium (20-mL samples) three times with an equal volume of ethyl acetate (1:1; v/v). Pool the organic fractions and concentrate them using a rotary evaporator and re-dissolve in 300 µL methanol for HPLC analysis.

A HPLC system (Thermo Separation Products Inc.), consisting of a Spectra System P2000 binary pump, a Spectra System AS1000 autosampler fitted with a 20 µL loop, and a Spectra System ultraviolet (UV)-visible optical scanning detector set at high-speed mode is used for capsaicinoids separation and detection. The detector signals are captured and processed by a PC1000 System Software (Thermo Separation Products, Inc.) loaded in an IBM 300PL personal computer. This software includes programs for system setup, control and monitoring of running samples, and quantitative and qualitative analyses. Separation of capsaicinoids in the samples is achieved using a Prodigy ODS2 5 column (250 × 4.6 mm internal diameter). Elute capsaicinoids using a gradient of 1 mM TFA and ACN as follows: (time [min], 1 mM TFA [%]): 0, 90; 5, 90; 10, 80; 25, 80; 35, 40; 55, 40; 60, 10; 65, 10; 70, 90; 80, 90. Maintain the rate flux at 1 mL/min, and the detector fixed at 230, 260, or 280 nm. Record the absorption spectra peaks between 200 and 365 nm at a rate of 10 spectra per second. The integration is set at 280, 260, or 230, as required.

3.4. Treatments With Precursors and Intermediates

The production of capsaicinoids in chili pepper cell suspensions can be enhanced by the addition of precursors and intermediates of the biosynthetic pathway. In order to test the effect of different precursors and intermediates, prepare freshly solutions of precursors and intermediates before the addition to the culture medium. Dissolve phenylalanine, vanillin, and vanillylamine in the MS culture medium described previously; cinnamic, p-coumaric, caffeic, and ferulic acids are first dissolved in a small volume of 1N KOH, adjusted to pH 5.7 with 0.1N HCl and then dissolved in MS culture medium. Prepare all solutions as 10 mM stocks, sterilize them by filtration in Millipore membranes (0.22 µm), and add them to the autoclaved culture medium to give a final concentration of 100 µM. This concentration of precursors does not cause significant growth alterations when added to the culture medium. Inoculate chili pepper cells at a final concentration equivalent to 5 mg dry weight (dw; 100°C/5 h) per mL in 125-mL Erlenmeyer flasks containing 50 mL of culture medium supplemented with each precursor. Incubate the cell cultures on a rotary shaker

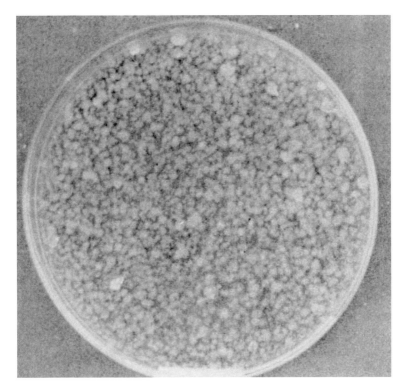

Fig. 3. Formation of chili pepper calli colonies from a cell suspension after 45 d of incubation.

(125 rpm) at $25 \pm 1°C$ under photoperiod (cool white fluorescent lamps; 16/8-h (light/dark) photoperiod; 50 µE/m²/s). In order to investigate the production of capsaicinoids during the culture cycle, collect the cell biomass from four individual Erlenmeyer flasks by filtration in Millipore membranes (0.22 µm) at 3-d intervals for 18 d. Fresh filtered cells are used for capsaicinoids extraction. Lyophilize the collected cells and freeze the filtered culture medium of each Erlenmeyer flask with liquid nitrogen. Store lyophilized cells and the frozen culture medium at $-70°C$ until use.

3.5. Isolation of Cell Clones

The isolation of chili pepper cell clones is recommended as a tool for detecting cells with different capacities to produce capsaicinoids. We have previously reported a technique for the isolation of variant cell clones *(4)*. Filtrate cell suspension cultures through a nylon mesh, and determine cell concentration of the suspension. This usually requires prior cell separation of the cells from cell aggregates with a treatment of 1 vol of culture with 2 vol of 8%

chromium trioxide (w/v), heating at 70°C for 10 min and then shaking vigorously for 10 min. This suspension is suitably diluted and counted on a cell counting slide. Adjust cell density to 1×10^5 with fresh medium, pour volumes (6–8 mL) of the cell suspension in Petri dishes and mix well with an equal volume of 2X semisolid MS culture medium containing 25 µM 2,4-D and 1.2% agar maintained at 40°C. Seal the Petri dishes with Vitafilm or an equivalent plastic wrap film and incubate the cultures under the same conditions described before for callus cultures. Cell colonies ready for analysis and subcultures are obtained after 45 d of incubation **(Fig. 3)**.

4. Notes

1. Different auxins have been tested for callus induction from chili pepper hypocotyl tissues *(2)*, but 2,4-D has been found better for friable callus production in the range of 6.25–50 µM. This auxin is also the best option for callus induction from other tissues, but the concentration of 2,4-D need to be adjusted in any case. α-Naphthaleneacetic acid (NAA) can also be used at the same levels as 2,4-D for callus formation.
2. The recommended concentration of Gelrite for semisolid culture medium preparation is 0.3%, but it has been observed that lower levels of this polymer can lead to the formation of more hydrated callus tissue (friable), which can be more suitable for cell suspension establishment.

References

1. Ochoa-Alejo, N. and Ramírez-Malagón, R. (2001) In vitro chili pepper biotechnology. *In Vitro Cell. Dev. Biol. Plant* **37,** 701–729.
2. Ochoa-Alejo, N. and García-Bautista, M. A. R. (1990) Morphogenetic responses in vitro of hypocotyls tissues of chili pepper (*Capsicum annuum* L.) to growth regulators. *Turrialba* **40,** 311–318.
3. Murashige, T. and Skoog F. (1962) A revised medium for rapid growth and bioassays with tobacco tissue cultures. *Physiol. Plant.* **15,** 473–497.
4. Salgado-Garciglia, R. and Ochoa-Alejo, N. (1990) Increased capsaicin in PFP-resistant cells of chili pepper (*Capsicum annuum* L.). *Plant Cell Rep.* **8,** 617–620.
5. Hall, R. D., Holden, M. A., and Yeoman, M. M. (1987) The accumulation of phenylpropanoid and capsaicin compounds in cell cultures and whole fruit of chilli pepper, *Capsicum frutescens* Mill. *Plant Cell, Tiss. Org. Cult.* **8,** 163–176.
6. Martínez-Juárez, V. M., Ochoa-Alejo, N., Lozoya-Glória, E., et al. (2004) Specific synthesis of 5,5'-dicapsaicin by cell suspension cultures of *Capsicum annuum* var. annuum (chili jalapeno Chigol) and their soluble and NaCl-extracted cell wall protein fractions. *J. Agric. Food Chem.* **52,** 972–979.

28

Isolation and Purification of Ribosome-Inactivating Proteins

Sang-Wook Park, Balakrishnan Prithiviraj, Ramarao Vepachedu, and Jorge M. Vivanco

Summary

Ribosome-inactivating proteins (RIPs) are cytotoxic *N*-glycosidases identified in plants, fungi, and bacteria. RIPs inhibit protein synthesis by virtue of their enzymatic activity, selectively cleaving a specific adenine residue from a highly conserved, surface-exposed, stem-loop (S/R loop) structure in the 28S rRNA of ribosomes. Some RIPs also exhibit a number of other enzymatic activities such as RNase, DNase, phospholipase, and superoxide dismutase (SOD). RIPs are considered to be plant defense-related proteins as they are able to inhibit the multiplication and growth of several pathogenic virus, fungi, and bacteria either alone or in conjugation with other defense-related proteins. The mechanism of inhibitory activity of RIPs against fungal pathogens seems to be by directly inhibiting fungal growth rather than depurinating host plant ribosomes and causing cell death as previously envisaged. This chapter describes the protocol used to isolate and purify RIPs from plant tissues.

Key Words: Antimicrobial activity; depurination; plant defense; ribosome-inactivating protein.

1. Introduction

Ribosome-inactivating proteins (RIPs) are cytotoxic *N*-glycosidases identified in plants, fungi and bacteria (**Table 1**). RIPs are largely divided into two classes: (1) type I RIPs consist of a single *N*-glycosidase domain; and (2) type II RIPs are chimero-RIPs constructed of an A-chain, functionally equivalent to a type-I RIP, which is attached to a sugar-binding B-chain lectin domain (*1*). RIPs have been proposed to inhibit protein synthesis by virtue of their enzymatic activity, selectively cleaving an adenine residue from a highly conserved,

---

[1]The first two authors contributed equally to this work.

From: *Methods in Molecular Biology, vol. 318: Plant Cell Culture Protocols, Second Edition*
Edited by: V. M. Loyola-Vargas and F. Vázquez-Flota © Humana Press Inc., Totowa, NJ

Table 1
List of Ribosome-Inactivating Proteins

Plant	Source	Name of RIP
Type I		
Phytolacca americana	Leaves	PAP
	Leaves (seasonally expressed)	PAP-I/II/III
	Seeds	PAP-S
	Roots	PAP-R
Phytolacca insularis		PIP
		PIP2
Phytolacca dioica	Seeds	PD-S2
	Leaves	PD-L1/2/3/4
Phytolacca dodecandra	Leaves	Dodecandrin
Mirabilis expansa	Roots	ME 1/2
Mirabilis jalapa	Leaves	MAP
Gelonium multiflorum	Seeds	Gelonin
Hura crepitans	Latex	RIP
Manihot palmata	Seeds	Mapalmin
Bryonia dioica	Leaves	Bryodin-L
	Roots	Bryodin
Citrullus colocynthis	Seeds	Colocin 1/2
Luffa cylindrica	Seeds	Luffin a
		Luffin b
Momordica charantia	Seeds	Momordin
Momordica cochinchinensis	Seeds	Momorcochin
Trichosanthes kirilowii	Roots	Trichosanthin
		TAP-29
	Seeds	Trichokirin
Hordeum vulgare	Seeds	barley RIP
Triticum aestivum	Germ	Tritin
Zea mays	Seeds	Corn RIP
Asparagus officinalis	Seeds	Asparin 1/2
Dianthus caryophyllus	Leaves	Dianthin 30/32
Lychnis chalcedonica	Seeds	Lychnin
Saponaria officianlis	Seeds	Saporin 5/6/9
Cinnamomin camphora	Seeds	Cinnamomin
		Camphrin
Trichosanthes anguina	Seeds	Trichoanguin
Iris hollandica	Bulbs	Iris RIP.A1/2/3
Volvariella volvacea	Fruiting Bodies	Volvarin
Sechuum edule	Seeds	Sechiumin
Hypsizigus marmoreus	Fruiting bodies	Hypsin

(continued)

Table 1 (Continued)
List of Ribosome-Inactivating Proteins

Plant	Source	Name of RIP
Type I		
Lyophyllum shimeji	Fruiting Bodies	Lyophyllin
Cucurbita pepo	Fruits	Pepocin
Sambucus nigra	Fruits	Nigritin
Type II		
Ricinus communis		Ricin
Abrus precatorius		Abrin
		Abrin II
Sambucus nigra	Bark	Nigrin b
		Basic Nigrin b
Sambucus sieboldiana	Bark	Sieboldin-b
Sambucus ebulus	Leaves	Ebulin 1
Momordica charantia	Seeds	B-momorcharin
Iris hollandica	Bulbs	Agglutinin b/r
Type III		
Hordeum vulgare	Leaves	JIP60
Zea mays	Seeds	Maize RIP

surface-exposed, stem-loop (S/R loop) structure in the 28S rRNA *(2)*. Thus, the site-specific deadenylation interrupts the interaction of elongation factors, EF1 and EF2, with the S/R loop, and blocks protein synthesis at the translocation step. Recently, Type-III RIPs have been identified as a single chain containing an extended carboxyl-terminal domain with unknown function; these are synthesized as precursors requiring proteolytic removal of an internal peptide for activity *(3)*.

1.1. Enzymatic Activity

RIPs are presently classified as rRNA N-glycosidases in the enzyme nomenclature (EC 3.2.2.22). Several studies have suggested that RIPs specifically cleave the N-C glycosidic bond of the adenine base in the tetra loop sequence (GAGA) located on the sarcin/ricin (S/R) loop of eukaryotic and prokaryotic ribosomes. The universally conserved adenine residue A4324 of the eukaryotic 28S rRNA (and A2660 in the prokaryotic 23S rRNA) has long been considered the only enzymatic target site for RIPs *(2,4)*, but several lines of evidence have recently revealed alternative substrates to the S/R loop. For instance, it has been shown that several RIPs can release adenine from multiple sites in rRNA *(5)*. Furthermore, RIPs have been found to target various nucleic acids, randomly removing adenine residues from single-stranded regions of nucleic acids and, to a lesser extent, guanine residues from wobble base-pairs in hairpin stems. This substrate recognition and enzymatic activity depends on the physical availability of nucleotides; denaturation of nucleic acid structures increases their interaction with RIPs *(5,6)*.

Apart from adenine and guanine glycosidase activity, RIPs exhibit a number of novel enzymatic activities such as RNase, DNase, phospholipase, and superoxide dismutase (SOD) activities *(7)*. These observations suggest that RIPs may possess dual biochemical activities and multiple biological roles, renewing our interest in the biological functions of RIPs because of the potential for diverse functions *in planta* of the possible primary/secondary roles of RIPs.

1.2. Antimicrobial Activity

RIPs are considered to be defense-related proteins *in planta* as RIPs are able to inhibit the multiplication and growth of several pathogenic organisms including virus, fungi, and bacteria. Despite differences in virus infection mechanisms, a number of RIPs show broad-spectrum antiviral activity against both plant and animal viruses including human immunodeficiency virus (HIV). In addition, several RIPs have shown direct inhibitory activity against the growth of various fungal and bacterial pathogens such as *Alternaria solani, Alternaria alternaria, Agrobacterium tumerfaciens, Agrobacterium rhizogenes* R100nal, *Botrytis cinerea, Fusarium oxysporum, Fusarium proliferatum, Mycosphaerella arachidicola, Neurospora crassa, Phycomyces blakesleeanus, Pythium irregulare, Physalospora piricola, Trichoderma reesei,* and *Verticillium dahliae, Bacillus subtilis, Rhizobium leguminosarum, Serratia marcescens, Pseudomonas syringae, Xanthomonas campestris* pv *versicatoria,* and *Erwinia carotovora (1)*.

Some RIPs, interestingly, show synergistic antifungal effects combined with other plant defense-related proteins *(8,9)*. The recently isolated root-specific PAP-H also shows traces of inhibitory activity against several fungal pathogens in the presence of other pathogenesis-related (PR) proteins such as chitinase and β-1,3-glucanase *(10)*. Facilitating this possible synergistic effect, disease resistance in transgenic tobacco plants is manifested by enhanced antifungal activity when the barley RIP is constitutively co-expressed with barley class II chitinase *(11)*. The combination of barley endosperm RIP with barley class II chitinase produces a significant increase in resistance to *Rhizotonia solani* in transgenic tobacco. Both transgenic proteins (RIP + chitinase) accumulate in the intercellular spaces of transgenic tobacco, and it was postulated that the cytotoxic effect of the barley RIP on fungal cells is enhanced by chitinase action resulting in an increased amount of RIP entering fungal cells. Maize RIP, b-32, under control of the potato wun1 gene promoter, also increases the tolerance of transgenic tobacco to *R. solani (12)*.

1.3. Classic Theory of Antimicrobial Mechanism

The defense mechanism of RIPs was explained previously by the so-called "suicide model" similar to the hypersensitive response (HR) *(13)*. The inhibitory mechanism of RIPs against microbial pathogens has been postulated based on the evidence and mechanistic information available from PAP and its isoform *(10)*. Based on the RI activity and the extracellular localization of PAP, it has been suggested that RIPs are synthesized in an inactive form, sequestered in the cell wall matrix, and re-enter the cytoplasm along with the pathogen at the infection sites. Thus, RIPs inhibit pathogen multiplication by inactivating host ribosomes and causing host cell death. However, the antimicrobial activity and mechanism of RIPs are gaining new attention through recent studies.

1.4. Revising the Working Model

As discussed, the widely accepted mechanism for antimicrobial action identifies host ribosomes as the target of RIP activity. However, no evidence has been discovered that would prove this suicidal mechanism. Several experiments with transgenic plants expressing RIPs also do not exhibit HR or other symptoms of spontaneous cell death in response to microbial infection although they are resistant to a wide range of pathogens. In addition, no pathogen has been reported to evoke the suicidal machinery in any plant system producing RIP(s).

Several independent studies suggest that the antimicrobial activity of RIPs can be independent from the enzymatic activity of RIPs, inactivating ribosomes of the plant host. The earliest demonstrations were made by Tumer and her colleagues, who assayed deletion mutants of PAP in transgenic tobacco plants *(9,13,14)*. Although PAPc (PAPW327stop) and PAPn (PAPG75D) were found to be unable to depurinate the S/R loop, they still conferred resistance against plant viral and fungal pathogens. Further experiments revealed that PAP is able to directly depurinate selective viral mRNA that has a 5' terminal m7GpppG cap; in contrast, PAP has no significant effect on uncapped mRNAs *(15,16)*. From these observations, the authors hypothesized that during viral infection, PAP may target selective viral RNAs by binding to the cap instead of inactivating host ribosomes. Recently, the inhibitory mechanism of RIPs against HIV replication that has generally been believed to relate to RI activity has also been readdressed *(17)*. Wang et al. *(17)* use site-directed mutagenesis approaches with TCS to demonstrate that the classical RI activity of TCS is not adequate to explain the anti-HIV action of TCS. In this study, an exception is revealed: TCS mutants with a C-terminal deletion or addition of amino acids retained almost all RI activity, but were devoid of anti-HIV activity. This result demonstrates that the C-terminus region of TCS is responsible for anti-HIV activity, and suggests that the defensive mechanism of RIPs against pathogens can be separated from RI activity.

As for the inhibitory mechanism of RIPs against fungal pathogens, more and more evidence has shown that RIPs can directly target and inhibit fungal growth rather than depurinating host plant ribosomes and causing cell death. The type-I RIPs—ME1, hypsin, lyophyllin, and hispin (*Benincasa hispida*)— show a direct inhibitory activity against an array of pathogenic and nonpathogenic fungi *(8–10,18,19)*. This fungal cytotoxicity has also been shown to be independent from RI activity *(20)*. Comparing three different RIPs—ME1, RTA, and saporin-S6 (isoform of saporin)—the latter two showed approx 10- to 50-fold higher enzymatic activity against isolated fungal ribosomes than did ME1. Nevertheless, ME1 showed higher inhibitory activity against fungal growth itself compared to RTA and saporin-S6. A study with transgenic tobacco plants expressing PAP and its mutants also demonstrates that the RI activity of RIP is not sufficient for host resistance to the fungal pathogen *(14)*. To further understand the antifungal mechanism of RIPs, Park et al. *(20)* labeled ME1 and saporin-S6 with NHS-fluorescein and the interaction between labeled RIPs and fungal cells was monitored. The labeled ME1 showed strong interaction with fungal hyphae, translocateed to the cell wall, and presumably penetrated into the cytosolic region, in contrast to the absence of interaction observed between labeled saporin-S6 and fungi. This result indicateed that some type-I RIPs, lacking any known carbohydrate-binding domain, are capable of interacting with the target cell surface. Taken together, these results indicate that: (1) the RI activity of RIPs does not adequately explain their defensive action against pathogens; (2) the defensive action of RIPs is independent from their enzymatic specificity; (3) some RIPs are able to recognize and target selective pathogens by possibly binding to specific cell membrane components; and, finally, (4) the defense mechanism of RIPs could proceed by directly targeting invading pathogens rather than host ribosomes.

1.5. Therapeutic Properties

The target specificity of monoclonal antibodies induced the development of immunotoxins, which deliver toxins, such as RIPs, to specific cells. PAP linked to specific antibodies has been shown to prevent the growth of leukemia cells *(21)*. Human CD19+ mixed lineage leukemia cells (RS4-11) proliferate in the hematopoietic tissues and other organs of mice with severe combined immunodeficiency in a manner similar to human acute leukemia. PAP linked to specific antibodies (anti-CD19-Pokeweed antiviral protein immunotoxin) selectively inhibit clonegenic RS4,11 cells in vitro. This immunotoxin is thus able to extend the life span of the mice inoculated with RS4,11 cells *(22)*.

Another relatively successful immunotoxin is TP3-PAP. This antibody (TP3) is reactive against an antigen on human and canine osteosarcoma. This tumor-associated antigen is expressed at very high levels on the surface membrane of human osteosarcoma cells *(23)*, and the antibody (IgG2b anti-P80, TP3) reacts with mesenchymal tumors like osteosarcomas, hemangiopericytomas, chondrosarcomas, malignant fibrous histiocytomas, and synovial cell sarcomas *(23)*. Studies with TP3 attached to PAP (TP3-PAP) show high activity against osteosarcoma cells in vitro, killing all the culture cells after 48 h.

Finally, studies with immunotoxins prepared by linking momordin 1, pokeweed antiviral protein from seeds (PAP-S), and saporin-S6 to the monoclonal antibodies 48-127, which recognize a glycoprotein (gp54) expressed in human bladder carcinomas, have shown that RIPs linked to those antibodies are effective against bladder tumor cell lines (T24) *(24)*. The same three RIPs linked to Ber-H2 monoclonal antibodies directed against CD30 antigen of human lymphocytes induced apoptosis in CD30 cell lines *(25)*. Furthermore, RIP gelanin was also shown to be effective on malarial parasites when linked to human transferin *(26)*. These studies indicate that RIPs could be very effective drugs when linked to proper targeting antibodies, and could thus be used as antitumor drugs.

1.6. Future Perspectives

Although RIPs have been studied for many years and their enzymatic and biological activities have been extensively investigated, very little is known about their role in plant biology. To elucidate RIP biology several questions will have to be addressed, such as how they gain access to their endogenous ribosomes substrates in vivo and how they interact with the translation machinery in the cell. The recent identification of RIP from the model system *Nicotiana tabacum* in corroboration with biotechnological approaches, such as mutagenesis and silencing, and genomic approaches, could be an effective way to bridge the gaps in our knowledge. Furthermore, the discovery of novel enzymatic activities could indicate some RIPs' dynamic participation in various cellular mechanisms in response to environmental cues. New insights into the defense mechanisms of RIPs and their additional nonribosomal substrates have also begun to increase our understanding of properties important for protecting organs and regulating cellular functions of host cells. Such varied approaches will likely accelerate our knowledge of the biological function of RIPs and establish their fundamental significance for medical applications in the next few years.

2. Materials

2.1. Plant Material

Candidate plant material is ideally grown under greenhouse or tissue culture conditions. Less fibrous material gives a higher RIP yield. Hairy root culture of some plants is also a good source of RIPs *(10)*.

2.2. Composition of Buffers

2.2.1. Plant Extraction Buffer (1 L)

1. 25 mM Phosphate buffer, pH 7.0.
2. 250 mM NaCl.
3. 10 mM Ethylenediaminetetraacetic acid (EDTA).
4. 5 mM Dithiothreitol (DTT).
5. 1 mM Phenylmethylsulfonyl fluoride (PMSF).
6. 1.5% [w/v] Polyvinylpolypyrrolidone (PVP).

2.2.2. Yeast Ribosome Extraction Buffer

1. 200 mM Tris-HCl, pH 9.0.
2. 200 mM KCl.
3. 25 mM MgCl$_2$.
4. 25 mM EGTA.
5. 200 mM Sucrose.
6. 25 mM β-mercaptoethanol.

2.3. Ribosome Depurination Assay (All Solutions Must Be Prepared With DEPC-Treated H$_2$O)

2.3.1. 7 M Urea/6% Polyacrylamide Gel

TBE solution (10X)	0.75 mL
Urea	3.15 g
Polyacrylamide (working solution)	1.5 mL
DEPC-treated H$_2$O	3.0 mL
TEMED	10 μL
APS (10%)	100 μL

2.3.2. Polyacrylamide (Working Solution; 50 mL)

Acrylamide	14.6 g
Bis-acrylamide	0.4 g

2.3.3. TBE (10X; 1 L)

Tris-base	108 g
Boric acid	55 g

EDTA	9.3 g
pH	8.2–8.4

2.3.4. Ethidium Bromide (0.5 mg/mL)

2.3.5. Sample Buffer

Formamide	10 mL
Xilemol cyanol	10 mg
Blomophenol blue	10 mg
0.5 M EDTA	20 µL
pH	8.0

2.3.6. RIP Buffer (2X; 10 mL)

1 M KCl	334 µL
0.5 M Tris	400 µL
1 M MgCl$_2$	200 µL
pH	7.2

Make up to 10 mL with DEPC-treated H$_2$O.

2.3.7. Extraction Buffer (2X; 100 mL)

1.5 M Tris-HCl, pH 8.8	2.2 mL
NaCl	0.83 g
0.5 M EDTA, pH 8.0	4 mL
10% Sodium dodecyl sulfate (SDS)	20 mL

Make up to 100 mL with DEPC-treated H$_2$O.

3. Methods

3.1. Protein Extraction From Plant Material

1. Plant materials are ground in liquid nitrogen along with acid-washed sand for maximum cell disruption.
2. The ground material is homogenized in three volumes of extraction buffer (*see* **Subheading 2.2.1.**) and centrifuged for 30 min at 10,000g.
3. The supernatant is decanted into a clean glass beaker and ammonium sulphate is added to a final concentration of 20% (w/v) with continuous stirring.
4. The mixture is left in the cold room for 1 h and centrifuged for 30 min at 10,000g.
5. The resulting supernatant is precipitated with increasing concentrations of 20–80% (w/v) ammonium sulphate and centrifuged at 14,000g for 30 min.
6. The pellet is suspended in 25 mM HEPES/NaOH, pH 8.0, containing 50 mM NaCl and then dialyzed against 25 mM HEPES/NaOH, pH 8.0, until it is free from the sulphate ions.
7. All extraction procedures are conducted at 4°C, and the ammonium sulphate fraction is stored at –80°C until use.

3.2. Purification of RIP

1. The total protein is fractionated using a cation-exchanger chromatography. Equilibration and loading are carried out using 25 mM NaH$_2$PO$_4$ buffer, pH 7.0, at a flow rate of 1 mL/min. The protein was fractionated with NaCl gradient, and further purified using gel filtration chromatography.
2. The gel filtration column is equilibrated with 25 mM HEPES/NaOH, pH 8.0, containing 100 mM NaCl. The basic protein is loaded on the column, and the proteins eluted with an isocratic gradient at a flow rate of 0.5 mL/min.
3. The fractions are assayed for RIP activity, and the active fraction is desalted using 5-kD cutoff ultrafiltration membranes (Millipore, Bedford, MA).

3.3. N-Glycosidase Activity Assay

3.3.1. Isolation of Yeast Ribosomes

Note: All buffers and solutions are prepared using DEPC-treated H$_2$O.

1. Yeast (*Saccharomyces cerevisiae*) cells are grown in yeast peptone dextrose (YPD) medium at 30°C overnight.
2. Yeast cells are harvested by centrifugation at 4000g for 20 min and washed several times.
3. The pellets are homogenized in extraction buffer (*see* **Subheading 2.2.2.**) under liquid nitrogen and centrifuged at 11,000g at 4°C for 20 min.
4. The supernatant is laid over a cushion of 10 mL sucrose (1 M sucrose, 20 mM KCl, and 5 mM MgCl$_2$ in 25 mM Tris-HCl, pH 7.6) in 70-Ti tubes (Beckman Coulter, Fullerton, CA) and centrifuged at 200,000g for 4 h at 4°C (L-70 Ultracentrifuge, Beckman Coulter).
5. The resulting pellet (ribosomes) is resuspended in 25 mM Tris-HCl buffer, pH 7.6, with 25 mM KCl and 5 mM MgCl$_2$ and stored at –80°C until use.

3.3.2. The Depurination Assay

Note: All buffers and solutions are prepared using DEPC-treated H$_2$O.

1. Preparation of reaction mixture (final volume: 100 µL).
    a. 10 µL of ribosome.
    b. 10 µL 10X RIP buffer.
    c. Approximately 10–20 µL RIP
    d. Make up to 100 µL with DEPC H$_2$O.
2. Incubate the mixture at either 30°C or 37°C for 30 min.
3. RNA is extracted using phenol/chloroform (1:1), and precipitated with ethanol.
4. The pelleted RNA is resuspended with 10 µL of DEPC H$_2$O.
5. Divide it into two parts (5 µL for aniline treatment, and 5 µL for control); 5 µL of RNA is pipetted into the new Eppendof tube for control. Store at –80°C until gel electrophoresis.

Ribosome-Inactivating Proteins

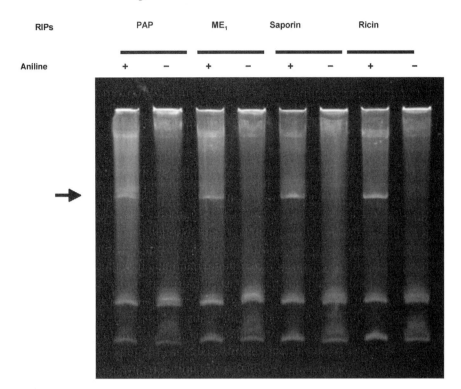

Fig. 1. Enzymatic activity of RIPs in vitro. Ribosomes were isolated from yeast and incubated with PAP (*Phytolacca americana*), ME1 (*Mirabilis expansa*), saporin (*Saponaria officinalis*), and ricin A-chain (*Ricinus communis*) as described in **Subheading 3.3.2.**, rRNAs were extracted, treated with aniline, separated on a 7 *M* 6% Urea polyacrylamide gel, and stained with EB. The presence (+) or absence (–) of aniline is denoted. The arrow shows the presence of the diagnostic 367-nucleotide cleavage product of rRNA.

6. Prepare aniline solution immediately before treatment.
    a. 400 µL DEPC H_2O.
    b. 50 µL Aniline.
    c. 60 µL Glacial acetic acid.
7. Carefully mix the 25 µL of aniline solution with 5 µL of RNA using a pipet.
8. Incubate the mixture on ice for 30 min, and RNA is precipitated with ethanol.
9. Add 10 µL sample buffer (2.3.5) into aniline-treated and control RNAs.
10. Incubate them at 65°C for 10 min.
11. RNAs are separated with 7 *M* urea/6% polyacrylamide gel.
12. The gel is visualized with ethidium bromide (**Fig. 1**).

Acknowledgments

Work reported in this communication was supported by a CAREER award from the National Science Foundation (MCB-0093014) to JMV, and by the Colorado State University Agricultural Experiment Station (JMV).

References

1. Nielsen, K. and Boston, R. S. (2001) Ribosome-inactivating proteins: a plant perspective. *Annu. Rev. Plant Physiol. Plant Mol. Biol.* **52**, 785–816.
2. Endo, Y. and Tsurugi, K. (1987) RNA *N*-glycosidase activity of ricin A-chain: mechanism of action of the toxic lectin ricin on eukaryotic ribosomes. *J. Biol. Chem.* **263**, 8735–8739.
3. Mundy, J., Leah, R., Boston, R., Endo, Y., and Stirpe, F. (1994) Genes encoding ribosome-inactivating proteins. *Plant Mol. Biol. Rep.* **12**, S60–S62.
4. Endo, Y., Mitsui, K., Motizuki, M., and Tsurugi, K. (1987) The mechanism of action of ricin and related toxic lectins on eukaryotic ribosomes: the site and the characteristics of the modification in 28S ribosomal RNA caused by the toxins. *J. Biol. Chem.* **262**, 5908–5912.
5. Barbieri, L., Gorini, P., Valbonesi, P., Castiglioni, P., and Stirpe, F. (1994) Unexpected activity of saporins. *Nature* **372**, 624.
6. Park, S. W., Vepachedu, R., Owens, R. A., and Vivanco J. M. (2004) The *N*-glycosidase activity of the ribosome-inactivating protein ME1 targets single-stranded regions of nucleic acids independent of sequence or structural motifs. *J. Biol. Chem.* **279**, 34,165–34,174.
7. Park, S. W., Vepachedu, R., Sharma, N., and Vivanco, J. M. (2004) Ribosome-inactivating proteins in plant biology. *Planta* **219**, 1093–1096.
8. Lam, S. K. and Ng, T. B. (2001) Hypsin, a novel thermostable ribosome-inactivating protein with antifungal and antiproliferative activities from fruiting bodies of the edible mushroom *Hyprizigus marmoreus*. *Biochem. Biophys. Res. Comm.* **285**, 1071–1075.
9. Lam, S. K. and Ng, T. B. (2001) First simultaneous isolation of a ribosome-inactivating protein and an antifungal protein from a mushroom (*Lyophyllum shimeji*) together with evidence for synergism of their antifungal effects. *Arch. Biochem. Biophys.* **393**, 271–280.
10. Park, S. W., Lawrence, C. B., Linden, J. C., and Vivanco, J. M. (2002) Isolation and characterization of a novel ribosome-inactivating protein from root cultures of pokeweed and its mechanism of secretion from roots. *Plant Physiol.* **130**, 164–178.
11. Jach, G., Gornhardt, B., Munday, J., et al. (1995) Enhanced quantitative resistance against fungal disease by combinatorial expression of different barley antifungal proteins in transgenic tobacco. *Plant J.* **8**, 97–109.
12. Maddaloni, M., Forlani, F., Balmas, V., et al. (1997) Tolerance to the fungal pathogen *Rhizoctonia solani* AG4 of transgenic tobacco expressing the maize ribosome-inactivating protein b-32. *Transgenic Res.* **6**, 393–402.
13. Tumer, N.E., Hudak, K., Di, R., Coetser, C., Wang, R., and Zoubenko, O. (1999) Pokeweed antiviral protein and its applications. *Curr. Top. Microbiol. Immunol.* **240**, 139–158.

14. Zoubenko, O., Uckun, F., Hur, Y., Chet, I., and Tumer, N. E. (1997) Plant resistance to fungal infection induced by nontoxic pokeweed antiviral protein mutants. *Nature Biotechnol.* **15,** 992–996.
15. Hudak, K. A., Dinman, J. D., and Tumer, N. E. (1999) Pokeweed antiviral protein accesses ribosomes by binding toL3. *J. Biol. Chem.* **274,** 3859–3864.
16. Hudak, K. A., Wang, P., and Tumer, N. E. (2000) A novel mechanism for inhibition of translation by pokeweed antiviral protein: depurination of the capped RNA template. *RNA* **6,** 369–380.
17. Wang, P., Zoubenko, O., and Tumer, N. E. (1998) Reduced toxicity and broad spectrum resistance to viral and fungal infection in transgenic plants expressing pokeweed antiviral protein II. *Plant Mol. Biol.* **38,** 957–964.
18. Vivanco, J. M., Savary, B. J., and Flores, H. E. (1999) Characterization of two novel type I ribosome-inactivating proteins from the storage roots of the Andean crop *Mibilis expansa. Plant Physiol.* **119,** 1447–1456.
19. Ng, T. B. and Parkash, A. (2002) Hispin, a novel ribosome-inactivating protein with antifungal activity from hairy melon seeds. *Protein Expres. Purification* **26,** 211–217.
20. Park, S. W., Stevens, N. M., and Vivanco, J. M. (2002) Enzymatic specificity of three ribosome-inactivating proteins against fungal ribosomes, and correlation with antifungal activity. *Planta* **216,** 227–234.
21. Ramakrishnan, S. and Houston, L. L. (1984) Inhibition of human acute lymphoblastic leukemia cells by immunotoxins: potentiation by chloroquine. *Science* **223,** 58–61.
22. Jansen, B., Uckun, F. M., Jaszcz, W. B., and Kersey, J. H. (1992) Establishment of a human t(4;11) leukemia in severe combined immunodeficient mice and successful treatment using anti-CD19 (B43)-pokeweed antiviral protein immunotoxin. *Cancer Res.* **52,** 406–412.
23. Bruland, O. S., Fodstad, O., Stenwig, A. E., and Pihl, A. (1988) Expression and characteristics of a novel human osteosarcoma-associated cell surface antigen. *Cancer Res.* **48,** 5302–5309.
24. Battelli, M. G., Polito, L., Bolognesi, A., Lafleur, L., Fradet, Y., and Stirpe, F. (1996) Toxicity of ribosome-inactivating protein-containing immunotoxins to a human bladder carcinoma cell line. *Int. J. Cancer* **65,** 485–490.
25. Bolognesi, A., Tazzari, P. L., Olivieri, F., Polito, L., Falini, B., and Stirpe, F. (1996) Induction of apoptosis by ribosome-inactivating proteins and related immunotoxins. *Int. J. Cancer* **68,** 349–355.
26. Surolia, N. and Misquith, S. (1996) Cell surface receptor directed targeting of toxin to human malaria parasite, *Plasmodium falciparum. FEBS Lett.* **396,** 57–61.

29

Catharanthus roseus Shoot Cultures for the Production of Monoterpenoid Indole Alkaloids

Elizabeta Hernández-Domínguez, Freddy Campos-Tamayo, Mildred Carrillo-Pech, and Felipe Vázquez-Flota

Summary

A protocol for the establishment of in vitro shoot cultures of *Catharanthus roseus* is described. Shoots can be maintained for more than 1 yr without evidence of tissue vitrification, disaggregation, or callus formation. Vindoline was the main alkaloid accumulated, reaching values similar to those found in leaves from field-grown plants, after a long period of culture. An induction methodology to reduce such waiting time is also presented.

Key Words: *Catharanthus roseus*; elicitation; in vitro shoot culture; methyl jasmonate; monoterpenoid indole alkaloids.

1. Introduction

The Madagascar periwinkle (*Catharanthus roseus*) produces more than 200 monoterpenoid indole alkaloids, some of them with pharmaceutical uses. This plant is the only commercial source of the cytotoxic dimeric alkaloids, vinblastine (VBL), and vincristine (VCR), which are widely used in treating various types of cancer, such as Hodgkin's disease and child leukemias. Both alkaloids are formed from the condensation of catharanthine and vindoline subunits in the leaves of mature *Catharanthus* plants, where they accumulate. The higheconomic value of these alkaloids, along with the low amounts recovered from *Catharanthus* plants, has prompted significant efforts to produce them through cell culture technology. Some of the strategies frequently used include the induction of cell lines from elite individuals, the systematic screening of cell lines for overproducing strains, the optimization of media composition, and the application of different elicitors, such as jasmonate, fungal homogenates, and different stress conditions *(1)*. However, even when undifferentiated cell cultures are capable of producing high levels of catharanthine and other alkaloids, they fail to accumulate vindoline *(2)*.

Extensive research has shown that the inability of cell cultures to produce vindoline is related to the lack of proper cell organization. In *Catharanthus* leaves at least two different cell types are required for vindoline formation. The enzymes involved in the early steps, which are common to vindoline and the rest of the *Catharanthus* alkaloids, are restricted to the upper epidermis, whereas those involved in the late reactions, which are exclusively committed to vindoline synthesis are limited to specialized cells, such as laticifers and idioblasts *(3,4)*.

In vitro shoot cultures, which can maintain a similar cell organization to that found in leaves, have been proposed as suitable system to produce vindoline *(5)*. Although vindoline concentration in in vitro shoots can compare to those in mature leaves, yield inconsistencies between different batches have been recorded *(6)*. Such differences may be related to the lack of coordination in the expression of the early and late biosynthetic steps *(2)*. These observations suggest that further research in the regulation of vindoline biosynthesis in these systems is required.

The first part of this chapter describes a protocol for the induction, maintenance, and characterization of a vindoline-producing *Catharanthus roseus* shoot culture. The second part presents a methodology for the induction of alkaloid induction is presented.

2. Materials

All media are prepared with deionized water and based on Murashige and Skoog's medium (MS) *(7)*.

1. 6-benzylaminopurine (BA), 1.0 mg/mL stock: dissolve 100 mg BA in a small volume of $0.1N$ HCl, and adjust volume to 100 mL with water.
2. Methyljamonate (MeJa) 100 mM stock: dilute 229 µL of MeJa (95%, $p = 1.030$ g/mL (Sigma-Aldrich Chemical Co., St. Louis, MO) in a total volume of 10 mL of dimethylsulfoxide (DMSO). Sterilize the solutions by filtration using 0.2 µm pore size nylon sterile membranes (Corning, NY).

3. Methods

3.1. Induction of Multiple Shoot Cultures

Shoot cultures are initiated from seedlings derived from immature embryos.

1. Unripe, siliques (approx 20-mm long) are collected from mature (6-mo-old) *C. roseus* (cv. Little Delicata) plants. Prior to seed extraction, siliques should be thoroughly washed with tap water and sterilized by subsequent immersion in 3%

sodium hypochlorite for 20 min and 70% ethanol for 5 min, and rinsed twice in sterile water.
2. Immature seeds are imbibed for 24 h in sterile water in the dark and then germinated in sealed jars containing semisolid MS media, without growth regulators or sucrose. Jars with seeds should be kept in the dark at 25°C for 5 d.
3. Five-day-old etiolated seedlings are then exposed to continuous white light (40 $\mu M/m^2/s$) for an additional 4 d at 25°C. Light-exposed seedlings are used for inducing the shoot cultures, as described next.
4. Seedlings (between 12 and 14 mm) with fully expanded cotyledons are collected. Radicles should be cut off with a scalpel and rootless seedlings (four per flask) are placed on semisolid MS media, supplemented with 3% sucrose and 1.0 mg/L of BA (1.0 mg/L BAP equals 4.34 mM).
5. Jars are kept at 25°C, either under continuous light or a 16-h photoperiod.

3.2. Shoot Formation and Maintenance

Shoots are formed from seedling explants within the first 10 d of culture. After 3 wk, an average of eight shoots per explant had been formed. Shoots at this stage were excised, and individually used as explants for the next rounds of subculture. In our laboratory, shoots had been maintained for up to 12 mo, without noticing callus or root formation. In contrast, shoots formed on 3.0 and 5.0 mg/L BA tend to form callus and roots after long culture periods (usually more than 6 wk). Shoots proliferated mainly from axial buds and took between 4 and 8 d to unfold (**Fig. 1**). The number of shoots formed per explant using 1.0 mg/L of BA is the same, either under continuous light or under a 16-h photoperiod. However, photoperiod shoots are smaller than those under continuous light. Therefore, shoots had been transferred to fresh semisolid MS media every 4 wk and maintained on 1.0 mg/L of BA, under continuous light. Under these conditions, a single shoot can form up to 10 new shoots after 3 wk *(8)*.

3.3. Induction of Shoot Cultures in Liquid Media

Significant damage could be inflicted to the shoots during subcultures on semisolid media. Liquid cultures may reduce tissue manipulation and damage. Liquid cultures are induced by transferring four clusters with approx 12 shoots each into 50 mL of MS media, contained in 250-mL flasks. Media was supplemented with 1.0 mg/L of BA and 30 g/L sucrose and cultures have been maintained at 25°C under continuous light on a rotary shaker at 50 rpm *(8)*. Slow agitation avoids tissue vitrification and disintegration, whereas a reduced volume prevents shoots from sinking and necrosis (**Fig. 2**).

3.4. Alkaloid Extraction and Quantification

Alkaloids are extracted and quantified as reported in *(9)*. Tissues are collected, frozen, and lyophilized prior to analysis.

Fig. 1. Aspect of *Catharanthus roseus* shoots formed from seedlings after 3 wk on Murashige and Skoog media supplemented with 1.0 mg/L of 6-benzylaminopurine.

1. Homogenize 100 mg of freeze-dried tissue (*see* **Note 1**) with 5.0 mL methanol and incubate at 50°C for 2 h. Remove debris is by filtration and evaporate the methanolic extract to dryness at reduced pressure.
2. Resuspend the residue in 2.0 mL of 2.5% sulfuric acid (v/v) and extract it 3× with 2 vol of ethyl acetate, discarding the organic phase each time.
3. Adjust the pH of the aqueous phase to 9.0 with ammonium hydroxide and extract it three times with 2 vol of ethyl acetate, keeping the organic phase each time.
4. Reduce to dryness the organic phase and resuspend the residue in 500 µL methanol. Recovery of ajmalicine, catharanthine, and vindoline should be higher than 95%.
5. Load 10–20 µL of the extract on silica gel 60 F254 chromatography plates (aluminum supported; Merck, Damstadt Germany) and separate alkaloids by thin layer chromatography (TLC), using a mobile phase of ethyl acetate with a drop of concentrated ammonium hydroxide (use 20 µL of ammonium hydroxide per 110 mL of ethyl acetate).

Production of Monoterpenoid Indole Alkaloids

Fig. 2. Shoots cultured in liquid media after a 28-d period under continuous light.

6. Identify alkaloids by comparing their R's values with authentic standards (ajmalicine: 0.562, catharanthine: 0.375 and vindoline: 0.25) and by their colored reaction to 1% ceric ammonic sulfate diluted in 85% phosphoric acid.
7. Quantify alkaloids by *in situ* densitometry. Plates are scanned at 280 nm using the absorbance reflection mode in a CS-930 densitometer Shimadzu Corporation (Kyoto, Japan).
8. Alternatively, filter the alkaloid extract and chromatograph it on a Hypersil 5 µm ODS column (200 × 2.1 mm Hewlett-Packard, Palo Alto, CA), with a guard column filled with RSiL C18 HL, using a HPLC system equipped with a 600E controller system and a 991 photodiode array detector (Waters, Milford MA), set up to scan from 210 and 350 nm. The eluent system consists of 42% acetonitrile in water (0.1% triethylamine), with a linear increase to 65% over a 30 min period *(10)*.

3.5. Jasmonate Induction of Alkaloid Synthesis

The formation of secondary metabolites by plant cell cultures is frequently enhanced by diverse stress factors, such as osmotic shock, excess of salts or heavy metals, and fungal homogenates, among others. These effects are fre-

Table 1
Alkaloid Contents in Different Culture Types of *Catharanthus roseus*

Tissue	Ajmalicine	Alkaloid (μg/g DW) Catharanthine	Vindoline
Shoots in semisolid culture (30 d)	40	150	2500
Shoots in liquid culture (28 d)	35	60	1931
Root cultures (14 d)	470	287	nd
Callus culture (12 d)	nd	264	nd
Leaf (3rd pair)	46	291	4600

nd, not detected.

quently mediated by chemical transducers, such as jasmonates *(10,11)*. The exposure of *Catharanthus* cell cultures to jasmonte activates at the transcriptional level, the synthesis of several alkaloids, but not that of vindoline *(12–14)*, confirming that the tissue organization is required for this process. Because shoot cultures respond to jasmonate elicitation, this system can be used to promote vindoline production *(8)*.

1. Start liquid cultures with 2.0 g of shoots in 50 mL of MS media, contained in 250-mL flasks.
2. Keep cultures under continuous light and low agitation (50 rpm) for 12 d. Then expose them to 0.1 mM MeJa (apply 50 μL of the stock solution under aseptic conditions).
3. Harvest tissues after 12 and 24 h of exposure, remove excess of media with tissue paper, frozen in liquid nitrogen and lyophilize.
4. Extract and quantify alkaloids as described previously (*see* **Note 2**).

4. Notes

1. Maximal vindoline accumulation occurs after long culture periods (over 25 d), reaching similar values to those found in leaves from plants grown in our nursery (**Table 1**).
2. Vindoline content in 12-d-old cultures is around 400 μg/g DW and it increases up to 4000 μg/g DW in response to 0.1 μM jasmonate for 12 h. Neither longer exposure, nor higher jasmonate concentrations resulted in any further increases. In 24-d-old cultures, vindoline content is around 2000 μg/g DW and no significant increase in response to jasmonate exposure can be observed.

References

1. Moreno, P. R. H., van der Heijden, R., and Verpoorte, R. (1995) Cell and tissue cultures of *Catharanthus roseus*: a literature survey II. Updating from 1988 to 1993. *Plant Cell Tissue Org. Cult.* **42**, 1–25.
2. De Luca, V., Balsevich, J., Tyler, R. T., Eilert, U., Panchuk, B. D., and Kurz, W. G. W. (1986) Biosynthesis of indole alkaloids: developmental regulation of the biosynthetic pathway from tabersonine to vindoline in *Catharanthus roseus*. *J. Plant Physiol.* **125**, 147–156.
3. St. Pierre B., Vázquez-Flota, F., and De Luca, V. (1999) Multicellular compartmentation of *Catharanthus roseus* alkaloid biosynthesis predicts intercellular traslocation of pathway intermediate. *Plant Cell* **11**, 887–900.
4. Vázquez-Flota, F., St-Pierre, B., and De Luca V. (2000) The light activation of vindoline biosynthesis does not involve cytodifferentiation in *Catharanthus roseus* seedlings. *Phytochemistry* **55**, 531–536.
5. Krueger, R. J, Carew, D. P., Lui, J. H. C., and Staba, E. J. (1982) Initiation, maintenance and alkaloid content of *Catharanthus roseus* leaf organ cultures. *Planta Med.* **45**, 56–57.
6. Hirata, K., Horiuchi, M., Ando, T., Miyamoto, K., and Miura, Y. (1990) Vindoline and catharanthine production in multiple shoot culture of *Catharanthus roseus*. *J. Ferm. Bioeng.* **70**, 193–195.
7. Murashige, T. and Skoog, F. (1962) A revised for rapid growth and bioassays with tobacco tissue cultures. *Physiol. Plant* **15**, 435–497.
8. Hernández-Domínguez, E., Campos-Tamayo, F., and Vázquez-Flota, F. (2004) Vindoline synthesis in *in vitro* shoot cultures of *Catharanthus roseus*. *Biotechnol. Lett.* **26**, 671–674.
9. Monforte-Gonzalez M., Ayora-Talavera, T., Maldonado-Mendoza, I., and Loyola-Vargas, V. M. (1992) Quantitative analysis of serpentine and ajmalicine in plant tissues of *Catharanthus roseus* and hyoscyamine and scopolamine in root tissue of *Datura stramonium* by densitometry in thin layer chromatography. *Phytochem. Anal.* **3**, 117–121.
10. Aerts, R. J., Gisi, D., De Carolis, E., De Luca, V., and Baumann, T. W. (1994) Methyl jasmonate vapour increases the developmentally controlled synthesis of alkaloids in *Catharanthus* and *Cinchona* seedlings. *Plant J.* **5**, 635–643.
11. Menke F. L. H., Parchmann, S., Mueller, M. J., Kijne J. W., and Memelink, J. (1999) Involvement of the octadecanoic pathway and protein phosphorylation in fungal elicitor-induced expression of terpenoid indole alkaloid biosynthetic genes in *Catharanthus roseus*. *Plant Physiol.* **119**, 1289–1296.
12. Van der Fits, L. and Memelink, J. (2000) ORCA3, a jasmonate-responsive transcriptional regulator of plant primary and secondary metabolism. *Science* **289**, 295–297.
13. Vázquez-Flota, F. and De Luca, V. (1998) Jasmonate modulates developmental- and light- regulated alkaloid biosynthesis in *Catharanthus roseus*. *Phytochemistry* **49**, 395–402.
14. Vázquez-Flota F., De Luca, V., Carrillo-Pech, M., Canto-Flick A., and Miranda-Ham M. L. (2002) Vindoline biosynthesis is transcriptionally blocked in *Catharanthus roseus* cell suspension cultures. *Mol. Biotechnol.* **22**, 1–8.

30

Methods for Regeneration and Transformation in *Eschscholzia californica*

A Model Plant to Investigate Alkaloid Biosynthesis

Benjamin P. MacLeod and Peter J. Facchini

Summary

Eschscholzia californica Cham. (California poppy) is a plant species that accumulates pharmacologically active alkaloids biosynthetically related to the morphinan alkaloids of *Papaver somniferum*. This, in combination with the relative ease with which it is propagated in vitro, makes it a key model for benzylisoquinoline biosynthesis. Transformation techniques are an important tool for these studies and for metabolic engineering attempts. *Agrobacterium* mediated transformation techniques for this model species have been developed in our lab and used for modulation of transcript levels relevant to the biosynthesis of these alkaloids. Here we describe the techniques used in our lab for production of transgenic callus, hairy root cultures, and whole plants.

Key Words: *Agrobacterium rhizogenes*; *Agrobacterium tumefaciens*; benzylisoquinoline alkaloids; California poppy; *Eschscholzia californica*; hairy roots; plant transformation protocol; somatic embryogenesis; transgenic callus.

1. Introduction

Eschscholzia californica Cham. (California poppy, also commonly cited as *E. californica* Cham.), a member of the Papaveraceae, accumulates benzylisoquinoline alkaloids derived from L-tyrosine. This class of alkaloids appears predominantly in higher plant species, most notably in the Ranunculales *(1)*. Medicinally important benzylisoquinolines used today include the commonly used narcotic, morphine, from the opium poppy, *Papaver somniferum* L., and the anti-plaque agent sanguinarine *(2)*, normally collected from bloodroot

(*Sanguinaria canadensis*), which also accumulates in *E. californica* (*3*). The pharmacological activities of these alkaloids may give some hint to the ecological role that the alkaloid plays, morphine could act as a feeding deterrent whereas sanguinarine is likely a phytoalexin (*4*). Supporting these roles, a number of the pathways involved in the biosynthesis of these alkaloids are regulated both developmentally (*5*) and in response to environmental factors such as biotic elicitors (*6*). A number of related regulatory networks are currently being mapped out at the molecular level in other alkaloid producing plant models (*7*).

Because *E. californica* produces benzylisoquinoline alkaloids biosynthetically related to the morphinan alkaloids of *P. somniferum*, information accumulated from this plant may be used for comparative studies in alkaloid biosynthesis and regulation. A sanguinarine-accumulating species, *E. californica* is an interesting model species for alkaloid biosynthesis as it shows various inducible responses in this pathway when cultured as a suspension (*8*) independent of the hypersensitive response (*9*). Additionally, this plant has proven to be easy to work with in vitro relative to *P. somniferum*.

Transformation of *E. californica* as with other plants used as models in secondary metabolism is a prerequisite to many approaches of study and is the primary tool to engineering secondary metabolism (*10*). The ability to transform plants allows for modulation of existing genes or introduction of foreign genes. Modulation of existing genes can be used in genetic selection systems to identify novel biosynthetic or regulatory genes (*11*), and to alter product accumulation, which although still a hit and miss approach to metabolic engineering, indirectly gives us a better understanding of how secondary metabolism is regulated in a plant (*12,13*). Introduction of novel genes may be used to report expression of promoters (*14*), to determine cellular localization of protein products (*15*), or to study protein–protein interactions in vivo (*16*). Novel gene introduction can also be used for metabolic engineering, resulting in the production of compounds normally not produced in the plant (*17*).

In our lab, *Agrobacterium* mediated transformation of *E. californica* was developed with many of these goals in mind. We used these methods to alter expression of genes involved in benzylisoquinoline alkaloid biosynthesis and as a tool to determine how these alterations affect expression of other biosynthetic genes (*12,13*). Hairy roots of California poppy have been produced using *Agrobacterium rhizogenes* (*18*), whereas *Agrobacterium tumefaciens* has been utilized for production of transgenic callus (*12*) and whole plants (*19*). Recently, another lab has also described the transformation of *E. californica* tissues (*20*).

1. *A. tumefaciens* mediated transformation of *E. californica* callus.
2. Regeneration of *E. californica* hairy roots transformed with *A. rhizogenes*.
3. Transformation of *E. californica* somatic embryos with *A. tumefaciens* for whole plant regeneration.

2. Materials
2.1. Agrobacterium *Preparation*

1. Electrocompetent *Agrobacterium* (*A. tumefaciens* GV3101 for transgenic callus or whole plants and *A. rhizogenes* R1000 for transgenic hairy roots).
2. Vector harboring gene of interest and kanamycin resistance.
3. Luri Bertani (LB) media with 50 mg/L kanamycin solidified with 1.5% (w/v) Bacto Agar in 100 × 15-mm Petri dishes.
4. AB media with 50 mg/L kanamycin solidified with 1.5% (w/v) Bacto Agar in 100 × 15-mm Petri dishes.
5. Standard B_5 *(21)* liquid media without sucrose.
6. Sterile working area, inoculation loop.
7. Electroporation apparatus.
8. Culture chamber (28°C).

2.2. E. californica *Explant Preparation*

1. *E. californica* seeds (store at 4°C when not in use to increase germination frequency).
2. 70% Ethanol.
3. 50% Bleach solution.
4. Detergent (e.g., Tween-20 or Triton X-100).
5. Gyratory shaker.
6. Sterile distilled water.
7. Sterile tissue or Kimwipes®.
8. Laminar flow hood, spatula, forceps, scalpel.
9. Germination media (standard B_5 media lacking sucrose, solidified with 0.3% (w/v) Gelrite in 100 × 15-mm Petri dishes) (*see* **Note 1**).
10. Growth chamber (25°C, 35 mmol/m^2/s light (Sylvania Gro-Lux Wide Spectrum), 16-h photoperiod).
11. Standard B_5 liquid media (optional).

2.3. A. tumefaciens-*Mediated Transformation of* E. californica *Callus*

1. Suspension of *A. tumefaciens* GV3101 prepared as in **Subheading 3.1**.
2. *E. californica* explants prepared as in **Subheading 3.2**.
3. Sterile tissue or Kimwipes.
4. Callus induction medium (standard B_5 media with 1 mg/L 2,4-dichlorophenoxyacetic acid (2,4-D) solidified with 0.8% (w/v) Phytagar in 100 × 15-mm Petri dishes) (*see* **Note 1**).
5. Callus induction medium supplemented with 70 mg/L paramomycin and 300 mg/L Timentin (store at 4°C, good for 1 wk at this temperature).
6. Laminar flow hood, forceps.
7. Growth chamber (25°C, no light).
8. Sterile distilled water.
9. Sterile distilled water supplemented with 300 mg/L Timentin (Timentin is unstable in solution, prepare fresh).

10. Suspension induction medium (standard B_5 media with 1 mg/L 2,4-D, 50 mg/L paramomycin, 300 mg/L Timentin, 1 g/L casein enzymatic hydrolysate) (only if suspension cultures required) (*see* **Note 2**).

2.4. Regeneration of Transgenic E. californica Hairy Roots With A. rhizogenes

1. Suspension of *A. rhizogenes* R1000 prepared as in **Subheading 3.1.**
2. *E. californica* explants prepared as in **Subheading 3.2.**
3. Sterile tissue or Kimwipes.
4. Hormone free solid media (standard B_5 media solidified with 0.8% (w/v) Phytagar in 100 × 15-mm Petri dishes) (*see* **Note 1**).
5. Laminar flow hood, forceps, scalpel.
6. Sterile distilled water.
7. Sterile distilled water supplemented with 300 mg/L Timentin (Timentin is unstable in solution, prepare fresh).
8. Hormone free solid B_5 media supplemented with 70 mg/L paramomycin and 300 mg/L Timentin (store at 4°C, good for 1 wk at this temperature).
9. Growth chamber (25°C, no light).
10. Hormone free liquid media supplemented with 70 mg/L paramomycin and 300 mg/L Timentin (if liquid root cultures are required) (*see* **Note 3**).
11. Growth chamber (25°C, 35 µmol/m^2/s light (Sylvania Gro-Lux Wide Spectrum), 16-h photoperiod) with gyratory shaker at 100 rpm.

2.5. Somatic Embryogenesis for Production of Transformed E. californica Plants

1. Suspension of *A. tumefaciens* GV3101 prepared as in **Subheading 3.1.**
2. *E. californica* explants prepared as in **Subheading 3.2.**
3. Sterile tissue or Kimwipes.
4. Primary callus induction media (standard B_5 media with 2 mg/L naphthaleneacetic acid (NAA) and 0.1 mg/L benzyladenine (BA) solidified with 0.8% (w/v) Phytagar in 100 × 15-mm Petri dishes) (*see* **Note 1**).
5. Growth chamber (25°C, no light).
6. Sterile distilled water.
7. Sterile distilled water supplemented with 300 mg/L Timentin (Timentin is unstable in solution, prepare fresh).
8. Laminar flow hood, forceps.
9. Primary callus induction media supplemented with 70 mg/L paramomycin and 300 mg/L Timentin.
10. Somatic embryo induction media (standard B_5 media with 1 mg/L NAA, 0.5 mg/L BA, 70 mg/L paramomycin, and 300 mg/L Timentin solidified with 0.8% (w/v) Phytagar in 100 × 15-mm Petri dishes.
11. Hormone free regeneration media (standard B_5 media with 70 mg/L paramomycin and 300 mg/L Timentin solidified with 0.8% (w/v) Phytagar in GA7 vessels.

12. Growth chamber (25°C, 35 µmol/m²/s light (Sylvania Gro-Lux Wide Spectrum), 16-h photoperiod).
13. Sterilized pots filled with moist vermiculite and covered with polyethylene bags.

3. Methods
3.1. Agrobacterium Preparation

1. Electroporate *Agrobacterium* strains (*see* **Note 4**) with a plasmid harboring gene of interest along with a kanamycin-selectable marker (*see* **Note 5**) and select for transformants on LB supplemented with 50 mg/L kanamycin (LB-Kan) by growing at 28°C for 2 d.
2. Streak a single colony selected on LB-Kan onto a plate with AB salts supplemented with 50 mg/L kanamycin (AB-Kan) (*see* **Note 6**) to confirm transformation and to obtain colonies for inoculation of liquid cultures.
3. Pick a single colony off the AB-Kan plates and inoculate 10 mL LB liquid supplemented with 50 mg/L kanamycin. Grow this culture overnight at 28°C on a gyratory shaker until mid to late log phase is reached.
4. Once an optical density (OD)600 of 0.5 to 1.0 is reached, collect the cells by centrifugation at 750*g* for 10 min. Decant the cell pellet and resuspend to an OD600 near 1.0 in liquid inoculation medium consisting of B_5 salts and vitamins. This cell suspension is ready to use immediately in any of the transformation protocols below.

3.2. E. californica *Explant Preparation*

1. Surface-sterilize seeds of *E. californica* (stored at 4°C when not in use) by rinsing first in 70% ethanol for 1 min. Rinse 70% ethanol from seeds with sterile distilled water. Decant and add enough 50% bleach solution (2.6% sodium hypchlorite) to cover seeds. To this, add a small amount of detergent (*see* **Note 7**) and shake seeds in a gyratory shaker for 20 min.
2. Rinse seeds with sterile distilled water for long enough to remove signs of detergent or bleach and then blot dry with sterile paper towel or tissue. Spread seeds on germination media (*see* **Note 8**) and germinate at 25°C under cool white fluorescent lights at 35 µmol/m²/s and set to a 16-h photoperiod.
3. Remove seedlings from germination media and de-root (*see* **Note 9**) in B_5 liquid media lacking sucrose 5–10 d after germination. De-rooted seedlings are then sliced longitudinally along the hypocotyl. The divided cotyledons may be used for each of the transformation procedures below (*see* **Note 10**).

3.3. A. tumefaciens-*Mediated Transformation of* E. californica *Callus*

1. Add suspension of *A. tumefaciens* GV3101 harboring the gene of interest on a kanamycin resistance encoding plasmid (*see* **Note 11**) as prepared in **Subheading 3.1.** to *E. californica* explants (*see* **Note 12**) as prepared in **Subheading 3.2.**
2. After immersion in the *A. tumefaciens* suspension for 15–30 min, blot the cotyledonary explants dry with sterile paper towel or tissue and transfer to callus induc-

tion media. Explants on callus induction media are incubated without light at 25°C.
3. After 2 d co-cultivation (*see* **Note 13**), rinse *E. californica* explants of *A. tumefaciens* 2× with distilled water. Then allow the explants to sit in a 100 mg/L Timentin solution (*see* **Note 14**) for 20–30 min. Blot the cleaned cotyledons dry with a sterile paper towel or tissue and transfer to callus induction media with 70 mg/L paramomycin and 300 mg/L Timentin. Incubate Petri dishes in growth chamber at 25°C in the dark (*see* **Note 15**).
4. To produce transgenic cell suspensions, transfer 0.5–1 g friable callus produced after 12 wk of incubation on callus induction medium with selection to 30 mL suspension induction medium in a 125-mL Erlenmeyer flask (*see* **Note 2**).

3.4. Regeneration of Transgenic E. californica *Hairy Roots* With A. rhizogenes

1. Add suspension of *A. rhizogenes* R1000 (*see* **Note 16**) as prepared in **Subheading 3.1.** to the *E. californica* explants (*see* **Note 12**) prepared in **Subheading 3.2.** The bacteria should harbor the gene of interest in a plasmid encoding kanamycin resistance.
2. After 15–30 min incubation at room temperature, cotyledonary explants are blotted dry on sterile paper towel and placed with the cut surface facing down on hormone-free solid media. Incubate plates at 25°C in the dark to allow *A. rhizogenes* to co-cultivate with the *E. californica* explants.
3. After 2-d co-cultivation at 25°C (*see* **Note 13**), remove cotyledonary segments and shake gently with sterile distilled water twice. Transfer rinsed explants to sterile distilled water containing 100 mg/L Timentin (*see* **Note 17**) and incubate for 20–30 min.
4. Transfer cleaned *E. californica* cotyledonary explants to hormone-free selection media supplemented with 70 mg/L paramomycin and 300 mg/L Timentin. Return explants to the dark at 25°C for root initiation (*see* **Note 18**).
5. Isolate roots initiated from wound sites after 4–5 wk and transfer to fresh hormone-free selection medium (*see* **Note 19**). Continue transfers of roots to hormone-free selection medium every 1–2 wk to obtain rapidly growing root cultures.
6. For liquid cultures of hairy roots, transfer 500 mg of rapidly growing hairy root culture to 40 mL hormone-free liquid media with 70 mg/L paramomycin and 300 mg/L^{-1} Timentin (*see* **Note 3**) in a 125-mL Erlenmeyer flask. Grow roots at 25°C under cool white fluorescent light at 35 µmol/m^2/s^1 and 16-h photoperiod on a gyratory shaker at 100 rpm. Continue to transfer cultures to fresh B_5 liquid media every 2 wk.

3.5. Somatic Embryogenesis for Production of Transformed E. californica *Plants*

1. Add suspension of *A. tumefaciens* GV3101 (*see* **Note 11**) as prepared in **Subheading 3.1.** to *E. californica* explants (*see* **Subheading 3.2.** and **Note 12**). The

bacteria should harbor the gene of interest in a plasmid encoding kanamycin resistance.
2. After incubation of wounded *E. californica* explants in *A. tumefaciens* for 15–30 min, blot them dry on using sterile paper towel. Transfer explants to primary callus induction medium (*see* **Note 20**). Incubate explants on this primary callus induction medium at 25°C in the dark for 2 d.
3. After 2-d of co-cultivation (*see* **Note 13**), rinse the explant tissue twice with sterile distilled water using gentle shaking and then transfer the explants to distilled water supplemented with 100 mg/L Timentin (*see* **Note 14**). Leave explants in this solution for 20–30 min and then blot dry using sterile paper towel or tissue.
4. Transfer clean explants to fresh primary callus induction medium supplemented with 70 mg/L paramomycin and 300 mg/L Timentin. Return explants to growth chamber set to 25°C without light for 4–5 wk to allow primary callus to form.
5. Transfer primary callus to somatic embryo induction medium. Return primary callus on this induction medium to a growth chamber at 25°C in the dark for 3–4 wk to allow somatic embryos to develop.
6. Transfer somatic embryos to hormone free regeneration media (*see* **Note 21**). These are left in a growth chamber at 25°C with 35 µmol/m^2/s light provided by cool fluorescent bulbs with a 16-h photoperiod for 3–4 wk.
7. Transplant putative transgenic plantlets (*see* **Note 22**) from GA7 vessels to pots with sterilized vermiculite covered in a polyethylene bag to keep humidity high and grow for 1 wk in a growth chamber set to 25°C with 35 µmol/m^2/s light provided by cool fluorescent bulbs at a 16-h photoperiod. After 1 wk, remove polyethylene bags and continue growth in the same chamber for another 3 wk. Once plants are acclimatized, transplant to regular soil and grow under standard greenhouse conditions.

4. Notes

1. Although Gelrite as a solidifying agent has advantages over Phytagar, including the convenience with which media can be removed from plant tissues, it cannot be used with paramomycin as a selection agent. We have noted that paramomycin selection is much less effective in the presence of Gelrite, possibly because of binding or precipitation of the paramomycin during media preparation.
2. To produce healthy suspension cultures, transfer the required amount of callus to liquid media and grow as described. On the first one to three transfer dates (generally transfer suspensions every 5–7 d for high-growth rates and reduced phenolic production) remove the flask from the gyratory shaker to allow the suspension to settle. Next, remove as much liquid media above the settled cells as possible and replace with the same volume of fresh media. In this way, a high density of cells is obtained which allows the suspension to cure the media. If slow growth rates are still a problem, it may be feasible to reduce selection pressure as long as the culture is monitored for transgenic cells as the reduction in selection pressure may allow a small population of wild-type cells surviving by living off the activity of transgenic cells to "take over" the culture. Once the

suspension has acclimatized and the growth rates have increased, continue transfers by adding 15 mL of suspension to 50 mL of fresh media in a 250-mL Erlenmeyer without letting cells settle.
3. Auxins may be added to this media to increase the growth rate of root cultures in suspension culture. We have noted that addition of NAA or indole butyric acid (IBA) at 0.5 or 1 mg/L increases growth rates but also increases the formation of callus *(18)*. The use of IAA at 1 mg/L on the other hand results in a similar increase in growth rates without concomitant formation of callus. An alternative is to reduce selection pressure after transferring the roots to liquid media, which may increase growth rates, but requires intermittent surveys to determine if wild-type tissues are present.
4. *A. rhizogenes* R1000, R1200rolD and 13333 have all been used successfully for production of transgenic hairy roots of *E. californica (18)* whereas *A. tumefaciens* GV3101 has been used to genetically transform both callus cultures *(12)* and whole plants *(19)* of *E. californica*.
5. Typically, the binary vector pBI121, which contains a 35S CaMV-driven *Npt*II gene is used for transformation experiments *(22)*. This vector along with kanamycin/paramomycin selection has a multiple cloning site for insertion of genes of interest. More recently we have used pCAMBIA series vectors containing an *Npt*II gene *(23)*.
6. The use of AB salts generally results in lower contamination rates, but LB may also be used and is easier to prepare. If the *Agrobacterium* strain being used has antibiotic resistance, the use of AB is less important. For example, GV3101 may be grown on media with gentamycin and rifampicin (*see* **Note 11**). Re-streaking colonies obtained after electroporation is important to ensure the colony picked does have the resistance conferred by the plasmid being used.
7. Detergents or "wetting agents" such as Tween-20 or Triton X-100 are used during sterilization of *E. californica* seeds to increase penetration of hypochlorite ions. It may be possible to obtain similar results using vacuum infiltration of a weak bleach solution.
8. B_5 media lacking sucrose is used for germination of *E. californica* to reduce growth of fungal or bacterial contaminants. The easiest way to plate *E. californica* seeds is to transfer a small mound (approx 50–80 seeds) to the middle of a 100 × 15 mm plate of germination media. Then, seeds can be spread evenly across the germination media using a flame-sterilized spatula. Germination rates are usually very high.
9. Seedlings processed for transformation tend to dry out after de-rooting and bisection of the hypocotyl. To avoid this, perform all seedling manipulations in a Petri dish with enough distilled water or B_5 media to cover the seedlings. This has the added advantage of being a solution where phenolics released by the wounded tissue can be collected for *Agrobacterium* stimulation (*see* **Note 12**). If older seedlings with true leaves are used, it is advantageous to remove them at this point as well. If these leaves are left, the seedlings tend to become entangled while rinsing explants of *Agrobacterium* after co-cultivation steps.

10. As an alternative to seedlings, wounded somatic embryos may be used as an explant source *(18)*. These somatic embryos may be produced on solid media *(29)* or as suspension cultures *(30)*.
11. *A. tumefaciens* GV3101 is equivalent to *A. tumefaciens* C58C1RifR-pMP90. This is a disarmed strain that carries chromosomal resistance to rifampicin *(26)* and the pMP90 Ti plasmid it carries encodes gentamycin resistance *(27)*. With this strain, all media can be supplemented with 25–50 µg/mL gentamycin and 10 µg/mL rifampicin.
12. The addition of *A. tumefaciens* or *A. rhizogenes* to the solution in which seedlings were excised (*see* **Note 9**) takes advantage of phenolics released by the wounded *E. californica* tissue, which should induce *Agrobacterium* virulence gene expression.
13. Two days after co-cultivation it is normal to have the explants partially or entirely covered in *Agrobacterium*. Occasionally the morphology of *Agrobacterium* at this point has been noted as unusual (e.g., discolored), but cultures of these unusual morphologies revert to normal morphologies if grown on LB or AB media. Explants after 2 d of co-cultivation also display unusual morphology including curled ends and hardening of the tissue. Removal of all *Agrobacterium* after co-cultivation has been noted as difficult with the strains used here. We have noted that silver nitrate added at 3–10 mg/L is helpful in reduction of *Agrobacterium* overgrowth after transfer of explants to selection media during production of hairy root cultures. Plates with this concentration added tend to turn dark, which can be minimized by adding filter-sterilized silver nitrate to the media at as low a temperature as possible and by storing plates in a cool, dark area.
14. Media containing Timentin (*see* **Note 17** for composition of Timentin) for reduction of *Agrobacterium* growth must be prepared fresh as it is heat sensitive in solution. Store Timentin as a powder at 4°C and add fresh to sterilized media. Storage of solutions of penicillins below 0°C is also not recommended, so prepararation of frozen concentrated stocks of Timentin is not advised. For pouring plates be sure to add the Timentin to the media at as low a temperature as possible. Plates prepared this way retain antibiotic activity for approx 1–2 wk if stored at 4°C.
15. Transfer *E. californica* callus at this point every 2–4 wk depending on how rapidly they discolor. We have also noted that *Agrobacterium* overgrowth on this media is much lower than in hairy root or somatic embryo transformations. This might be owing to the presence of 2,4-D.
16. *A. rhizogenes* R1000 is an armed *Agrobacterium* strain that has the same chromosomal background as *A. tumefaciens* C58 *(28)* along with the complete Ri plasmid of *A. rhizogenes* A4 (pRiA4b) *(24)* that includes virulence and tumor-inducing regions *(25)*.
17. Timentin is 30:1 (w:w) mixture of disodium ticarcillin and clavulanic acid.
18. Root initials may be visible as early as 2–3 wk after transfer to hormone-free selection media. Usually the roots are not large enough for excision and transfer as in **Subheading 3.4.5.**, until 4–5 wk have passed. Leaving the explants on hor-

mone-free selection media longer at this point may be advantageous for increased root growth rate. Transfer of explants to fresh hormone-free selection media every 2–4 wk reduces chance of *Agrobacterium* overgrowth.
19. Roots are usually reddish in color and covered in whitish root hairs. More than one root often emerges from a single explant and it is likely that each of these roots is to the result of an individual transformation event. The original explant if kept on hormone-free selection media for long enough often begins to grow as a large lump of callus covered in root initials. Occasionally, somatic embryos may spontaneously emerge from hairy roots. This tends to become more frequent with longer culture periods especially if transfer to fresh media is performed less often.
20. The hormones used for primary callus induction differ from those used in production of non-transgenic *E. californica* somatic embryos *(29)*. The use of 2,4-D in primary callus induction media as is used in nontransgenic somatic embryogenesis is counterproductive to *Agrobacterium* mediated transformation (*see* **Note 15**). It may be possible to start explants on primary callus induction media as described in **Subheading 3.5.2.** and then transfer to primary callus induction media as described for non-transgenic *E. californica* somatic embryogenesis *(29)* to reduce *Agrobacterium* overgrowth and still obtain transformed primary callus.
21. Alternatively, if transgenic selection seems certain enough, transfer to B_5 supplemented with 300 mg/L Timentin and solidify with Gelrite rather than Phytagar to make removal of media for transfer to vermiculite easier (*see* **Note 1**).
22. There are numerous ways to test for presence of transgene in these plants, which will not be dealt with in detail here. In our own work with California poppy we have commonly used PCR or a GUS reporter to indicate presence of a transgene.

References

1. Jensen, U. (1995) Secondary compounds of the Ranunculiflorae. *Plant Syst. Evol.* (Suppl.) **9**, 85–97.
2. Dzink, J. L. and Socransky, S. S. (1985) Comparative in vivo activity of snaguinarine against oral microbial isolates. *Antimicrob. Agents Chemother.* **27**, 663–665.
3. Tomé, F., Colombo, M. L., and Califiroli, L. (1999) A comparative investigation on alkaloid composition in different populations of *Eschscholtzia californica* Cham. *Phytochem. Anal.* **10**, 264–267.
4. Cline, S. D. and Coscia, C. J. (1988) Stimulation of sanguinarine production by combined fungal elicitation and hormonal deprivation in cell suspension cultures of *Papaver bracteatum*. *Plant Physiol.* **86**, 161–165.
5. Kamo, K. K., Kimoto, W., Hsu, A. F., Mahlberg, P. G., and Bills, D. D. (1982) Morphinane alkaloids in cultured tissues and redifferentiated organs of *Papaver somniferum*. *Phytochem.* **21**, 219–222.
6. Mahady, G. B. and Beecher, C. W. (1994) Elicitor-stimulated benzophenanthridine alkaloid biosynthesis in bloodroot suspension cultures is mediated by calcium. *Phytochem.* **37**, 415–419.

7. Menke, F. L. H., Parchmann, S., Mueller, M. J., Kijne, J. W., and Memelink, J. (1999) Involvement of the octadecanoid pathway and protein phosphorylation in fungal elicitor-induced expression of terpenoid indole alkaloid biosynthetic genes in *Catharanthus roseus*. *Plant Physiol.* **119,** 1289–1296.
8. Schumacher, H. M., Gundlach, H., Fiedler, F., and Zenk, M. H. (1987) Elicitation of benzophenanthridine alkaloid synthesis in *Eschscholtzia* cell cultures. *Plant Cell Rep.* **6,** 410–413.
9. Roos, W., Dordschbal, B., Steighardt, J., Hieke, M., Weiss, D., and Saalbach, G. (1999) A redox-dependent, G-protein-coupled phospholipase A of the plasma membrane is involved in the elicitation of alkaloid biosynthesis in *Eschscholzia californica*. *Biochim. Biophys. Acta* **1448,** 390–402.
10. Sato, F., Hashimoto, T., Hachiya, A., et al. (2001) Metabolic engineering of plant alkaloid biosynthesis. *Proc. Natl. Acad. Sci. USA* **98,** 367–372.
11. van der Fits, L., Hilliou, F., and Memelink, J. (2001) T-DNA activation tagging as a tool to isolate regulators of a metabolic pathway from a genetically non-tractable plant species. *Transgenic Res.* **10,** 513–521.
12. Park, S. U., Yu, M., and Facchini, P. J. (2002) Antisense RNA-mediated suppression of benzophenanthridine alkaloid biosynthesis in transgenic cell cultures of California poppy. *Plant Physiol.* **128,** 696–706.
13. Park, S. U., Yu, M., and Facchini, P. J. (2003) Modulation of berberine bridge enzyme levels in transgenic root cultures of California poppy alters the accumulation of benzophenanthridine alkaloids. *Plant Mol. Biol.* **51,** 153–164.
14. Ouwerkerk, P. B. and Memelink, J. (1999) Elicitor-responsive promoter regions in the tryptophan decarboxylase gene from *Catharanthus roseus*. *Plant Mol. Biol.* **39,** 129–136.
15. Saslowsky, D. and Winkel-Shirley, B. (2001) Localization of flavonoid enzymes in *Arabidopsis* roots. *Plant J.* **27,** 37–48.
16. Panicot, M., Minguet, E. G., Ferrando, A., et al. (2002) A polyamine metabolon involving aminopropyl transferase complexes in *Arabidopsis*. *Plant Cell* **14,** 2539–2551.
17. Bohmert, K., Balbo, I., Kopka, J., et al. (2000) Transgenic *Arabidopsis* plants can accumulate polyhydroxybutyrate to up to 4% of their fresh weight. *Planta* **211,** 841–845.
18. Park, S. U. and Facchini, P. J. (2000) *Agrobacterium rhizogenes*-mediated transformation of opium poppy, *Papaver somniferum* L., and California poppy, *Eschscholzia californica* Cham., root cultures. *J. Exp. Bot.* **51,** 1005–1016.
19. Park, S. U. and Facchini, P. J. (2000) *Agrobacterium*-mediated genetic transformation of California poppy, *Eschscholzia californica* Cham., via somatic embryogenesis. *Plant Cell Rep.* **19,** 1006–1012.
20. Lee, J. and Pedersen, H. (2001) Stable genetic transformation of Eschscholzia californica expressing synthetic green fluorescent proteins. *Biotechnol. Prog.* **17,** 247–251.
21. Gamborg, O. L., Miller, R. A., and Ojima, K. (1968) Nutrient requirements of suspension cultures of soybean root cells. *Exp. Cell Res.* **50,** 151–158.

22. Jefferson, R.A., Kavanagh, T.A., Bevan, M.W. (1987) GUS fusions: β-glucuronidase as a sensitive and versatile gene fusion marker in higher plants. *EMBO J.* **6,** 3901–3907.
23. Roberts, C. S., Rajagopal, S., Smith, L.A., et al. pCAMBIA binary vectors: A comprehensive set of modular vectors for advanced manipulations and efficient transformation of plants by both *Agrobacterium* and direct DNA uptake methods. Accessed on 5, 22, 2005. Available at: www.cambia.org/pCAMBIA_vectors.html; Genbank accession numbers: AF234290 to AF234316.
24. Van Larebeke, N., Engler, G., Holsters, M., et al. (1974) Large plasmid in Agrobacterium tumefaciens essential for crown gall-inducing ability. *Nature* **252,** 169–170.
25. Simoens, C., Alliottel, T. H., Mendel, R., et al. (1986) A binary vector for transferring genomic libraries to plants. *Nucleic Acids Res.* **14,** 8073–8090.
26. Goodner, B., Hinkle, G., Gattung, S., et al. (2001) Genome sequence of the plant pathogen and biotechnology agent *Agrobacterium tumefaciens* C58. *Scien1ce* **294,** 2323–2328.
27. White, F. F., Taylor, B. H., Huffman, G. A., Gordon, M. P., and Nester, E. W. (1985) Molecular and genetic analysis of the transferred DNA regions of the root-inducing plasmid of *Agrobacterium rhizogenes. J. Bacteriol.* **164,** 33–44.
28. White, F. F. and Nester, E. W. (1980) Relationship of plasmids responsible for hairy root and crown gall tumorigenicity. *J. Bacteriol.* **144,** 710–720.
29. Park, S. U. and Facchini, P. J. (2000) High-efficiency somatic embryogenesis and plant regeneration in California poppy, *Eschscholzia californica* Cham. *Plant Cell Rep.* **19,** 421–426.
30. Park, S. U. and Facchini, P. J. (2001) Somatic embryogenic cell suspension cultures of California poppy, *Eschscholzia californica* Cham. *In Vitro Cell. Dev. Biol. Plant* **37,** 35–39.

APPENDIX A

The Components of the Culture Media

Víctor M. Loyola-Vargas

1. Introduction

A medium is defined as a formulation of inorganic salts and organic compounds (apart from major carbohydrate sources and plant growth regulators) used for the nutrition of plant cultures (*1*). The success in the technology and application of plant tissue culture is greatly influenced by the nature of the culture medium used. A better understanding of the nutritional requirements of cultured cells and tissues can help to choose the most appropriate culture medium for the explant used.

Most of our knowledge of the nutrition of plant cultures comes from the solutions developed for the hydroponic culture of intact plants during the last part of the 19th and beginning of the 20th centuries. The major changes made in the composition of the media since the early days have been the introduction of ammonium as nitrogen source, together with the higher amount of nitrate and potassium, as well as the use of organic additives mainly vitamins, and amino acids.

Plant tissue culture provide major (*macro-*), minor (*micro-*), a carbon source and trace amounts of certain organic compounds, notably vitamins, amino acids and plant growth regulators (**Tables 1–8**). In general, the tissue culture medium must contain the 16 essential elements for plant growth (*2*).

The most important difference among media may be the overall salt level. There seems to be basically three different media types by this classification: high salt (e.g,. Murashige and Skoog [MS] medium) (**Table 1**), intermediate level (e.g., Nitsch and Nitsch) (**Table 6**), and low salt media (e.g., White) (**Table 5**).

Table 1
Murashige and Skoog (5) Media Composition

	mg/L	mM
Macroelements		
NH_4NO_3	1650	20.60
KNO_3	1900	18.80
$CaCl_2 \cdot 2H_2O$	440	3.00
$MgSO_4 \cdot 7H_2O$	370	1.50
KH_2PO_4	170	1.25
Na_2EDTA	37.30	0.10
$FeSO_4 \cdot 7H_2O$	27.80	0.10
Microelements	mg/L	µM
H_3BO_3	6.20	100.0
$MnSO_4 \cdot 4H_2O$	22.30	100.0
$ZnSO_4 \cdot 7H_2O$	8.60	30.0
KI	0.83	5.0
$Na_2MoO_4 \cdot 2H_2O$	0.25	1.0
$CuSO_4 \cdot 5H_2O$	0.025	0.1
$CoCl_2 \cdot 6H_2O$	0.025	0.1
Organic components	mg/L	µM
myo-Inositol	100.0	555.10
Nicotinic acid	0.5	4.06
Pyridoxine·HCl	0.5	2.43
Thiamine·HCl	0.1	0.30
Glycine	2.0	26.60
Sucrose	30,000	87.64 mM
pH	5.7–5.8	

Researchers quickly found that the addition of "complexes" to the basic medium frequently resulted in successful growth of the tissues and organs. Some of these complexes have included green tomato extract, coconut milk, orange juice, casein hydrolysate, and yeast and malt extract (2).

Table 2
Linsmaier and Skoog (7) Media Composition

	mg/L	mM
Macroelements		
NH_4NO_3	1650	20.60
KNO_3	1900	18.80
$CaCl_2 \cdot 2H_2O$	440	3.00
$MgSO_4 \cdot 7H_2O$	370	1.50
KH_2PO_4	170	1.25
Na_2EDTA	37.30	0.1
$FeSO_4 \cdot 7H_2O$	27.80	0.1
Microelements	mg/L	μM
H_3BO_3	6.20	100.0
$MnSO_4 \cdot 4H_2O$	22.30	100.0
$ZnSO_4 \cdot 4H_2O$	8.60	30.0
KI	0.83	5.0
$Na_2MoO_4 \cdot 2H_2O$	0.25	1.0
$CuSO_4 \cdot 5H_2O$	0.025	0.1
$CoCl_2 \cdot 6H_2O$	0.025	0.1
Organic components	mg/L	μM
myo-Inositol	100.0	555.10
Thiamine·HCl	0.4	1.20
Sucrose	30,000	87.64 mM
pH	5.6	

It is very important when choosing a medium to take into account that some of the components of the culture media are not just nutriments and can have a very deep influence on the growth of the cultures in the differentiation process. Another important fact is that vigorous colonies of callus tissue required more nutriments than slowly growing ones, while the situation can be reverse for other nutriments.

Table 3
Gamborg et al. *(8,9)* Media Composition

	mg/L	mM
Macroelements		
$(NH_4)_2SO_4$	134	1.0
KNO_3	2528	25.0
$CaCl_2·2H_2O$	150	1.0
$MgSO_4·7H_2O$	250	1.0
$NaH_2PO_4·H_2O$	150	1.1
Na_2EDTA	37.30	0.1
$FeSO_4·7H_2O$	27.80	0.1
Microelements	mg/L	µM
H_3BO_3	3.0	50.0
$MnSO_4·H_2O$	10.0	60.0
$ZnSO_4·7H_2O$	2.0	7.0
KI	0.75	4.5
$Na_2MoO_4·2H_2O$	0.25	1.0
$CuSO_4·5H_2O$	0.025	0.1
$CoCl_2·6H_2O$	0.025	0.1
Organic components	mg/L	µM
myo-Inositol	100.0	555.10
Nicotinic acid	1.0	8.12
Pyridoxine·HCl	1.0	4.86
Thiamine·HCl	10.0	30.0
Sucrose	20,000	58.42 mM
pH	5.5	

It is also important to pay attention to a number of inaccuracies and errors that have appeared in several widely-used plant tissue cultures basal medium formulations *(3,4)*, such as the exact hydration and molar equivalence of iron and its chelating agent. Inconsistencies exist in popular commercial prepara-

Table 4
Phillips and Collins *(10)* Media Composition

	mg/L	mM
Macroelements		
NH_4NO_3	1000	12.5
KNO_3	2100	20.8
$CaCl_2 \cdot 2H_2O$	600	4.1
$MgSO_4 \cdot 7H_2O$	435	1.8
KH_2PO_4	325	2.4
$NaH_2PO_4 \cdot H_2O$	85	0.6
$FeSO_4 \cdot 7H_2O$(EDTA)	25.0	0.1
Microelements	mg/L	µM
H_3BO_3	5.0	82.0
$MnSO_4 \cdot H_2O$	15.0	90.0
$ZnSO_4 \cdot 7H_2O$	5.0	17.5
KI	1.0	6.0
$Na_2MoO_4 \cdot 2H_2O$	0.4	1.7
$CuSO_4 \cdot 5H_2O$	0.1	0.4
$CoCl_2 \cdot 6H_2O$	0.1	0.4
Organic components	mg/L	µM
myo-Inositol	250.0	1400.0
Pyridoxine·HCl	0.5	2.43
Thiamine·HCl	2	6.0
Sucrose	25,000	73 mM
pH	5.8	

tions for several common basal media, in addition to those found in the primary literature. Even the primary literature can have some mistakes *(5)* or the same name is used to designate the same medium. There have been many different versions presented in print by White et al. *(4)*.

Table 5
White *(11,12)* Media Composition

	mg/L	mM
Macroelements		
KNO_3	80	0.79
$Ca(NO_3)_2 \cdot 4H_2O$	300	1.27
KCl	65	0.87
$CaCl_2 \cdot 2H_2O$	440	3.00
$MgSO_4 \cdot 7H_2O$	720	2.92
$NaH_2PO_4 \cdot H_2O$	19	0.13
Na_2SO_4	200	1.40
$Fe_2(SO_4)_3$	2.5	0.006
Microelements	mg/L	μM
H_3BO_3	1.5	24.2
$MnSO_4 \cdot 4H_2O$	7.0	31.3
$ZnSO_4 \cdot 7H_2O$	3.0	10.4
KI	0.75	4.5
MoO_3	0.0001	0.007
$CuSO_4 \cdot 5H_2O$	0.001	0.004
Organic components	mg/L	μM
Nicotinic acid	0.5	4.0
Pyridoxine·HCl	0.1	0.5
Thiamine·HCl	0.1	0.3
Glycine	3.0	40.0
Sucrose	20,000	58.42 mM
pH	5.5	

Minor variations in medium composition can determine the success or failure of certain protocols. The excesses of chelating agents, although small, may influence micronutrient availability. Investigators should examine original papers carefully and compare them with commercial formulations when seeking details on a given nutrient or medium.

Table 6
Nitsch and Nitsch (13) Media Composition

	mg/L	mM
Macroelements		
NH_4NO_3	720	9.0
KNO_3	950	9.4
$CaCl_2$	166	1.5
$MgSO_4 \cdot 7H_2O$	185	0.75
KH_2PO_4	68	0.5
Na_2EDTA	37.2	0.1
$FeSO_4 \cdot 7H_2O$	27.8	0.1
Microelements	mg/L	µM
H_3BO_3	10	161.7
$MnSO_4 \cdot 4H_2O$	25	112.1
$ZnSO_4 \cdot 7H_2O$	10	42.8
$Na_2MoO_4 \cdot 2H_2O$	0.25	1.0
$CuSO_4 \cdot 5H_2O$	0.025	0.1
Organic components	mg/L	µM
myo-Inositol	100.0	555.10
Nicotinic acid	5.0	40.6
Pyridoxine·HCl	0.5	2.43
Thiamine·HCl	0.5	1.50
Glycine	2.0	26.60
Folic acid	0.5	1.1
Biotin	0.05	0.2
Sucrose	20,000	58.42 mM
pH	5.5	

In relation to the Kao and Michayluk medium, the presence of vitamin-free casamino acid and coconut water is essential for the culture of protoplasts, but they are not necessary for the culture of cells. This medium is one of the more complex of all the media used in plant tissue culture. It is used primarily for the growth of very low-cell density cultures, as well protoplasts in liquid media (**6**).

Table 7
Schenk and Hildebrandt (14) Media Composition

	mg/L	mM
Macroelements		
KNO_3	2500	24.72
$MgSO_4 \cdot 7H_2O$	400	1.63
$NH_4H_2PO_4$	300	2.60
$CaCl_2 \cdot 2H_2O$	200	1.36
Na_2EDTA	20	0.053
$FeSO_4 \cdot 7H_2O$	15	0.054
Microelements	mg/L	µM
$MnSO_4 \cdot H_2O$	10.0	59.17
H_3BO_3	5.0	80.86
$ZnSO_4 \cdot 7H_2O$	1.0	3.47
KI	1.0	6.02
$CuSO_4 \cdot 5H_2O$	0.2	8.00
$Na_2MoO_4 \cdot 2H_2O$	0.1	0.41
$CoCl_2 \cdot 6H_2O$	0.1	0.42
Organic components	mg/L	µM
myo-Inositol	1000.0	5500.6
Nicotinic acid	5.0	40.6
Thiamine·HCl	5.0	14.8
Pyridoxine·HCl	0.5	2.43
Sucrose	30,000	58.42 mM
pH	5.9	

Reference

1. George, E. F. (1993) Plant Propagation by Tissue Culture. Part 1. The Technology, Second ed., Exegetics Limited, Great Bretaña.
2. Conger, B. V. (1980) *Cloning Agricultural Plants Via In Vitro Techniques,* CRC Press, Boca Raton, Florida.
3. Wallace, R. J. (1992) Rumen microbiology, biotechnology and ruminant nutrition: The application of research findings to a complex microbial ecosystem. *FEMS Microbiol. Lett.* **100,** 529–534.
4. Singh, M. and Krikorian, A. D. (1981) White's standard nutrient solution. *Ann. Bot.* **47,** 133–139.
5. Murashige, T. and Skoog, F. (1962) A revised medium for rapid growth and bioassays with tobacco tissue cultures. *Physiol. Plant.* **15,** 473–497.
6. Kao, K. N. and Michayluk, R. (1975) Nutritional requeriments for growth of *Vicia hajastana* cells and protoplasts at a very low population density in liquid media. *Planta* **126,** 1095–1100.

Table 8
Kao and Michayluk (6) Media Composition

	mg/L	mM
Macroelements		
NH_4NO_3	600	7.49
KNO_3	1900	18.80
$CaCl_2 \cdot 2H_2O$	600	4.08
$MgSO_4 \cdot 7H_2O$	300	1.21
KH_2PO_4	170	1.25
Sequestrene® 330Fe	28	—
Microelements	mg/L	µM
H_3BO_3	3.00	48.5
$MnSO_4 \cdot H_2O$	10.00	59.2
$ZnSO_4 \cdot 7H_2O$	2.00	7.0
KI	0.75	4.5
$Na_2MoO_4 \cdot 2H_2O$	0.25	1.0
$CuSO_4 \cdot 5H_2O$	0.025	0.1
$CoCl_2 \cdot 6H_2O$	0.025	0.1
Vitamins	mg/L	µM
myo-Inositol	100.0	555.10
Nicotinamide	1.0	8.19
Pyridoxine·HCl	1.0	4.86
Thiamine·HCl	1.0	3.00
Calcium D-pantothenate	1.0	4.20
Folic acid	0.4	0.90
p-Aminobenzoic acid	0.02	0.15
Biotin	0.01	0.04
Choline chloride	1.00	7.16
Riboflavin	0.20	0.53
Ascorbic acid	2.00	11.35
Vitamin A	0.01	0.03
Vitamin D_3	0.01	0.02
Vitamin B_{12}	0.02	0.01
Organic acids	mg/L	µM
Sodium pyruvate	20.0	181.8
Citric acid	40.0	208.2
Other sugars and sugar alcohols	mg/L	µM
Fructose	250.0	1.38
Ribose	250.0	1.66
Xylose	250.0	1.66

(Continued)

Table 8 *(Continued)*
Kao and Michayluk *(6)* Media Composition

	mg/L	μM
Other sugars and sugar alcohols		
Mannose	250.0	1.38
L-amino acids	mg/L	mM
All are used at a concentration of 0.1 mg/L except:		
Glutamine	5.6	38.3
Alanine	0.6	6.7
Nucleic-acid bases	m/L	mM
Adenine	0.10	0.74
Guanine	0.03	0.20
Thymine	0.03	0.24
Uracil	0.03	0.27
Hypoxanthine	0.03	0.22
Xanthine	0.03	0.19
Other	mg/L	mM
Vitamin-free casamino acid	250.0	—
Coconut water	20.0 mL/L	—
Sucrose	20 g/L	58.40 mM
Glucose	10 g/L	55.49 mM
pH	5.6	

Note: This medium is filtered-sterilized.

7. Linsmaier, E. M. and Skoog, F. (1965) Organic growth factor requirements of tobacco tissue cultures. *Physiol. Plant.* **18**, 100–127.
8. Gamborg, O. L., Miller, R. A., and Ojima, K. (1968) Nutrient requirements of suspension cultures of soybean root cells. *Exp. Cell Res.* **50**, 151–158.
9. Gamborg, O. L., Murashige, T., Thorpe, T. A., and Vasil, I. K. (1976) Plant tissue culture media. *In Vitro Cell. Dev. Biol. Plant* **12**, 473–478.
10. Phillips, G. C. and Collins, G. B. (1979) *In vitro* tissue culture of selected legumes and plant regeneration from callus cultures of red clover. *Crop Sci.* **19**, 59–64.
11. White, P. R. (1943) *A Handbook of Plant Tissue Culture,* Science Press Printing, Lancaster, Pennsylvania.
12. White, P. R. (1963) *The Cultivation of Animal and Plant Cells, Second ed.*, Ronald Press, New York.
13. Nitsch, J. P., and Nitsch, C. (1969) Haploid plants from pollen grains. *Science* **163**, 85–87.
14. Schenk, R. U., and Hildebrandt, A. C. (1972) Medium and techniques for induction and growth of monocotyledonous and dicotyledonous plant cell cultures. *Can. J. Bot.* **50**, 199–204.

APPENDIX B

Plant Biotechnology and Tissue Culture Resources on the Internet

Victor M. Loyola-Vargas

1. General Resources
- American Ag-Tec Potato Mini-Tuber Production (http://www.ag-tec.com/potato.htm)
- Biotechnology Information Center, National Agricultural Library (http://www.nal.usda.gov/bic)
- Center for Plant Biotechnology Research, Tuskegee University (http://www.tuskegee.edu/Global/story.asp?S=1122588)
- Soybean Tissue Culture and Genetic Engineering Center (http://www.cropsoil.uga.edu/homesoybean/)
- Cevie asbl (http://www.cevie.com/)
- Conifer Biotech Working Group (http://www.nal.usda.gov/bic/Newsletters/cbwg.news/)
- Ethical, Legal, and Social Aspects (ELSA) of biotechnology (http://www.cordis.lu/elsa/src/prj-biot.htm)
- European Plant Embryogenesis Network (EPEN) (http://epen.tran.wau.nl/)
- Fruit Tree Research Institute, Italy (http://www.agora.stm.it/htbin/wwx?fi^C.Damiano)
- Istituto Sperimentale per la Frutticoltura; Sezione di Propagazione (http://www.agora.stm.it/htbin/wwx?fi^C.Damiano)
- Iowa State University Biotechnology (http://www.biotech.iastate.edu/)
- Iowa State University Plant Transformation Facility (http://www.biotech.iastate.edu/Research1998/ServicesFacilities.html#ptf)

From: *Methods in Molecular Biology, vol. 318: Plant Cell Culture Protocols, Second Edition*
Edited by: V. M. Loyola-Vargas and F. Vázquez-Flota © Humana Press Inc., Totowa, NJ

- Kew Gardens Micropropagation News on the Internet (http://www.rbgkew. org.uk/science/micropropagation/bgmnews.html)
- Laboratoire de Biotechnologie et Amélioration des Végétaux (http://www.univ-brest.fr/UFR/SCIENCES/ISAMOR/BioTech.html)
- Listserv: PLANT-TC (http://plant-tc.coafes.umn.edu/listserv/)
- National Center for Biotechnology Information (http://www.ncbi.nlm.nih.gov/)
- National Botanical Research Institute (http://www.nbri-lko.org/)
- National Research Council of Canada/Plant Biotechnology Institute (http://www.pbi.nrc.ca/)
- NCGR-Corvallis Information—Tissue Culture (http://www.ars-grin.gov/ars/PacWest/Corvallis/ncgr/tc.html)
- Oklahoma Plant Transformation Facility (http://www.ptf.okstate.edu/)
- Plant Hormones—Long Ashton Research Station (http://www.plant-hormones.info/)
- Plant Tissue Culture Research at the University of Minnesota (http://plant-tc.coafes.umn.edu/)
- Plant Tissue Culture for Home Gardeners (http://www.une.edu.au/~agronomy/AgSSrHortTCinfo.html)
- Plant World Explorations (http://members.ozemail.com.au/~mhempel/)
- Research activities at the LG Molekulargenetik (http://www.lgm.uni-hannover.de/LGM%20English/index.htm)
- Succulent Tissue Culture (http://www.succulent-tissue-culture.com/)
- Institute for Plant Genomics and Biotechnology (http://ipgb.tamu.edu/)
- Texas A&M Plant Tissue Culture Information Exchange (http://aggie-horticulture.tamu.edu/tisscult/tcintro.html)
- The Sweet Potato Biotechnology Program at Tuskegee University (http://aggie-horticulture.tamu.edu/tisscult/tuskegee/tuskegee.html)
- USDA Biotechnology Permit Information (http://www.aphis.usda.gov/brs/)

2. Micropropagation

- Alan Bickell Orchids (http://web.idirect.com/~orchids/index.html)
- Asymbiotic Micropropagation Of Orchids From Seeds
- In Vitro Propagation—Embryo Culture of Cephalanthera rubra
- Kitchen Culture Kits, Inc. (http://www.kitchenculturekit.com/)
- Micropropagation Unit at Royal Botanic Gardens Kew (http://www.rbgkew.org.uk/ksheets/pdfs/k1microprop.pdf)
- Oglesby Plant Laboratories (http://www.oglesbytc.com/)
- Orchid Species Culture (http://www.orchidculture.com/)
- Plant Tissue Culture for Home Gardeners http://www.une.edu.au/~agronomy/AgSSrHortTCinfo.html)
- The Micropropagation of Aroids (http://www.aroid.org/horticulture/tculture.html)
- Tissue culture in the homa kitchen (http://www.labs.agilent.com/botany/public_html/cp/slides/tc/tc.htm)

- Tropica Tissue Culture Laboratory—Aquarium Plants (http://192.38.244.204/go.asp?article=193)

3. Databases

- AGRICOLA (http://agricola.nal.usda.gov/help/aboutagricola.html)
- Angiosperm DNA C-Values Database (http://www.rbgkew.org.uk/cval/database1.html)
- BIC Biotechnology Bibliographies and Resource Guides (http://www.nal.usda.gov/bic/Biblios/)
- Bibliography on Tissue Culture, National Agricultural Library
- Internet Directory for Botany (http://www.botany.net/IDB/botany.html)
- Terminology Associated With Cell, Tissue, and Organ Culture, Molecular Biology And Molecular Genetics
- The WWW Virtual Library—BioScience Resources
- The WWW Virtual Library—Plant Science Resources

4. Books

- Agritech Publications (http://agritechpublications.com/)
- Recent Advances in Plant Tissue Culture. Vol.1. Regeneration, Micropropagation and Media 1988–1991 (http://www.agritechpublications.com/rec1book.htm)
- Recent Advances in Plant Tissue Culture. Vol. 2. Secondary Metabolite Production 1988–1993 (http://www.agritechpublications.com/rec2book.htm)
- Recent Advances in Plant Tissue Culture. Vol. 3. Regeneration and Micropropagation:Techniques, Systems, and Media 1991–1995 (http://www.agritechpublications.com/rec3book.htm)
- Recent Advances in Plant Tissue Culture. Vol. 4. Microbial Contamination of Plant Tissue Cultures (http://www.agritechpublications.com/rec4book.htm)
- Recent Advances in Plant Tissue Culture. Vol. 5. New Techniques and Systems for Growth, Regeneration, and Micropropagation 1995–1997 (http://www.agritechpublications.com/rec5book.htm)
- Recent Advances in Plant Tissue Culture. Vol. 6. Regeneration and Micropropagation: Techniques, Systems, and Media 1997–1999 (http://www.agritechpublications.com/rec6book.htm)
- Regeneration and Micropropagation: Techniques, Media, and Applications 1999–2002. Vol. 7 of Recent Advances in Plant Tissue Culture (http://www.agritechpublications.com/rec7book.htm)
- Microbial "Contaminants" in Plant Tissue Cultures: Solutions and Opportunities 1996–2003. Vol. 8 of Recent Advances in Plant Tissue Culture (http://agritechpublications.com/rec8book.htm)

5. Editorials

- Annual Reviews (http://arjournals.annualreviews.org/;jsessionid= iBIj52VDf_R8)
- Cambridge University Press (http://www.cup.cam.ac.uk/)
- CRC Press (http://www.crcpress.com/)

- Elsevier (http://www.elsevier.com/wps/find/homepage.cws_home)
- Exegetics
- Plant Culture Media. Vol 1. Formulations and Uses, 1999 (http://agritechpublications.com/media1.htm)
- Plant Culture Media. Vol 2. Commentary and Analysis, 1999 (http://agritechpublications.com/media1.htm)
- Humana Press (http://www.humanapress.com/Index.pasp)
- Kluwer (http://www.kluweronline.com/)
- Springer-Verlag (http://www.springeronline.com/sgw/cda/frontpage/0,0,0-0-0-0-EAST,0.html)
- Timber Press (http://www.timberpress.com/books/index.cfm)
- Vedams (https://www.vedamsbooks.com/index.htm)

6. Journals and Newsletters

- Agricell Report (http://www.agritechpublications.com/)
- American Journal of Botany (http://www.amjbot.org/current.shtml)
- Biologia Plantarum (http://www.kluweronline.com/issn/0006-3134/)
- Biotechnology Letters (http://www.kluweronline.com/issn/0141-5492/contents)
- Biotechnology Process (http://pubs3.acs.org/acs/journals/toc.page?incoden=bipret)
- Botanic Gardens Micropropagation News (http://www.rbgkew.org.uk/science/micropropagation/bgmnews.html)
- Canadian Journal of Botany (http://pubs.nrc-cnrc.gc.ca/cgi-bin/rp/rp2_desc_e?cjb)
- Crop Science (http://crop.scijournals.org/)
- Current Opinion in Biotechnology (http://www.current-opinion.com/jbio/about.htm?jcode=jbio)
- Current Opinion in Cell Biology (http://www.current-opinion.com/jcel/about.htm?jcode=jcel)
- Current Opinion in The Plant Biology (http://www.current-opinion.com/jpbl/about.htm?jcode=jpbl)
- Electronic Journal of Biotechnology (http://www.ejbiotechnology.info/)
- Euphytica (http://www.kluweronline.com/issn/0014-2336/)
- In Vitro Cellular and Developmental Biology—Plant (http://www.cabi-publishing.org/Journals.asp?SubjectArea=Pla&Subject=Biotechnology%2C+Plant+Breeding+and+Genetic+Resources&PID=23)
- Journal of Experimental Botany (http://www3.oup.co.uk/jnls/list/exbotj/adrates/)
- Journal of Phytophatology (http://www.blackwell-synergy.com/servlet/useragent?func=showIssues&code=jph)
- Journal Plant Physiology (http://www.elsevier-deutschland.de/jpp)
- Nature Biotechnology (http://www.nature.com/biotech/)
- New Phytologist (http://www.blackwell-synergy.com/servlet/useragent?func=showIssues&code=nph)
- Physiologia Plantarum (http://www.blackwell-synergy.com/servlet/useragent?func=showIssues&code=ppl)

Appendix B

- Plant Biotechnology Journal (http://www.blackwell-synergy.com/servlet/useragent?func=showIssues&code=pbi)
- Plant Breeding (http://www.blackwell-synergy.com/servlet/useragent?func=showIssues&code=pbr)
- Plant Cell (http://www.plantcell.org/)
- Plant Cell and Enviroment (http://www.blackwell-synergy.com/servlet/useragent?func=showIssues&code=pce)
- Plant Cell Physiology (http://pcp.oupjournals.org/)
- Plant Cell Reports (http://springerlink.metapress.com/app/home/journal.asp?wasp=5hggyjwrqldn1dkt9t0m&referrer=parent&backto=linkingpublicationresults,1:100383,1)
- Plant Cell, Tissue, and Organ Culture (http://www.kluweronline.com/issn/0167-6857/contents)
- Plant Growth Regulation (http://www.kluweronline.com/issn/0167-6903/)
- Plant Journal (http://www.blackwell-synergy.com/journals/issuelist.asp?journal=tpj)
- Plant Molecular Biology (http://www.kluweronline.com/issn/0167-4412/contents)
- Plant Physiology (http://www.plantphysiol.org/contents-by-date.0.shtml)
- Planta (http://springerlink.metapress.com/app/home/journal.asp?wasp=53flumtrmrdradhwrh5u&referrer=parent&backto=linkingpublicationresults,1:100484,1)
- Theoretical and Applied Genetics (http://springerlink.metapress.com/app/home/journal.asp?wasp=43wrc22qrg3uxh80vbft&referrer=parent&backto=subject,136,144;)
- Trends in Biochemical Sciences (http://www.sciencedirect.com/science?_ob=JournalURL&_cdi=5180&_auth=y&_acct=C000059356&_version=1&_urlVersion=0&_userid=2974920&md5=6b7bf54a429a339ce0e7775f612aa38a)
- Trends in Biotechnology (http://www.sciencedirect.com/science?_ob=JournalURL&_cdi=5181&_auth=y&_acct=C000059356&_version=1&_urlVersion=0&_userid=2974920&md5=b193df7f479ce8d9bee711a193c4c92d)
- Trends in Cell Biology (http://www.sciencedirect.com/science?_ob=JournalURL&_cdi=5182&_auth=y&_acct=C000059356&_version=1&_urlVersion=0&_userid=2974920&md5=7a8600c9b4d5bd24f73a3c9110d9a9e5)
- Trends in Genetics (http://www.sciencedirect.com/science?_ob=JournalURL&_cdi=5183&_auth=y&_acct=C000059356&_version=1&_urlVersion=0&_userid=2974920&md5=7d205e09f5c512f5a2d6da9b773b6e94)
- Trends in Plant Science (http://www.sciencedirect.com/science?_ob=JournalURL&_cdi=5185&_auth=y&_acct=C000059356&_version=1&_urlVersion=0&_userid=2974920&md5=d89d0b0b12068fabbac190b1a237867c)

7. PTC Inside the Classroom

- Access Excellence (http://www.accessexcellence.org/)
- Cloning Plants By Tissue Culture (http://www.jmu.edu/biology/biofac/facfro/cloning/cloning.html)
- Introduction To Plant Breeding—Crop Tissue Culture (Course Notes From AGRONOMY 815 at the University of Nebraska) Plant Cell Protoplasts (http://agronomy.unl.edu/815/cropt.htm)
- Plant Tissue Culture: An Alternative For Production Of Useful Metabolite (http://www.fao.org/docrep/t0831e/t0831e00.htm)
- Plant Tissue Culture (http://academy.d20.co.edu/kadets/lundberg/ptc.html)

8. Societies

- American Phytopathological Society (www.apsnet.org)
- American Society for Horticultural Science (http://www.ashs.org/)
- American Society of Plant Physiologists (http://www.aspb.org/membersonly.cfm?CFID=1213590&CFTOKEN=96692329)
- Botanical Society of America (http://www.botany.org/)
- International Association for Plant Tissue Culture and Biotechnology (IAPTC&B) (http://www.genetics.ac.cn/iaptcb.htm)
- International Plant Propagators' Society (http://www.ipps.org/Default.asp)
- Society for In Vitro Biology (http://www.sivb.org/)
- The American Orchid Society (http://orchidweb.org/)
- The International Society For Plant Molecular Biology (http://www.uga.edu/ispmb/)

9. Design and Layout of a Micropropagation Facility

- Design and Layout of a Micropropagation Facility (http://aggie-horticulture.tamu.edu/tisscult/microprop/facilities/microlab.html)

Index

A

Abscisic acid, spruce embryo maturation, 94, 95, 97, 193, 194
Agave species,
  micropropagation,
    contamination control, 168, 169, 174, 175
    cutting of explants, 170
    explant culture, 170, 175, 176
    field testing, 172, 173
    large-scale production, 173
    materials, 167, 168, 173, 174
    meristematic tissue extraction, 168
    multiplication, 170, 171, 176, 177
    pre-adaptation, 171, 172, 177
    quality control, 173, 177, 178
    rooting, 172
    shoot growth, 171
    soil transplantation, 172
  reproduction, 166, 173
  tissue culture applications, 166
  uses, 165
Agrobacterium-mediated transformation, *see* California poppy; Maize; Petunia
Albinism, spectinomycin induction, 220, 224, 228, 229
Allium sativum, *see* Garlic
Aloe, *see* Chinese aloe
Aluminum tolerance,
  aluminum concentration assay in culture medium, 307, 308, 311
  cell suspension culture for tolerant line production, 306, 307, 311, 312
  detoxification pathways, 305, 306
  phospholipase C assay,
    incubation conditions and calculations, 310, 312
    materials, 307, 311
    reaction mix, 310, 312
    sample preparation, 308, 309
    substrate stock solution preparation, 309, 312
  toxicity,
    economic impact, 305
    phospholipase C signaling, 305, 306
Amylase, isozyme analysis in cybrids, 228
Analysis of variance (ANOVA),
  continuous data analysis, 149, 150, 156
  dosage or concentration treatments, 153, 154
  orthogonal contrasts, 152
  treatment means analysis, 150
ANOVA, *see* Analysis of variance
Automated Temporary Immersion Reactor (RITA), micropropagation, 122, 123

B

Bacanora, *see Agave* species
Bacillus thuringiensis, chloroplast transformation of genes for pest control, *see* Chloroplast transformation
Benzylisoquinoline alkaloids, *see* California poppy
Biolistics, *see* Chloroplast transformation; Wheat
BioMINT™ bioreactor,
  micropropagation,
    addition of chemicals, 128
    contamination control, 127, 128
    cutting and grading, 127

385

equipment, 123, 124, 128
filling, 125, 126, 128
incubation, 126
media changing, 127
platform operation, 126, 128
sterilization, 125
Brassicacae family,
  cybrid production,
    albino line production, 224, 228, 229
    cytogenetic studies, 228
    DNA analysis, 226, 227
    isozyme analysis, 227, 228
    materials, 221–223
    plant regeneration and hybrid
      selection, 226
    protoplast isolation and fusion,
      224–226, 229, 230
    seedling culture, 223, 224
  oilseed rape chloroplast
    transformation, *see* Chloroplast
    transformation
  somatic hybridization applications,
    220, 229, 230

C

California poppy,
  Agrobacterium-mediated
    transformation,
    bacteria preparation, 361, 364
    callus transformation, 361, 362, 365
    explant preparation, 361, 364
    materials, 359–361, 363
    overview, 358
    somatic embryogenesis of
      transformed plants, 362, 363,
      365, 366
  hairy root regeneration with
    Agrobacterium rhizogenes,
    362, 365, 366
  medicinal uses of benzylisoquinoline
    alkaloids, 357, 358
Callus culture,
  capsaicin production, 330, 334

characteristics, 51, 52
friable embryogenic callus, *see*
  Cassava
garlic,
  clove disinfestation, 63
  growth kinetics, 66–68
  initiation and maintenance, 63,
    64, 68
  plant regeneration, 64, 66
growth measurements,
  dry-weight determination, 54
  fresh-weight determination, 54
  materials, 53, 54
  overview, 52, 53
  parameters,
    doubling time, 57, 58
    growth index, 56, 57
    specific growth rate, 57
historical perspective, 11
Capsaicin,
  applications, 327
  biosynthesis, 327, 328
  production in cell cultures,
    callus culture, 330, 334
    capsaicinoid extraction,
      separation, and quantification,
      330–332
    cell clone isolation, 333, 334
    enhancement with precursors and
      intermediates, 332, 333
    materials, 329
    suspension culture, 330
Cassava,
  indirect somatic embryogenesis,
    friable embryogenic callus,
      culture and maintenance, 104–106
      plant regeneration, 105, 106
      greenhouse transfer, 105, 107
      leaf explant isolation and culture,
        104, 106
      materials, 103, 104
      overview, 102, 106
    uses, 101

Catharanthus roseus, see Madagascar periwinkle
Cayenne pepper, *see* Capsaicin
CCP, *see* Critical control point
Chili pepper, *see* Capsaicin
Chinese aloe,
  medicinal use, 181
  micropropagation,
    bacteria-free explant establishment, 183
    materials, 181, 182
    rationale, 179–181
    rooted plantlet transplantation, 184
    rooting, 184
    shoot induction, 184
Chloroplast transformation,
  Bacillus thuringiensis genes, 287
  biolistic transformation of oilseed rape chloroplasts,
    bombarded tissue culture, 291
    bombardment, 290, 297, 298
    bottlenecks, 298, 299
    chloroplast expression vector construction, 290, 291, 296, 297
    DNA coating of microcarriers, 289, 290
    evaluation of transformants,
      insecticide resistance bioassay, 294, 295
      polymerase chain reaction and Southern blot, 294
    gold microcarrier preparation, 289, 296
    homoplasty of transformants, 291, 293
    materials, 287, 288
    tissue preparation, 288
  historical perspective, 286
  overview of approaches, 286, 287
  rationale and advantages, 285, 286
Classroom, Internet resources, 384
Clonal propagation, historical perspective, 17

Coconut,
  breeding programs, 131
  photoautotrophic plant culture,
    acclimatization ex vitro, 138, 139
    autotrophic plantlet establishment, 138
    carbon dioxide concentration, 134, 135
    embryo germination, 136, 138
    exogenous sucrose, 133, 134
    growth room conditions, 138
    light intensity, 134
    materials, 136, 139
    overview, 131–133
    photosynthetic capacity optimization, 135, 136
Coconut water, embryo culture, 11
Coffee, direct somatic embryogenesis, 111–116
Conifers, *see* Spruce
Contamination, tissue culture,
  Agave micropropagation control, 168, 169, 174, 175, 177, 178
  aseptic culture establishment, 45, 46
  BioMINT™ bioreactor control, 127, 128
  diagnosis, 37–39
  Hazard Analysis Critical Control Points management, 46, 48
  identification of pathogens and biological contaminants,
    critical control point biomarkers, 38, 40, 48
    genomic fingerprinting, 41–44
    kits, 40
    metabolomic fingerprinting, 44
    microscopy, 39, 40
    proteomic fingerprinting, 44
    selective culture media, 40
    serological diagnostics, 40, 41
  indexing of pathogens, 44, 45
  overview, 36
  prospects for study and management, 47, 48

Corn, *see* Maize
Critical control point (CCP),
  biomarkers, 38, 40, 48
Cryopreservation, *see* Encapsulation-vitrification,
Cybrids, *see* Brassicacae family
Cytoplasmic male sterility,
  transmission, 219, 220

D–E

Denaturing gradient gel electrophoresis
  (DGGE), pathogen
  identification, 44
DGGE, *see* Denaturing gradient gel
  electrophoresis
Echinacea,
  protoplast isolation, culture, and
    plant regeneration,
    applications, 211, 212
    culture, 213, 214, 216, 217
    isolation, 213
    materials, 212
    plant material preparation, 212,
      213, 216
    plant regeneration, 214, 215
  uses, 211
ELISA, *see* Enzyme-linked
  immunosorbent assay
Embryo culture, historical perspective,
  10, 11
Encapsulation-vitrification,
  embryogenic cell suspension
    maintenance, 80
  freezing and rewarming, 81, 82
  materials, 79, 80
  principles, 78, 79
  survival, regrowth, and plant
    regeneration, 82–84
  technique, 80–83
  unloading, 82, 83
Endangered species,
  micropropagation, *see* Chinese aloe;
    Pacific yew

  preservation, 6, 7
Enzyme-linked immunosorbent assay
  (ELISA),
  pathogen identification, 43
  serological diagnostics, 40, 41
Eschscholzia californica, *see* California
  poppy
Esterase, isozyme analysis in cybrids, 228
Evans Blue, viability assay, 73, 74

F–G

Flow cytometry, interspecific hybrids,
  259, 261
Friable embryogenic callus, *see* Cassava
Gamborg media, composition, 372
Garlic, somatic embryogenesis
  induction,
  callus culture,
    clove disinfestation, 63
    growth kinetics, 66–68
    initiation and maintenance, 63, 64, 68
    plant regeneration, 64, 66
  materials, 60–62
Gene gun, *see* Chloroplast
  transformation; Wheat
Genomic fingerprinting, pathogens, 41–44
Growth measurements,
  callus culture,
    dry-weight determination, 54
    fresh-weight determination, 54
    materials, 53, 54
    overview, 52, 53
  parameters,
    doubling time, 57, 58
    growth index, 56, 57
    specific growth rate, 57
  suspension culture,
    materials, 53, 54
    overview, 52, 53
    packed cell volume, 55, 56
    settled cell volume, 55
Guard cell protoplasts,
  guard cell features, 233, 234

history of study, 234
isolation, culture, and plant
regeneration from *Nicotania glauca*,
callus culture, 245
isolation, 242–244, 247
materials, 236, 238–241, 246, 247
overview, 235, 236
plant growth, 241, 242, 246, 247
plant regeneration, 247, 248
primary culture and colony
formation, 244, 245, 247
signal transduction studies, 235, 236

H–L

HACCP, *see* Hazard Analysis Critical Control Points
Hairy roots, *see* California poppy
Hazard Analysis Critical Control Points (HACCP), contamination management, 46, 48
Henequin, *see Agave* species
Historical perspective, plant tissue culture, 9–20
Internet resources, plant biotechnology and tissue culture, 379–383
Interspecific hybrids, *see Primula* interspecific hybridization
Jasmonate, alkaloid induction in Madagascar periwinkle shoot cultures, 353, 354
Kao and Michayluk media, composition, 377, 378
Kinetin, historical perspective, 11
Knop's solution, historical perspective, 12
Large-scale propagation, overview, 4, 5
Leaf disk, *see* Petunia
Lesquerella fendleri, *see* Brassicacae family
Linsmaier and Skoog media, composition, 371
Linum, embryo rescue from nonviable seed crosses, 10
Lycopersicon esculentum, *see* Tomato

M

Madagascar periwinkle,
vinblastine production, 349
vincristine production, 349
vindoline production from shoot cultures,
alkaloid extraction and quantification, 351–353
jasmonate induction, 353, 354
overview, 349, 350
shoots,
culture initiation, 350, 351
formation and maintenance, 351
induction in liquid media, 351
Maize, *Agrobacterium*-mediated transformation with binary vector,
Agrobacterium preparation,
culture, 318, 320
master plate, 318
strain, 318
working plate, 318
confirmation, 320
embryo preparation, 319
infection and co-cultivation of embryos, 319, 320
materials, 316–318, 320
overview, 315, 316
regeneration of T0 plants, 319
selection of putative transgenic events, 319
Metabolic engineering, overview, 7, 8
Metabolomic fingerprinting, pathogens, 44
Micropropagation,
Agave species, *see Agave* species
BioMINT™ bioreactor,
addition of chemicals, 128
contamination control, 127, 128
cutting and grading, 127
equipment, 123, 124, 128
filling, 125, 126, 128
incubation, 126
media changing, 127

platform operation, 126, 128
sterilization, 125
endangered plants, *see* Chinese aloe;
Pacific yew
Internet resources, 380, 381, 384
liquid medium, 122
semi-solid medium, 121, 122, 128
stages, 128
temporary immersion, 122, 123
Molecular beacons, pathogen identification, 41
Monoterpenoid indole alkaloids, *see* Madagascar periwinkle
Murashige and Skoog media, composition, 370

N–P

Nitsch and Nitsch media, composition, 375
NT-1 cells, *see* Tobacco
Oilseed rape, *see* Chloroplast transformation
Orychophragmus violaceus, *see* Brassicacae family
Pacific yew,
  micropropagation,
    bacteria-free explant establishment, 182
    materials, 181, 182
    rationale, 179, 180
    rooted plantlet transplantation, 183
    rooting, 183
    shoot induction and elongation, 182, 183
  taxol,
    applications, 180
    needles in production, 180
Packed cell volume (PCV), determination, 55, 56, 67–69
Papever somniferum, history of in vitro fertilization, 13
Passiflora edulis, *see* Passionfruit

Passionfruit,
  breeding programs, 202
  leaf protoplast isolation, culture, and plant regeneration from *Passiflora edulis*,
    applications, 202
    culture, 206, 208
    isolation, 205–208
    materials, 202–204
    plant regeneration and characterization, 206–208
    seedling growth, 203, 207
    species, 201, 202
Pathogens, *see* Contamination, tissue culture
PCR, *see* Polymerase chain reaction
PCV, *see* Packed cell volume
Peroxidase, isozyme analysis in cybrids, 228
Petunia,
  advantages as model system, 265, 266
  Agrobacterium-mediated transformation of leaf disks,
    analysis of transformants, 270, 271
    inoculation of leaf disks, 268–271
    materials, 266, 267
    overview, 266
    seeding germination, 268, 270
    transformed shoot recovery, 269–271
Philips and Collins media, composition, 373
Phospholipase C (PLC),
  activity assay,
    incubation conditions and calculations, 310, 312
    materials, 307, 311
    reaction mix, 310, 312
    sample preparation, 308, 309
    substrate stock solution preparation, 309, 312
  aluminum toxicity signaling, 305, 306
Picea, *see* Spruce
PLC, *see* Phospholipase C

Polymerase chain reaction (PCR),
    Agrobacterium T-DNA detection in
        transformants, 270, 320
    chloroplast transformant analysis,
        294
    cybrid DNA analysis, 226, 227
    pathogen identification, 41–44
Polynomial contrasts, statistical
        analysis, 161, 162
Poppy, see California poppy
Primula interspecific hybridization,
    artificial crossing,
        crossing, 256, 259, 260
        emasculation, 256, 259
        materials, 255
    overview, 253–255
    rescue culture,
        abnormal seedlings, 257
        embryos and immature seeds,
            256, 257, 260
        flow cytometry, 259, 261
        medium preparation, 256
        random amplified polymorphic
            DNA analysis, 258, 259, 261
Proteomic fingerprinting, pathogens, 44
Protoplasts, see Brassicacae family;
    Echinacea; Guard cell
        protoplasts; Passionfruit

R

Random amplified polymorphic DNA
        (RAPD) analysis,
    cybrid DNA, 226, 227
    interspecific hybrids, 258, 259, 261
    pathogen identification, 42
RAPD analysis, see Random amplified
        polymorphic DNA analysis
Restriction fragment length
        polymorphism (RFLP), cybrid
        DNA analysis, 226, 227
RFLP, see Restriction fragment length
        polymorphism

Ribosome-inactivating proteins (RIPs),
    antimicrobial activity and
        mechanism, 338–341
    enzymatic activities, 337, 338
    N-glycosidase activity assay, 344, 345
    isolation and purification,
        materials, 342, 343
        protein extraction, 343
        purification, 344
    prospects for study, 341
    therapeutic properties, 340, 341
    types and classification, 335–337
RIPs, see Ribosome-inactivating
        proteins
RITA, see Automated Temporary
        Immersion Reactor

S

Schenk and Hildebrandt media,
        composition, 376
Sisal, see Agave species
Softwoods,
    clonal propagation rationale, 188
    distribution, 187, 188
    spruce, see Spruce
    uses, 187
Somatic embryogenesis,
    direct versus indirect, 102, 111, 112
    induction, see California poppy;
        Cassava; Coffee; Garlic; Spruce
Somatic hybridization, see Brassicacae
        family
Southern blot,
    Agrobacterium T-DNA detection in
        transformants, 270
    chloroplast transformant analysis, 294
Spectinomycin, plastid ribosome
        deficiency induction, 220, 224,
        228, 229
Spruce,
    development, 88, 189
    history of culture, 88

organogenesis,
   materials, 189, 194
   mature explants, 192, 194
   seedling explants,
      explant preparation and culture initiation, 189, 190, 194
      rooting, 191, 194
      shoot development and multiplication, 190, 191, 194
   somatic embryogenesis in suspension cultures,
      conversion, 96, 97, 193, 195
      development, 93, 94, 193
      induction,
         culture, 92, 192–194
         sterilization of seeds, 91, 92, 97
      maintenance, 92, 93, 193
      materials, 88, 189, 190
      maturation,
         abscisic acid, 94, 95, 97, 193, 194
         desiccation of embryos, 95, 97
         morphology, 95, 96
      overview, 88–91
   uses, 187–189
Statistical analysis, plant biotechnology,
   binomial data analysis, 148, 155
   continuous data analysis, 149, 150, 156
   count data analysis, 148, 149, 156
   dosage or concentration treatments, 153, 154, 159–163
   importance, 145, 146
   multiple comparison and multiple range tests, 150, 151
   orthagonal contrasts, 151, 152, 157–159
   petunia culture model, 146, 147, 154, 155
   software, 146, 147
   standard error, 152, 153, 159
   treatment means analysis, 150
Sugar beet, guard cell protoplast culture, 235, 236

Suspension cultures,
   aluminum-tolerant line production, 306, 307, 311, 312
   capsaicin production, 330
   characteristics, 52
   growth measurements,
      cell density measurement,
         cell counting, 56, 58
         chromium trioxide disaggregation of cell clusters, 56
         hydrolytic enzyme disaggregation of cell clusters, 56, 58
      materials, 53, 54
      overview, 52, 53
      packed cell volume, 55, 56
      parameters,
         doubling time, 57, 58
         growth index, 56, 57
         specific growth rate, 57
      settled cell volume, 55
   spruce, *see* Spruce
   viability assays,
      Evans Blue assay, 73, 74
      FDA assay, 74
      materials, 72–74
      microscopic assay, 74
      MTT/TTC assay, 73, 74

T

Taxus, *see* Pacific yew
Temporary Immersion System, micropropagation, 122, 123
Tequila, *see* *Agave* species
Tobacco,
   NT-1 cell suspension maintenance,
      materials, 61–63
      packed cell volume determination, 67–69
      viability assay, 68, 69
   protoplasts, *see* Guard cell protoplasts

Index

ribosome-inactivating proteins, 341
Tomato, viability assays in suspension cultures,
  Evans Blue assay, 73, 74
  FDA assay, 74
  materials, 72–74
  microscopic assay, 74
  MTT/TTC assay, 73, 74
Tree tobacco, *see* Guard cell protoplasts
Triticum aestivum, *see* Wheat

V

Viability assays,
  Evans Blue assay, 73, 74
  FDA assay, 74
  materials, 72–74
  microscopic assay, 74
  MTT/TTC assay, 73, 74
  NT-1 cell suspensions, 68, 69
  overview, 71, 72
Vinblastine, *see* Madagascar periwinkle
Vincristine, *see* Madagascar periwinkle

W

Wheat,
  Agrobacterium-mediated transformation, 274
  crop loss, 274
  domestication, 273
  particle bombardment for transformation,
    culture, selection, and regeneration, 279, 281
    DNA-coated gold particle preparation and delivery, 277–279, 281
    embryo culture, 276, 277, 280, 281
    materials, 274–276, 280
    rationale, 274
    validation of transformation, PAT assay, 279, 280
    reporter assay for transient expression, 279
White media, composition, 374
White spruce, *see* Spruce